Design Theory '88

AF683380

WITHDRAWN
CAL POLY SLO
KENNEDY LIBRARY

Design theory '88

DATE DUE

DEC 07 1990
NOV 27 REC'D
JUN 30 1991
MAY 20 REC'D
JUN 30 1992
APR -3 REC'D
NOV 01 1992
OCT 12 REC'D
JUN 30 1994
APR 27 REC'D
JAN 05 1995
DEC 05 REC'D
DEC 09 1996
NOV 12 REC'D

GAYLORD PRINTED IN U.S.A.

CALIFORNIA POLYTECHNIC STATE UNIVERSITY LIBRARY

Sandra L. Newsome W.R. Spillers
Susan Finger
Editors

Design Theory '88

Proceedings of the 1988 NSF Grantee Workshop
on Design Theory and Methodology

With 97 Illustrations

Springer-Verlag
New York Berlin Heidelberg
London Paris Tokyo

Sandra L. Newsome
Department of Psychology
Rensselaer Polytechnic Institute
Troy, New York 12180-3590
U.S.A.

W.R. Spillers
Department of Civil Engineering
Rensselaer Polytechnic Institute
Troy, New York 12180-3590
U.S.A.

Susan Finger
Robotics Institute
Carnegie Mellon University
Pittsburgh, Pennsylvania 15213
U.S.A.

Library of Congress Cataloging-in-Publication Data
NSF Grantee Workshop on Design Theory and Methodology (2nd : 1988 : Troy, N.Y.)
Design theory '88 : proceedings of the 1988 NSF Grantee Workshop on Design Theory and Methodology / Sandra L. Newsome, W.R. Spillers, Susan Finger [editors].
p. cm.
1. Engineering design--Congresses. I. Newsome, Sandra Lynn. 1949- . II. Spillers, William R., 1934- . III. Finger, Susan IV. Title.
TA174.N77 1988
620'.00425--dc19 89-5994

Printed on acid-free paper.

© 1989 by Springer-Verlag New York, Inc.
All rights reserved. No part of this book may be translated or reproduced in any form without written permission from the publisher (Springer-Verlag, 175 Fifth Avenue, New York, New York 10010, U.S.A.), except for brief excerpts in connection with reviews or scholarly analysis. Use in connection with any form of information storage and retrieval, electronic adaptation, computer software, or by similar or dissimilar methodology now known or hereafter developed is forbidden.
The use of general descriptive names, trade names, trademarks, etc. in this publication, even if the former are not especially identified, is not to be taken as a sign that such names, as understood by the Trade Marks and Merchandise Act, may accordingly be used freely by anyone.

Text provided by the editors in camera-ready form.
Printed and bound by Edwards Brothers, Inc., Ann Arbor, Michigan.
Printed in the United States of America.

9 8 7 6 5 4 3 2 1

ISBN 0-387-96976-4 Springer-Verlag New York Berlin Heidelberg
ISBN 3-540-96976-4 Springer-Verlag Berlin Heidelberg New York

This book is dedicated to
THE STUDENTS OF THE WORKSHOP
who will determine the future of Design Theory

FOREWORD

In 1984, Nam Suh, who was then the Assistant Director for Engineering at the National Science Foundation (NSF), created the Design Theory and Methodology Program. Among his goals in creating this program were to develop a science of engineering design and to establish design as an accepted field of engineering research. From 1984 to 1986 this program was directed by Susan Finger; from 1986 to the present Jack Dixon has been the director. The program itself has covered a broad range of disciplines, from chemical engineering to architecture, and a broad range of research paradigms, from psychological experiments to mathematical models.

The present volume is based on the second NSF Grantee Workshop on Design Theory and Methodology, called *Design Theory '88*, which was held June 2–5, 1988 at Rensselaer Polytechnic Institute in Troy, NY, USA. It is, however, not strictly a proceedings since it includes some material that was not presented at a the Workshop and since it omits some papers and discussions that were presented at the Workshop. At the Workshop, invited speakers presented overviews of six different research areas based on summaries submitted in advance by the grantees of the Design Theory and Methodology Program. Since most of the speakers were not supported under the NSF program they brought fresh views to it. The other papers in this book were submitted directly to this volume and were not presented at the Workshop.

This volume begins with some remarks by Roland Schmitt who gives an industrial perspective on the problems of engineering design today. His paper is followed by the research papers, grouped into five categories. The last section of the volume addresses issues of research methodology.

There are five categories of research papers:

- Cognitive and Social Aspects of the Design Process. Papers in this section focus on human cognitive and social processes involved in engineering design.
- Computational Models of Design Processes. Computer simulations of design decisions are addressed in this section.
- Environments, Languages, Representations and Data Bases for Design. Formal or mathematical methods of describing form, function, ... are described in this section.

- Analysis to Support Design. Papers in this section focus on mathematical techniques that are used either in the evaluation of designs or in support of the earlier stages of the design process.
- Design for Manufacturing and the Life Cycle. This section is devoted to research on the inclusion of life–cycle concerns, such as manufacturing or maintenance, within the design process.

These groupings are arbitrary and many alternative groupings are possible. For example, a grouping of papers that have a behavioral science perspective would include not only papers from the first category but also papers in both Computational Models of Design and the section on Environments, Language, Representations, and Data Bases for Design. Mathematical concepts would also serve as a dimension of similarity uniting papers from Computational Models of Design, Environments, Languages, Representations, and Data Bases for Design with papers in Analysis to Support Design. Still a third dimension of similarity, computer tools to facilitate design activities, would include some papers from all five categories.

We would like to thank all of those who helped make the Workshop a success. In particular, we have benefited from the delightful administrative assistance of Linda Jaynes and Susan Daniels, both of whom are graduate students at RPI. A number of anonymous reviewers provided comments on the papers included in this book, and we are grateful for their time and effort. We deeply appreciate the work that Jack Dixon, the present program director for Design Theory and Methodology at NSF, put into the organization of the Workshop. Finally, we acknowledge the pleasure of working with Springer–Verlag and Dr. Julia Abrahams, our editor.

S. L. Newsome
W. R. Spillers
S. Finger

Troy and Pittsburgh
September 1988

CONTENTS

* Invited Paper

PART 1

THE INDUSTRIAL PERSPECTIVE

The following is the text of the Keynote Address given by Roland Schmitt at the Workshop. Dr. Schmitt, who became the 16th President of Rensselaer Polytechnic Institute in March of 1988, was for many years a Senior Vice President of the General Electric Company and director of their Research and Development Center. He has furthermore had direct access to federal decision making through his position as Chairman of the National Science Board, the policy making body of the National Science Foundation. There is no one more qualified to describe the industrial world of design and we are delighted to be able to include his remarks in this volume.

the editors

DESIGN-CENTERED INNOVATION

Dr. Roland W. Schmitt
President
Rensselaer Polytechnic Institute, Troy, NY 12180

Being competitive is easy. Just introduce a product with the best performance, the highest quality and reliability and the lowest cost! There's nothing new here. What's new, in the competitive world economy today, is how to do these things. The new answer, and the theme of my remarks today, is a shift from technology-centered innovation on the one hand or cost-centered innovation on the other to design-centered innovation.

Let me say what I mean by some of those terms, then discuss some examples, and finally tell you what I mean by a shift to design-centered innovation. In the powerfully simplified model I will adopt, there are only a few generic approaches to creating innovative products. One is to aim for a totally new capability in technology--to invent the transistor, or the laser, or xerography, or the electronic digital computer, for example. In these cases, the initial challenge is to put together something that works. You rely on the technological uniqueness to create a market. There need be little initial concern for cost if you are truly supplying something that conventional technology cannot match. This is technology centered-innovation. It is the area where, up to now, Europe and the United States have excelled.

Consider a few examples of innovations carried out successfully in the U.S. and Europe--such as the high-bypass turbofan aircraft engine... the microprocessor... computer software for word processing and automated spreadsheets.... and, indeed, the entire computer industry, from mainframes to personal computers. Perhaps the classic example of what might be termed the "Western approach" to engineering excellence is the VAX computer, developed by the Digital Equipment Corporation. The designer, Gordon Bell, set out to provide a single computer architecture for many different computing styles. Combining an innovative approach (in terms of a wide range of instructions and a large address space) with a conservative approach to design and hardware, his team succeeded in developing a product that was still dynamically evolving and popular a decade after its introduction, an incredible longevity in the computer world.

Another major category, cost-centered innovation, typically comes somewhat later in the life of a product. Once many companies are able to build a high-technology product, the emphasis shifts from technology to cost advantages. We see the learning curves that result in a sharp decrease in the price of the product with each doubling of its cumulative production. This has been most spectacularly evident in the electronic industry, but we should remember that most older products went through that stage earlier.

The country that has especially excelled at cost-centered innovation is Japan. I should qualify that slightly by saying that in many cases Japanese engineers have improved products not only by cutting costs, but by enhancing their features--for example, by building superior quality into small automobiles, or shrinking the size as well as the cost of compact disk audio players. Another classic case is the microwave oven. An engineer for the New Japan Radio Company (Keisha Ogurua) totally redesigned the magnetron, the oven's microwave source, to simultaneously achieve technical superiority and lower cost. More recently, a team of engineers at JVC, a subsidiary of Matsushita, also achieved both technological innovation-the ability to put two hours of programming on a single tape-and substantial cost reduction by their development of the VHS format videocassette recorder. But it is the cost-centered innovation of Japan that is most impressive, and the prime example is the semiconductor industry.

An example that illustrates even more clearly the Japanese approach is the success of the company Fanuc in the area of numerical controls. A Fanuc engineer told one of my colleagues in GE part of that company's secret. The difference between U.S. and Japanese approaches, he said, is the difference between a snake and a wall. Both U.S. and Japanese companies start a project aiming at roughly the same cost and performance targets. But the U.S. engineers, in their obsession with performance, allow the characteristics of the product to snake upward in both performance and cost, until they have purchased performance improvements at a cost the customer is unwilling to pay.

The Japanese company, by contrast, begins the project by erecting a cost wall. The wall is rigidly enforced, and the engineer is challenged to build the best product he can without crossing the wall. The result often cannot match the peak performance of the American rival. But usually it surpasses the American product in the total package of performance, quality, and cost. One might label this the Eastern approach to innovation.

Product development need not be either cost or technology centered. In some cases, unfortunately, it is neither! This is essentially what happened in some major U.S. industries in the 1960s and 1970s. Believing that they had a captive market, and that the technology had matured and was not susceptible to much improvement, companies backed off from pushing either technology or costs. The automobile industry settled into a rut of unimaginative designs, steadily inflating costs, and declining quality standards. The fragmented machine tool industry failed to put much effort into either the development of new technology or cost-conscious manufacturing. The steel industry was slow to adopt innovations. The consumer electronics industry, with just a couple of exceptions, failed to either adopt a cost-centered approach to established products such as television, or a technology-centered approach to new ones such as video cassette recorders.

One can get another perspective on the "Eastern" and "Western" approaches to innovation by looking at an interesting intermediate case, that of the revolutionary medical diagnostic method CAT scanning. This was a European invention, developed by Godfrey Hounsfield in the laboratories of the British company EMI. It was further developed by the two companies that have subsequently become the world leaders in the area, General Electric and Siemens.

More recently, the Japanese entered the picture, and showed how the Eastern approach can be effective even on such a technologically sophisticated product. When Yokogowa of Japan formed a joint venture with GE in 1982, they knew little about designing medical systems. But, with access to GEs technology, within four years the joint venture had totally redesigned GE's high performance CAT scanner, producing a model "equal in performance to the original, but easier to use and at half the development costs" (R. Bowen, GE-Medical Systems, in IEEE Spectrum, May 87, p. 46).

That and many other similar instances had led some of us in the U.S. to ask the question- "why can't we achieve all of this ourselves? Why couldn't we simultaneously achieve the technology advance that our high-performance scanner represented and the cost-conscious approach to engineering the Japanese practice? The Japanese appear to be asking themselves a similar question - why not retain excellence at cost-centered innovation while picking up some of that Western method of technology pioneering? There is good evidence that the Japanese are attempting to do just that. If you look at the hottest pioneering area right now, high temperature superconductors, they are right out there at the forefront with a large numnber of first rate research efforts.

In the United States and in Europe, I see many efforts to approach to problem from the other side. Many of us are coming to recognize that competitiveness has two parts. First, innovation - the embodiment of new technology in novel features, improved performance, unique functions of new products and processes. This provides an outlet for the inventive spirit, the pioneering urge of the engineer. Second, execution - the low cost manufacture of high quality, reliable products that are attractive to customers. This provides maximum value and satisfaction to the user and customer.

In our effort to get both of these at once, many in the United States were initially tempted to take a technology centered approach. Recognizing that the problem was not in the R&D laboratory but on the manufacturing floor, our answer was initially a new high-tech approach to manufacturing. To reduce costs, replace people with robots. If the robots aren't able to do some of the things that the people could - like fitting together parts that are slightly out of tolerance - then develop smarter robots!.

We have learned in the past decade or so that this technology-centered approach is not sufficient. Certainly technology has provided many opportunities for cost reduction and faster development of innovative products. I might point, for example, to the use of computer aided design, analysis, and simulation.

But in many cases those tools have not lived up to their full potential. We have learned that the reasons for these disappointments are not in the tools themselves, but in the way we use them.

The problem has been labeled by one of my former colleagues in GE (S. Holmes, GE-MABG) "functional parochialism." We have cut up the process of product development into neat little compartments - R&D, marketing, design, manufacturing, quality control. The people in each compartment have sought to do that particular job as well as possible, drawing on the best available technology. But the outcome of the whole process has been considerably less optimized than each of the parts.

Performance has not been optimum because we have carried specialization too far. A certain degree of specialization is inevitable, and even desirable. After all, nobody can be an expert in every phase of innovation. In order to give inventors enough freedom to make truly original inventions, they have to have some degree of isolation from day-to-day problems.

But in the United States, and perhaps in Europe as well, many of us have overdone this specialization. We have taken it to the point where the R&D people, the marketing people, and the manufacturing engineers are scarcely able to communicate with each other. The marketing expert gives little thought to the potential of technology when he does his market surveys. The R&D team gives little thought to the marketplace, or the problems the manufacturing engineer will have in producing the new product. As a result we end up with innovations that are too costly to make... or with low cost manufacture of exotic features that nobody wants... or with the best designs of an obsolete technology.

The answer to these problems is to step back from that functional overspecialization, and put the center of the innovation process where it belongs. That center should be in the design process. That's where you decide what the product is going to be. And, in fact, if you look at each of those successful innovations I referred to earlier, whether it is the microwave oven, or the VCR or the VAX, in each case you find an outstanding designer and an outstanding, clearly articulated design concept at the center. But design centered innovation is more than just a matter of putting a gifted designer in charge. The other crucial part of the idea is integrating invention, marketing and manufacturing right at the beginning of the product cycle. You might think of these people as looking over the shoulders of the designer.

I suggest that this idea of design-centered innovation is the common denominator toward which both East and West are headed. It is being sought today not only by companies such as Yokogowa and Fanuc, but also by such well established Western companies as GE, Ford and Siemens. Ford's resurgence in recent years has been built around the design centered innovation of the Taurus and Sable automobiles, and equally great innovations in the organization of the development process that bring together marketing, design, and manufacturing people early in the process. Siemens became aware of the fact that its marketing, engineering, and manufacturing people were working in isolation, and created a new post of manager of design with the specific responsibility of overcoming that isolation.

The "looking over the shoulder" role takes the form of fully shared responsibility. At crucial stages in this design process - at the stage of approving design specifications, milestone design reviews, and at releasing the drawings, for example - each member of the team must sign off and accept full joint responsibility. This full responsibility puts the burden on each person to do his homework all the way through. He can no longer say, "I did my part all right, it was those other guys who messed up."

The basic idea then, is pretty simple. Put design at the center of the process... put the R&D, design, marketing, and manufacturing people together as a team at the start of the project... and insist on full shared responsibility. But the implications of this idea may be very large indeed. In the United States, for example, it's causing us to look more closely at the way we manage technology. Do our organizational setups and management styles make functional parochialism inevitable? It's also causing us to take a new look at the way we educate engineers. Are we educating them in the image of their professors, with too much emphasis on research, and not enough on design and manufacturing? Can we build into engineering education some of that interdisciplinary emphasis and design-centered attitude that will be needed in the industrial world? Is there a science of design that can be as useful in putting things together as the traditional sciences are in understanding their parts?

Those possible wider implications are still uncertain, and matters of the future. But let me sum up at the argument I have arrived at based on the present situation.

Innovation to date has tended to have two main forms. Technology centered innovation has been the preferred form for the West - for the U.S. and Europe. We have sought to achieve competitiveness through forefront innovative products or processes. Our rivals in Japan have tended more toward cost-centered innovation - using technology to develop intrinsically low cost designs, or lower cost productive methods for making established products.

As the world gets more competitive, both East and West are recognizing that neither approach is sufficient by itself. Both are moving toward the idea of design-centered innovation. In part, this is a natural evolution of the way the best innovations have always happened. Look behind most successful high-technology products and you see the hand of a strong and creative design team.

The new element today is to have all the important participants - marketing, engineering, and manufacturing - looking over that designer's shoulder from the beginning, making their concerns heard throughout the process, and sharing full responsibility at each key milestone.

Design centered innovation is not a cure all. It can degenerate into design-by-committee, where responsibility is diffused and initiative is stifled. It must be coupled with the strong leadership that has always enabled individuals to put their personal creative stamp on the product. Done correctly it can result in designs that are both innovative and amenable to low cost manufacturing... that offer the right balance between features, performance and cost... and that combine the best in advanced product technology with the best in cost-driven manufacturing technology.

PART 2

COGNITIVE AND SOCIAL ASPECTS OF THE DESIGN PROCESS

DESCRIPTIVE MODELS OF COGNITIVE ASPECTS OF THE ENGINEERING DESIGN PROCESS

Gary Perlman
Computer and Information Science
The Ohio State University
Columbus, OH 43210

Abstract

In this paper, I discuss some similarities and differences among a collection of research reports on NSF research on the engineering design process. I compare the methods of gathering information on which the researchers draw their conclusions. I discuss why I think hypertext is a promising basis for technical assistance in the design process and how more controlled experimental methodology may provide more insights into the design process.

Introduction and Overview

Dixon (1988) describes a descriptive cognitive theory of design as follows:

> An essential ingredient in a descriptive cognitive theory is an active intelligent designer.... A complete design system in a descriptive cognitive theory of design requires relating the meaningful operationally defined variables from each of the following:
> - The person or persons;
> - The problem;
> - The organizational environment;
> - The design environment, ..., and;
> - Time.

Some goals of a descriptive cognitive theory include providing a sound theoretical basis for prescriptions about design personnel (e.g., educational and knowledge backgrounds) and organizational and technical support (e.g., processes and tools).

In this paper, I will summarize parts of progress reports on Descriptive Cognitive Models of Design Processes, part of the NSF program in Design Theory and Methodology. My own orientation differs from the researchers' in that I am a software designer in practice and a cognitive psychologist by training. Most of the research described here is in mechanical engineering, although there are many similarities between mechanical design and software design. Design of any kind is a difficult creative process, and measuring and modeling the variables in Dixon's list will be correspondingly difficult. Any cognitive design system -- one that includes human input -- must acknowledge human cognitive capabilities by augmenting those capabilities and compensating for human limitations.

My strategy in this paper is to find some common themes in the research, both in the methods used to gather information on which conclusions are based, and in the conclusions. Then I will discuss some areas that may help the researchers in their future work.

Review of Work by NSF Grantees

For each of five research teams, denoted by the last names of the authors, I have listed points about how data were collected and analyzed, and some of their major conclusions. To fully understand the research done, it is necessary to read the reports because the following summaries are condensed.

Data Collection and Analysis

The information gathered by the researchers ranged from studying existing experimental research to conducting structured interviews or analysis of protocols. None of the research teams base their conclusions on controlled experimental data, which may be because they are primarily looking for new hypotheses, not trying to test them. Some of the data are collected from small samples, primarily because of the high cost of data collection and analysis (e.g., video protocol analysis can be extremely time consuming), but this has dictated trying to generalize from anecdotes.

Newsome Spillers Vosburgh

- survey of existing studies

Papalambros

- interdisciplinary teams in time-limited competition
- teams keep a timed log of approaches and decisions
- students evaluate software support tools

Peterson Hagel Nadler Chignell

- structured interviews of five expert designers
- focus on successful real-world projects
- designers chosen by peers in several engineering specialties
- gathered "retrospective timeline scenarios"
- time oriented analysis of people, tasks, information, events

Ullman Dietterich

- five subjects (three professional designers, two graduate students)
- two problems (both simple but realistic)
- protocol analysis fit to detailed task model (TEAM)

Waldron Waldron Owen

- observations of six experts (three company presidents, three professors), and some graduate students
- protocol analysis fit to systemic representational model

CONCLUSIONS BY RESEARCHERS

The following are some common conclusions by the researchers. Designers gather information from a variety of sources and draw heavily on past experience with problems with similar design requirements. Designers work through a series of stages. Papalambros' Designer tool passively guides students through these. Insights from other researchers (e.g., Ullman *et al*, Waldron *et al*) may help advance such a tool by providing more detailed information about stages. The high cost of protocol analysis might be reduced by a technological assistant (OCS, 1987). It is not clear how conscious designers are of these stages, and whether teaching them or guiding designers through them improve design. The final design specification (e.g., a drawing) does not contain information on the design process. Designers need access to different versions of the design to be able to determine why design decisions were (not) made.

Newsome Spillers Vosburgh

- must empirically verify the utility of design support technology

Papalambros

- sophisticated methods may not be practical for time-constrained simple problems
- feedback from designers can improve tools
- insights are gained from interdisciplinary design teams

Peterson Hagel Nadler Chignell

- few multi-purpose process aids available

- need better organizational communication
- need more support for conceptual design
- design is relatively unstructured
- difficult to text hypotheses using retrospective case studies

Ullman Dietterich

- TEAM model has great descriptive power
- detailed data representation required for intelligent CAD

Waldron Waldron Owen

- experts use functional terms, naive subjects use structural descriptions of problems
- expert designers work from designs in their repertoires and adapt them or choose others until constraints are satisfied
- systemic representation allows different views of design information (e.g., decisions) that is usually lost

Commentary

With the possible exception of Waldron *et al*, these are not cognitive models in the sense that they do not provide explanations, in cognitive terms, how or why designers work the way they do. If the goal is to understand how people design, then these studies are severely lacking because they provide few if any insights into cognitive processes. However, if the goal is to improve design by aiding the design process, then the most promising approach seems to be to build and refine design-support tools. This is the approach taken by Papalambros. For tools to support people, there should be a theoretical basis for tool development based on human cognitive capabilities and limitations. Because our knowledge of human cognition is limited, empirical evaluation must be used to validate the utility of support tools and to refine them. This suggests a need for collaborative efforts between engineers and psychologists, the sort of multidisciplinary approach taken by Papalambros.

Technical Issues

I will now suggest an area where technical support may aid design: *hypertext*. In general, technical aids to the cognitive design process must augment human capabilities while compensating for human weaknesses. The information demands on the human designer, in gathering relevant information, in comparing and discussing alternatives, and in tracking design decisions, are severe, even in small design problems. Hypertext, or more generally, hypermedia systems can provide a representational mechanism for all this information. (The ideas in hypertext systems correspond closely to those of object-oriented databases.) A good background source on hypertext can be found in Conklin (1987).

Briefly, hypertext systems can represent complex networks of information in which *nodes* of information can be connected by *labeled arcs* to other nodes. Nodes contain data, which might include, but are not limited to text, graphics, and attributes. In the context of design, nodes can represent requirements, constraints, justifications, physical parts, and versions of the same (Smolensky *et al*, 1987; Conklin & Begeman, 1977; Perlman & Moorhead, 1988). For example, in an assembly, one part may be constrained to be part of a larger composite part, and the design of the composite part may be motivated by certain requirements, but constrained by previous design decisions, perhaps in another assembly or in another part in the same assembly. Reasons for not choosing certain design alternatives can also be represented, and only presented when requested, thereby avoiding the problem of information overload. In representational power, hypertext subsumes systemic networks (see Waldron *et al*).

However, hypertext introduces many possible human interface problems which must be evaluated (see the arguments by Newsome *et al*). Critical to hypertext usability is an access and display system to selectively locate and display those aspects of the design network of interest to the designer. For an example of a system that supports engineering design, see GECI (1988), which allows designers to work at different levels of detail.

With a hypertext representation of design information, it may be possible to automate aspects of design and provide intelligent assistance. For example, the effects of changes in requirements or in the addition of constraints can be anticipated by finding all parts of a design that depend on it. More ambitiously, analogies between designs may be detected (by finding common sub-networks of representations) and suggested to designers.

Hypertext can represent the information in the conceptual design phase, which is not well supported by tools (Peterson *et al*), and may provide more flexibility than the fixed process outlines in Papalambros' experimental tool. A history of discussions among members of a design team can be recorded and included in the overall design representation (see Malone *et al*, 1987). Furthermore, hypertext allows simultaneous representation of multiple timelines (Peterson *et al*), and can provide a representation of mechanical design data for expert CAD, a design history tool (Ullman *et al*).

Methodological Issues

If the benefits of methods are untestable, then there is no proven basis for following so-called "good design practices." In contrast with Dixon, I think that with empirical evaluation of technological or methodological support for design, we should be looking for prescriptions, at least along the way to deeper understanding. We must qualify prescriptions with the caveat that our understanding is incomplete.

I am motivated to answer the following related questions. *Given two adequate designs, how can differences in designs be explained? Given two designs, one with advantages over another, can the differences be explained by or predicted from the backgrounds of the designers? the tools used? Can this information be used to direct design education?*

The studies in the NSF reports do not try to control for many variables in the design process. Referring to Dixon's breakdown in the introduction, both the *persons* changed and the *problems* changed, so any design protocol or interview had many variables changing at once. This complexity makes drawing conclusions difficult or impossible. What might be more efficient is to introduce some control while still maintaining a natural environment.

	Problem one	Problem two	Domain two
Designer 1 (expert)	11	12	1*
Designer 2 (novice)	21	22	2*

Figure 1. Designer and Problem Variables.

Two variations on paradigms in the NSF studies come to mind. Each allows new designs to be viewed as variations on others. Both the attributes of the designer and the problem can be systematically manipulated. For example, in Figure 1 differences between solution 11 and 21 can be attributed to differences between designers (e.g., perhaps level of expertise or experience with designing a certain type of artifact). Across rows, similar designs by the same designer may be compared both within a domain (11 vs 12) or across domains (11 vs 1*).

While **holding person variables constant**, we can vary the problem, and see how the same designer attacks different problems. Similar problems should produce highly comparable designs, while dissimilar problems, possibly moving to a different design domain, might show transfer of higher level design strategies.

While **holding problem variables constant**, we can vary the persons, and see how different designers attack the same problem. The persons may differ in the amount of their design experience or domain expertise. They may also differ in their familiarity with the organizational and design environments. For example, Papalambros' study showed different teams working on same problem in competition, however, it offered no conclusions based on the successes/failures of teams based on team differences or use of tools. My own experience (Perlman, 1988) with having different student teams design software with similar design goals provided me with some insights into how comparable designs can differ. Students had expertise with different operating systems (design environments) and had different motivations for developing skills in areas that would be useful for them with their employers (organizational environment). Beyond these major differences, many design differences could be attributed to adapting to the constraints imposed by the software tools most readily available.

Acknowledgements

I thank Beth Adelson and Sandra Newsome for their suggestions. This paper was written while I was a visiting scientist at the MIT Sloan School of Management.

References

References from the Design Theory Workshop:

Dixon, J. R. "**On research methodology towards a scientific theory of engineering design.**" To appear in Artificial intelligence for engineering design, analysis, and manufacturing (AI EDAM), Vol. 1, No. 3, June 1988.

Newsome, S. L., Spillers, W. R., & Vosburgh, A. M. "**Quantitative evaluations of human factors savings in CAD workstations.**" Proceedings of the Third International Conference on Civil and Structural Engineering Computing (Civil-Comp '87). 105-108.

Papalambros, P., Y. "**The design laboratory: Interdisciplinary research and education.**"

Peterson, J. G., Hagel, M., Nadler, G., & Chignell, M. H. "**Aids to the design process based on techniques used by expert designers.**"

Ullman, D. G. & Dietterich, T. G. "**Progress in understanding the process of mechanical design.**"

Waldron, M. B., Waldron, K. J, & Owen, D. H. "**Use of systemic theory to represent the conceptual mechanical design process.**"

Other References:

Conklin, J. (1987) "**Hypertext: An introduction and survey.**" IEEE Computer, 20, 17-41.

Conklin, J. & Begeman, M. L. (1987) "**gIBIS: A hypertext tool for team design deliberation.**" Proceedings of Hypertext '87, ACM, New York.

GECI - Groupe d'Etudes et Conseils en Ingenierie (1988) "**Hyperdoc -More intelligent documentation.**" System shown at RIAO '88, Boston. Address: Tour Winterthur, Cedex 19 92085 Paris La Defense, FRANCE.

Malone, T. W., Grant, K. R., Turbak, F. A., Brobst, S. A., & Cohen, M. D. (1987) "**Intelligent information-sharing systems.**" Communications of the ACM, 30:5, 390-402.

OCS (1987) (Observational Coding System Tools). Exhibit at the 1987 meeting of the Human Factors Society, New York, NY. *This system allows developers to code and analyze video protocols. As you watch a video, you see various events, like errors, grunts, drawing, etc. OCS Tools lets you sit at a PC and code these events by typing key codes you define. The codes are stored and can be used to search through the video to specified event types. A data analysis program produces summaries of the event codes.* Contact: Triangle Research Collaborative, Box 12167, RTP, NC 27709 (919) 549-9093

Perlman, G. (1988) "**User interface development.**" Software Engineering Institute, Carnegie-Mellon University, Pittsburgh.

Perlman, G. & Moorhead, A. H. (1988) "**Applying hypertext methods for the effective utilization of standards.**" Proceedings of COMPSTAN '88, IEEE, New York. 55-59.

Smolensky, P., Bell, B. Fox, B., & Lewis, C. (1987) "**Constraint-based hypertext for argumentation.**" Proceedings of Hypertext '87, ACM, New York.

META-DESIGN: REFLECTIONS ON A GRADUATE COURSE IN DESIGN THEORY AND METHODOLOGY

Alice M. Agogino, P.E., Principal Investigator, Associate Professor
Jonathan Cagan and Michael J. Molezzi, Graduate Research Assistants
Mechanical Engineering Department , University of California, Berkeley, CA 94720

Abstract

Motivated by the directive of the NSF Division of the same name, a graduate course in Design Theory and Methodology was offered in the Fall 1987 Semester at the University of California at Berkeley. Participating students took an active role in the design of the class itself, including allocating resources and defining an appropriate grading scheme. This paper summarizes the conclusions of a meta-design team composed of the principal investigator, three students in the class and an outside student involved in the manufacture and redesign of the major class project. The team not only studied the design process followed by students involved in the class projects, but also the design of the class itself. Results from the introspective and retrospective analysis of the meta-design team are compared to the results of other researchers and provide hypotheses for formal testing in future experiments. Recommendations on computational aids to improve the design process are proposed.

Introduction

Hierarchical levels of design and cognitive processes were recurring themes in the Design Theory and Methodology class described in this paper. At the top level was the design of the class itself and the research goals behind it [2,13,15,24]. The course content and projects ranged from creative and conceptual design [1,7,16], to functional design [26-27], to design for manufacturability and detailing [6,9,33] and finally, failure analysis and risk assessment of the final product [18,25] (references corresponding to each topic were used as reading material for the course.) A schedule of the topics covered in the class are included in Appendix A. A follow-up student project was initiated during the Spring 1988 Semester in order to redesign and manufacture the final product and study the complete design cycle.

Although not always apparent to the students, the meta-design theme was used to force introspection by the students and faculty into the design process, at all levels. We were both researchers and participants in the design process. All design projects had "clients" that could provide feedback and evaluate the final product. Feedback from the meta-design team observing the design process and class experiment was intended to provide input to the principle investigator in identifying promising research directions for developing useful design methods, theories, and computational architectures for *engineering* design.

Graduate status and an undergraduate course in design were the only prerequisites and reading assignments were selected from references [1-36]. The course requirements consisted of various individual exercises and two class projects. Only the class projects will be described in this paper: (1) logo design and (2) design of a computer workstation for wheelchair users.

Logo Design

In order to stimulate creative thought and allow the students to practice suggested approaches to conceptual design described in the literature, the first class project was the design of a new logo for the Berkeley Expert Systems Technology Laboratory and the associated Intelligent Systems Research Group [2]. The class was informed that a desirable logo should communicate at least two of the following concepts: (1) mechanical systems, (2) AI - artificial intelligence and (3) UC Berkeley. Elegance and simplicity were preferred to detailed "noisy" designs. The logo was required to be of a form that would be usable on stationary, journal publications and promotional literature. The principal investigator was the final judge and the students could ask her opinion of any intermediate concepts.

The anecdotal experience involved in this humble design project illuminates the benefits of antagonistic duality and synergism in fostering creativity in design teams. Because this was the first class project and most of the students were first-year graduate students, very few of the students knew each other very well. The principal investigator did not force collaboration but subtly tried to encourage it. One woman student concentrated on artistically drawing "pretty" letters associated with the Lab and UC Berkeley. Her logo was criticized by her male desk neighbor as not having enough "content". His logo showed a human head with a large gear instead of a brain. He argued that his logo captured the concepts of AI and mechanical systems. She rebuffed by exclaiming "gear head!" In spite of their different perspectives, the two students decided reluctantly to collaborate. The result is the logo that was most liked by the members of the Lab and forms the basis of the logo that was finally adopted (Fig. 1). The initial antagonism eventually was redirected and led to creative synergy in the design process.

BERKELEY EXPERT SYSTEMS
TECHNOLOGY LABORATORY
Intelligent Systems Research Group

Figure 1. UC Berkeley Logo Designed by Class

Computer Workstation for Wheelchair Users

The major design project for the class was the design of a computer workstation for wheelchair disabled users. Although the table must be versatile enough to accommodate a variety of users with a range of disabilities, one quadriplegic student was identified as the primary user of the final product (this undergraduate student was concurrently performing research in the Lab). The project incorporated the creative, innovative, and routine levels of design. The final designs of the class were then further synthesized and redesigned by a graduate student not in the class (co-author, M. Molezzi) and then manufactured as a finished product. The rest of this section discusses the design project and the next section summarizes the results of the synthesis and redesign toward a finished product.

The class was divided into three groups. Most of the class joined one of the two design teams while three class members (including co-author J. Cagan) participated in a class meta-design team which observed the design process in order to make descriptive generalizations and recommend further studies and normative approaches to automate and enhance the design process. The two design teams were allowed to pick their own identifying titles so as to avoid the subconscious biases associated with names like: "one and two" or "A and B". They picked "The Breakfast Club" (they often met early in the morning before class) and "The Dirty Dozen" (based on the number of team members). We will refer to them as BC and DD for short.

During the course of the project, the students "went meta" and redesigned the grading systems of the course to encourage both competition and cooperation. The resulting scheme included three grades for the project, equal weights on the (1) class, (2) team and (3) individual contribution grades. One BC student felt the need to redefine the design problem in a more empathetic fashion in order to motivate the project:

> *Imagine that you are the same person you are right now except that you do not have full use of the muscles in your arms and legs. You could be a student, an engineer, a teacher or in a number of other active professions. Due to a spinal cord injury, cerebral palsy, birth defects, or other medical conditions, thousands of students, engineers (or potential engineers), paraplegic and quadriplegic students are in this situation.*
>
> *Imagine that you are enrolled in ME290P: Design Theory and Methodology and you have to use the UNIX™ computer system for the first assignment. You've paid your computer fees, the computer room is wheelchair accessible but when you get to the computer terminal you discover that the terminal workspace is just not suitable or comfortable to work from a wheelchair. The table is too low, the keyboard is hard to access and the screen is too far away and difficult to adjust -- the overall situation is frustrating. You can only manage by letting someone else work the terminals for you and you look over his/her shoulder. You are dependent on the goodwill of a fellow student and you have lost much of the cognitive benefits of interactive computer graphics and design. You wish someone, maybe even yourself, would design and build a computer workspace more suited to disabled students in wheelchairs.*

The two design teams differed in their management styles and approaches. The BC team refused to identify a leader (instead they elected a "communication coordinator"), practiced consensus decision making, and pursued only one major design concept in depth. They tended to agree with each other and thus competing ideas were not pursued. They were internally motivated, sincerely hoped to build a useful product and were not concerned with competition from the other team. The DD team, on the other hand, had a strong team leader who practiced a hierarchical decision making style. He was highly respected and allowed team members to voice disagreement, which they frequently did to the disutility of the project (retrospective and introspective conclusions by the DD team at the end of the semester). The factionalism within the larger DD team was further compounded by its division into two subgroups, each pursing alternate competing design concepts in parallel. The DD team members were highly competitive and were keenly interested in observing the other team. An academic version of industrial espionage was reported. In spite of the grading scheme the students designed to encourage some cooperation between teams (to improve the overall class grade) no cross-team cooperation was observed.

The class meta-design team concentrated on three aspects of design: the design process, optimization in design, and the use of language and vocabulary in design. Two concerns for the team were to observe the design process without affecting it as is desired in protocol analysis [34] and ethnographic research, and to have the designers retrospectively report the design process via case studies [36]. Although the meta-design team was present during the group sessions, it did not participate or intervene at all. Thus, it was difficult to draw out the complete chain of thought processes of the designers, and data in the form of conversation, sketches, and notes, were difficult to analyze.

Design Process

Although initial brainstorming was performed by both groups, each group quickly settled on one or two design concepts to pursue, greatly limiting the possibilities of the final design. These final concepts were influenced by the individual goals of the two teams. The BC team desired a working product by the end of the semester; this led to decisions to purchase and modify as necessary any off-the-shelf components. The DD team, on the other hand, wanted a least-cost original design by the end of the semester; designing each major component from scratch from inexpensive parts. Each team developed its own series of design constraints. Some constraints were based on the client (i.e., professor and target disabled user) and placed limitations on cost, design time, and location of final product. Each team decided on the same number of system degrees of freedom independently (eg., screen tilt, height adjustment, etc.). Sketches of some of the designs of the two teams are provided in Appendix B.

Both teams studied design theories [15-16,19-22,28,31], but neither followed an organized identifiable, *normative* design procedure. We offer two possible interpretations: (1) good design requires unstructured creative activity and opportunistic decision making or (2) the teams could have been more efficient and could have derived a better design had they followed a good design process. The meta-design team suggested an experiment consisting of two teams; one trained in design theories and consciously utilizing those theories whenever possible, and the other team naive to published design theories and allowed to intuitively define their approach to design. The quality and efficiency of the final design product and process of the two teams could then be analyzed to give evidence for or against a given design theory. Although an interesting suggestion, the practical difficulties and ethical concerns are apparent.

The general design observations of the meta-design team are listed in Table 1. The observation that major design decisions are made very early in the design process were also concluded in the study by Ullman, Stauffer and Dietterich [34] as are observations (3), (5) and (8). Because the Ullman et al. study used protocol analysis on individual designers, they provide no conclusions concerning design team behavior and thus can not be used for comparison to the other meta-team observations. The retrospective study of the design team described in Waldron and Waldron [36], however, report results similar to observations (2), (4), (5) and (8). Although not derived from controlled test conditions, the observations of the meta-design team provide a resource of hypotheses for formal testing in future experiments [13].

Table 1. Meta-Design Team Observations

(1) Major design decisions are made very early in the design process.
(2) Only one or two concepts are pursued at a time.
(3) Decisions made by a team are sometimes forgotten and rehashed. (The members of the team that did this claimed that this was a process failure due to improper recording of decisions.)
(4) Within the groups, design work is divided into sub-tasks.
(5) Designers move alternatively from detailed design to global design.
(6) Some important decisions were made by the groups on impulse with no apparent justification.
(7) Team organization and global goals affect the fundamental design.
(8) Effective use of communication tools such as blackboards, CAD systems, prototypes, and working environment influence both the global and detailed design.
(9) Team personalities influence the design process. Teams that try to understand one another and support each others views, tend to work harmoniously together in order to create a successful product. When team members clash, the group as a whole suffers, hurting the efficiency of the design process and possibly the quality of the final product.

Design Optimization

The meta-design team considered two aspects of optimization in design; one is the optimization of the actual design product, the other is how to optimize the design process. Although each team studied formal optimization techniques in the course [3-4], neither team was concerned about mass-production and thus neither team performed any type of detailed formal optimization. However, both teams performed a kind of qualitative optimization [4], beginning with conceptual designs based on qualitative goals and few design constraints and moving to a final, highly constrained detailed product. In general, the initial brainstorming phase has few constraints and the resulting concepts define and thus restrict the design space from which the final design will be derived. Brainstorming is the global search phase and is often much too brief; if design concepts are not created, the appropriate vocabulary may not exist for other designs to be derived. As new constraints are added to the design space and numerous designs are regarded as being infeasible or suboptimal, a local search takes place over one or two designs to home in on a final product. This transition between global and local search is a dynamic one and numerous qualitative aspects of the design are considered.

By observing the design teams, the meta-team proposed an optimized group structure for design teams of more than four or five members. The team should have a team leader to organize and control the design process. The first meeting should be a social, bonding experience to encourage positive group dynamics. However, after the initial meeting, distractions should be removed and communication tools should be utilized to concentrate on the appropriate design activity.

Design Language

Language is an important part of human action, interaction, and intelligence. It is imperative to have a design language which is robust, efficient and yet not restrictive [11,12,32]. A well-structured communication context enables designers to avoid needless confusion and design iterations, and encourages the expression and creation of new designs. On the other hand, a restrictive or biased design language can, in fact, limit the design possibilities. For example, the principal investigator made the mistake of using the expression "computer table" to describe the design project early on in the course. Although she corrected herself and used words like "orientation device" and "computer workstation" so as to allow designs that did not resemble "tables", the students stuck with the table form rather than concentrating on the desired function (relative positioning of the user and computer) and creating alternative forms. (During the initial brainstorming alternate forms were suggested but quickly discarded, such as magnetically floating computers and hanging or cantilevered workstations).

The meta-design team observed the naive design language used among the design teams. The naive language of the BC group appeared to be adequate for most of the design project, due to the cultural homogeneity of the group and its relatively small size. The larger DD team required more formal language structures than the smaller BC team to ensure meaningful communication. This was due to its larger size and the multicultural nature of the group. At times verbal communication worked best; at other times a visual sketch was most effective. Imprecision in the vocabulary led to confusion in communicating design concepts. These observations suggest important considerations for design automation. Intelligent CAD system must provide a sufficient and consistent vocabulary to communicate with multiple users. *Language biasing* must be minimized [29].

Design Automation

The meta-design team synthesized their observations and made a series of recommendations for improving current CAD technology. Some of these ideas would be trivial to automate; others will require further research and development.

- Record-keeping: Ideas and decisions should be systematically "recorded" to prevent a design team from reinventing their own wheel (recording design histories and design intent). In addition, this would be useful in backtracking or redesign, easing the retreat from a dead-end idea to the nearest feasible design path. Hypertext concepts may be useful here.

- Checklist: CAD can include reminders to define goals and constraints completely, and to consider all the necessary aspects of the design. This could help ensure legally defensible consideration of safe design choices.

- Design Languages: Much of this software will be based on a language designed (or selected) to include the relevant design concepts and symbols. Current research in object-oriented data structures and feature-based design works towards these goals [5,12].

- Creativity: Brainstorming and innovation in design can be stimulated by flashing ideas [1] (random or linked) onscreen. Software could help the designer play with existing designs by modifying them (at random or optimally-directed [8]).

- Education: Design situations based on case studies and common applications can help speed novices along the learning curve [17,19-20]. Development of additional and more recent case studies are recommeneded. Knowledge-based approaches are appropriate here.

Research on Design Synthesis and Redesign

Compartmentalized design is a common occurrence in industrial product design. One or more groups are responsible for a specific phase of product design, then all results and information are passed on to the next group, and so on. In the case of the computer workstation design as discussed in the previous section, two separate and complete designs (the BC team produced a working prototype and the DD team developed a CAD simulation and detailed drawings) were passed on to a single student responsible for final product development and manufacture. Access was given to all methods and sources used by the original design groups as documented in each student's design notebooks. The purpose of this portion of the project was to synthesize a final product design from the original design proposals, allowing additional input from the new designer, and documenting the methods used for this type of redesign. In this case, innovative and routine levels of design dominate. The synthesis and redesign process will be the focus of the following sections.

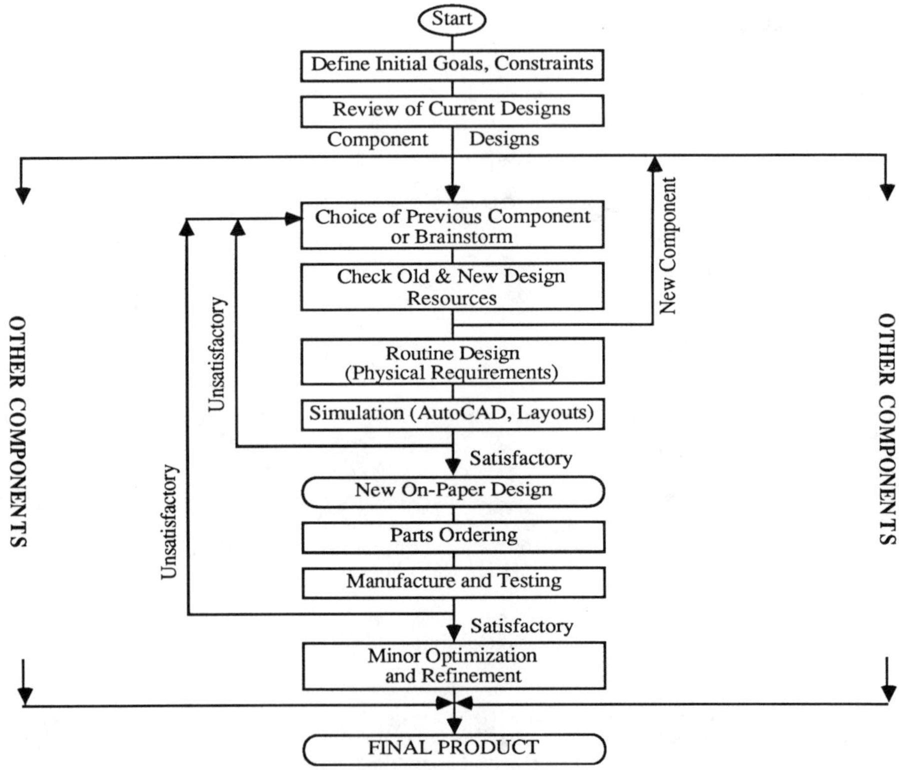

Figure 2. Design Synthesis and Redesign Flowchart

Methodology

A simplified flowchart outline of the methodology used is shown in Fig. 2. The goals and constraints were first defined, then a review of previous and updated information performed. The redesigner (co-author M. Molezzi) then decided that the overall design efficiency would be best served by breaking up the design into separate hardware components (terminal stand, keyboard platform, base table design, etc.). This allowed for work scheduling in parallel, minimized the effects of delays in any one component, and balanced the level of detailing between components at each stage of the global design.

Each component design began with a choice of either a previous design or a new innovative idea. This choice was then analyzed through a review of the design resources (the reasons and knowledge upon which design choices were made) listed by the previous designers and any new resources discovered by the redesigner. This review, although repetitive, was considered necessary to assure all relevant data had been gathered. This step involved a review of the design notebooks, discussions with possible users of the final product, checks on the availability and prices of parts, and so on. At this point, it was found that a review of design resources could present additional design requirements and options, such as the need for a new or additional component. This created a new component design path as indicated by a return loop in Fig. 2. For example, a discussion with a quadriplegic computer user for the purpose of reviewing the original designs revealed that the use of a simple motorized turntable to hold books and reference materials would be more useful than providing stationary desktop space, an idea not discussed in either of the original designs.

Evaluation of the design functionality was then performed through the use of routine numerical design and simulation. Depending on the results of each evaluation, the component was either accepted as the final paper design, or rejected, resulting in an iterative loop back to a new component choice.

The process then proceeded to the manufacturing stage. Again, the components were each evaluated while being built, resulting in the next process branch. In an ideal design process, the paper design and simulation is final, eliminating the need for prototype testing. In the realistic case, prototype testing reveals the need for small adjustments or parameter optimization to improve performance. In the worst case, testing of the prototype can sometimes reveal infeasibilities or flaws not apparent in the paper design. This requires an iterative loop back to the component design selection stage. Fortunately, this cost- and time-consuming scenario did not occur in the design discussed here.

Redesign Conclusions

A large portion of the redesign was actually an overlap of tasks previously performed by other design groups. Although we concluded that most of the overlap was necessary, the detailed physical design data from the previous proposals were not. The redesign primarily made use of conceptual component designs and the design resources provided by the previous groups. The detailed physical designs completed by the previous groups were either not used (unselected components) or completely redone (selected and modified components). A sketch of the final design is provided in Appendix C (based predominantly on the BC design concept).

The documentation of previous design intent, resources and references was found to be vital in the derivation of the final design. Access to this information allowed the new designer to get a quick understanding of the design requirements and, as stated above, allowed for the discovery of many new possibilities. This reinforces the notion that "design intent "and "design histories" will need to be captured for effective intelligent CAD systems.

Conclusions

In this paper we describe a graduate design course as an educational vehicle and experimental laboratory to complement a research program in design theory and methodology. Although not always apparent to the students, the meta-design theme was used to force introspection by the students and faculty into the design process, at all levels. We were both researchers and participants in the design process. In addition to the obvious goal of enhancing the design curriculum, the class experiment provided valuable input to the principle investigator in identifying promising research directions for developing useful design methods, theories, and computational architectures for *engineering* design. The observations from the meta-design team listed in Table 1 provide hypotheses concerning the design process for use in formal testing in future experiments. The need for rapid prototyping tools and intelligent CAD software for both teaching design and performing design was painfully obvious, as has been documented by previous design researchers and educators [10,24]. Recommendations for intelligent CAD design tools and research needs included use of hypertext for record keeping of design histories, formal checklists for life cycle design considerations, design languages, creativity tools, and knowledge-based approaches.

Guest lecturers from the UCB School of Architecture [28], the UCB Department of Computer Sciences [30], the UCB Department of Industrial Engineering and Operations Research [14], the San Francisco State University Engineering Design Center, and industry [23] enhanced the interdisciplinary aspects of the course. However, the principle investigator recognizes that this is insufficient and an Engineering College-wide course will be pursued for future offerings, with increased industrial participation.

Acknowledgements

The authors would like to thank David Ernest, Muriel Strand, Keith Yokum, the ME290P (Design Theory and Methodology) students and participating faculty for their contributions to this study. The authors gratefully acknowledge NSF for partial support of this course through Grant #DMC-8451622, supplemental funds for disabled researchers.

References

1. Adams, J.L. (1987), *Conceptual Blockbusting: Guide to Better Ideas*, The Portable Stanford Series, Stanford Alumni Association, Stanford, CA.
2. Agogino, A.M. (ed.) (1988), "Research Summaries of the Berkeley Expert Systems Technology Laboratory," Spring 1988, 5136 Etcheverry Hall, University of California, Berkeley, CA 94720.
3. Agogino, A.M., Almgren, A.S. (1987), "Techniques for Integrating Qualitative Reasoning and Symbolic Computation in Engineering Optimization", *Engineering Optimization* Vol. 12 (2), Sept./Oct 1987., pp. 117-135.
4. Agogino, A.M. (1987), "AI in Computer-Aided Design: Qualitative Reasoning and Symbolic Computation," *Proceedings, NSF Workshop on Design Process* (M.B. Waldron, ed.), Berkeley, CA, February 8-10, pp. 189-214.
5. Agogino, A.M., Guha, R. (1987), "Object-oriented Data Structures for Reasoning about Functionality, Manufacturability, and Diagnosability of Mechanical Systems", Working Paper 87-0901-P, 5136 Etcheverry Hall, Mechanical Engineering Department, University of California, Berkeley, CA 94720. (Prepared for the 1988 NSF Workshop on Features in Design and Manufacturing, University of California at Los Angeles, Feb. 26-28, 1988; based on an invited presentation at the 1987 ASME Design Automation Conference, Boston, MA, August 29, 1987.)
6. Agogino, A.M., Nour-Omid, O, Imaino, W., Wang, S.S. (1987), "Decision-Analytic Methodology for Cost-Benefit Evaluation of Diagnostic Testers," Working Paper 87-0804-0, Rev. 1, August 1987, 5136 Etcheverry Hall, Mechanical Engineering Department, University of California, Berkeley, CA 94720. (Accepted for publication, with revision, in *Transactions of the IIE*).
7. de Bono, E. (1970), *Lateral Thinking: Creativity Step by Step*, Harper & Row, Publishers, New

York, NY.
8. Cagan, J., Agogino (1988), A.M., "Innovative Design of Mechanical Structures from First Principles," in press: *AI EDAM*, Vol 1 (3).
9. Chryssolouris, G., Wright, K., Wesley, C. (1986), "Manufacturing Systems Control with Aid of Value/Utility Functions," ASME Preprint, Winter Annual Meeting, Anaheim, CA, Dec. 7-12, 1986.
10. Conway, L. (1981), "The MPC Adventures: Experiences with the Generation of VLSI Design and Implementation Methodologies," *Second Caltech Conference on VLSI*, Pasadena, CA.
11. Cutkosky, M.R., Tenenbaum, J.M. (1987), "Report of the Working Group on Design Languages," *Proceedings, NSF Workshop on Design Process* (M.B. Waldron, ed.), Berkeley, CA, February 8-10, pp. 313-324.
12. Dixon, J.R., Cunningham, J.J., Simmons, M.K. (1988), "Research in Designing with Features," *IFIP WG 5.2 Workshop on Intelligent CAD*, Cambridge, MA, Oct. 1987, North-Holland Publishers.
13. Dixon, J.R. (1988), "On Research Methodology Towards a Scientific Theory of Engineering Design", (in print) *AI EDAM*, V. 1 (3).
14. Dreyfus, H.L and S.E. Dreyfus (1986), *Mind over Machine*, The Free Press, Macmillan, Inc., N.Y.
15. Felder, R.M. (1987), "On Creating Creative Engineers," *Engineering Education*, Jan. 1987, pp. 222-227.
16. French, M.J., (1983) *Conceptual Design for Engineers*, Second Edition, Springer-Verlag, NY.
17. Fuchs, H.O. and R.F. Steidel, Jr. (eds.), *10 Cases in Engineering Design*, Longman Group, Ltd., London, 1973.
18. Gibson-Harris, S. (1986), "Expecting the Unexpected: Hazard Recognition in Machine Design," ASME Preprint, Winter Annual Meeting, Anaheim, CA, Dec. 7-12, 1986.
19. IEEE (1987), "A Special Report: The Global Automobile," *Spectrum*, Vol. 24, No. 5, pp. 31-78.
20. IEEE (1987), "A Special Report: On Good Design," *Spectrum*, Vol. 24, No. 10, pp. 25-72.
21. Koen, B.V. (1985), *Definition of the Engineering Method*, ASEE Publications, Suite 200, 11 Dupont Circle, Washington, D.C. 20036.
22. Koen, B.V. (1987), "Generalization of the Engineering Method to the Universal Method," *Engineering Education*, Jan. 1987, pp. 214-221.
23. Macfarlane, J., Donath, M. (1988), "The Automated Symbolic Derivation of State Equations for Dynamic Systems," to appear in *Proceedings of the IEEE Conference on Artificial Intelligence Applications*, San Diego, CA, March 1988.
24. Papalambros, P.Y. (1987), "Design Research in an Experimental Environment," Presented at the International Congress on Planning and Design Theory Conference, ASME, New York. (Copies of the paper can be obtained from the author at The University of Michigan, Department of Mechanical Engineering and Applied Mechanics, Ann Arbor, Michigan 48109, USA.)
25. Petroski, H. (1985), *To Engineer is Human: The Role of Failure in Successful Design*, St. Martin's Press, New York, NY.
26. Rinderle, J.R., Suh, N.P. (1982), "Measures of Functional Coupling in Design," *ASME Transactions*, Journal of Engineering for Industry, Vol. 100, No. 2, May 1978.
27. Rinderle, J.R. (1987), "Function and Form Relationships: A Basis for Preliminary Design," *Proceedings, NSF Workshop on Design Process* (M.B. Waldron, ed.), Berkeley, CA, February 8-10, pp. 295-310.
28. Rittel, H.W.J., "The Reasoning of Designers," School of Architecture, University of California, Berkeley, CA 94720.
29. Russell, S.J. (1986) "Preliminary Steps Toward the Automation of Induction", *Proceedings of AAAI-86*, Philadelphia, PA August 11-15, pp. 447-484.
30. Sequin, C.H. (1983), "Managing VLSI Complexity: An Outlook, *"Proceedings of the IEEE*, Vol. 71, No. 1, Jan. 1983, pp. 149-166.
31. Shigley, J.E. (1977), *Mechanical Engineering Design*, McGraw-Hill, 3rd Edition.
32. Stiny, G. (1981), "An Introduction to Shape and Shape Grammars," *Environment and Planning*, B, No. 7, pp. 343-351.
33. Suh, N.P., Bell, A.C., Gossard, D.C. (1978), "On an Axiomatic Approach to Manufacturing Systems," *ASME Transactions*, Journal of Engineering for Industry, Vol. 100, No. 2, May 1978.
34. Ullman, D.G., Stauffer, L.A. Dietterich, T.G. (1987), "Preliminary Results of an Experimental Study of the Mechanical Design Process", *Proceedings, NSF Workshop on Design Process* (M.B. Waldron, ed.), Berkeley, CA, February 8-10, pp. 145-188.
35. Waldron, M., ed. (1987) *Proceedings, NSF Workshop on Design Process* (M.B. Waldron, ed.), Berkeley, CA, February 8-10, 1987.
36. Waldron, K.J., Waldron, M.B. (1987) "A Retrospective Study of a Complex Mechanical System Design", *Proceedings, NSF Workshop on Design Process* (M.B. Waldron, ed.), Berkeley, CA, February 8-10, 1987, pp. 109-144.

Appendix A. Class Schedule

2 weeks	Introduction, theories of design
2 weeks	Conceptual design, creativity, problem formulation and analysis
5 weeks	Quantitative and systematic methods (qualitative and quantitative optimization), multiple objective methods, graph-theoretic methods, design abstractions)
3 weeks	Design decomposition and synthesis (design for manufacturability, marketability, reliability, etc., case studies, systems modeling)
3 weeks	Design automation (CAD, graphics, artificial intelligence and knowledge-based systems)

Appendix B. Sketches of Class Designs; (a) BC and (b) DD Scissor Table Design

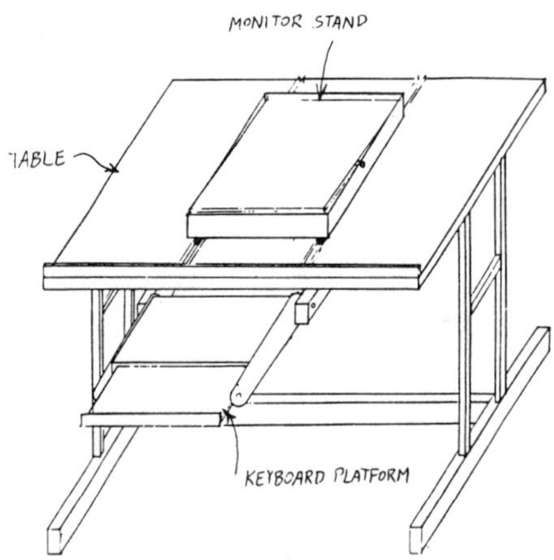

(a)

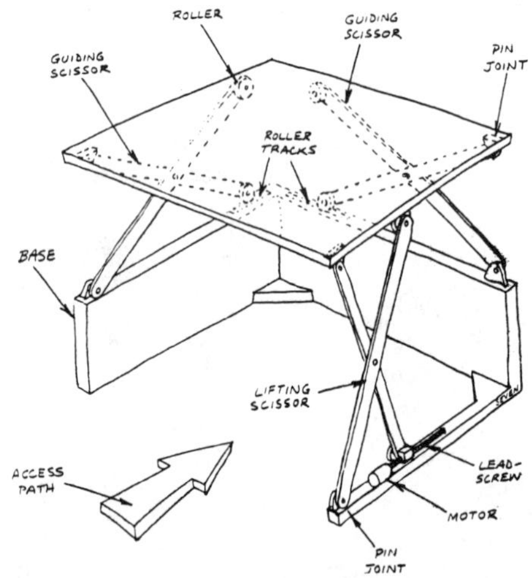

(b)

Appendix C. Sketch of Final Design

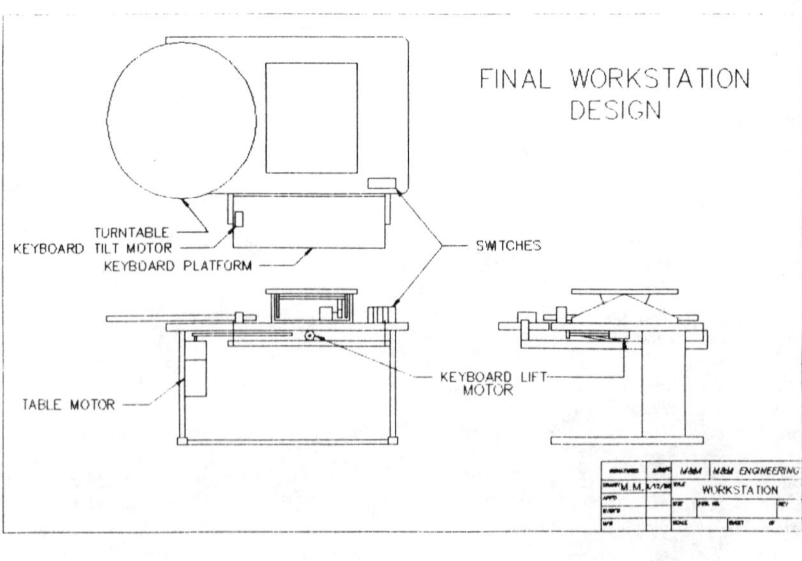

Design Theory and Methods -- An Interdisciplinary Approach

Donald A. Schon and Louis L. Bucciarelli
Massachusetts Institute of Technology
Cambridge, Massachusetts 02139

Designing is different from problem solving, though to design one must solve problems. Designs are not always works of art, though designing demands the exercise of creativity. And though the artifacts of engineers and of architects stand alone, ideally expressing something akin to an organic unity, most designs are the productions of groups of individuals with different interests and responsibilities. Our premise in researching design theory and methods is that the nature of designing requires an interdisciplinary approach.

Members of the Design Theory and Methods Group at MIT[1] are all concerned, in one way or another, with the nature of design inquiry and designers' knowledge, but they are diverse in the disciplines they represent, the methods they use and the substantive design domains they choose to examine.

They choose examples from architecture, environmental design, planning, and engineering. Their projects range over the development of computer models of designing, the evaluation of computer environments for designing, reconstruction of the inner logic of particular architectural designs, the use of games for

---

[1] Founding members of the Design Theory and Methods Groups, and contributors to the earlier Design Studies publication were Stanford Anderson, Aaron Fleisher, Mark Gross, John Habraken, Gary Hack, William Porter, Patrick Purcell, Edward Robbins, and Donald Schon from MIT's School of Architecture and Planning, and Louis Bucciarelli from MIT's School of Engineering. Other faculty members who later joined the group: James Anderson, Ranco Bon, Stephen Ervin, Joseph Ferreira from Architecture and Planning, Jeanne Bamberger from the Music Section, John Whiteman from Harvard's Graduate School of Design, and Gabriella Goldschmidt, a visiting scholar from the Technion, Israel, who has participated in the group's research since 1986.

experimentation in design theory and methodology, and participant/observation study of design activity within an engineering firm. Ideas of interest to the members of the group include the possible existence of a "language" of designing, design as search within a field of constraints, functions of metaphor in design discourse, and the place of uncertainty and ambiguity in the "process world" of design organizations.

In the years that have elapsed since its founding, the Design Theory and Methods Group has continued its research, aided by the support from the National Science Foundation and the National Endowment for the Arts. As a result of the continued exchange of ideas (participants have an opportunity to discuss their work weekly at a Design Research Seminar), connections among members' research have become clearer and some common language has developed. At the same time, perspectives and methods of study have remained diverse.

What follows is a report on the group's progress -- a brief summary of the current status of each of six research ventures and an elaboration of some shared questions, insights, and methods. Issues of inquiry and knowledge remain central. The possible computability of aspects of designing has become an increasingly important focus of attention, even for authors from whom computational models or assistants are not primary goals.

Mark Gross, Stephen Ervin, James Anderson, and Aaron Fleisher continue their development of a "Constraint Manager," a computer environment for designing based on a model of designing as search within field constraints. In [1] "Constraints: Knowledge Representation in Design," they give a succinct description of the conceptual architecture of the Constraint Manager, illustrate some of its existing applications, and describe at greater length its evolving use in constructing and reading diagrams and its possible linkage to a system for retrieving and manipulating video references.

The Constraint Manager functions as a flexible environment in which designers may construct and manage their own knowledge bases. In an allied effort, Patrick Purcell works to evaluate computer environments as defined by particular software/hardware packages and explores how they might be adapted and further developed to assist designers. Reference [2] is a report on comparative study of four existing "expert systems" for designers, each a "shell" that has been equipped with low level design information -- for example, information useful for cafeteria planning and layout. His purpose is to compare the comparative utility of these shells in constructing knowledge based design systems.

A third article [3] in the same volume of Design Studies describes John Habraken's and Mark Gross's "Concept Design Games," a conceptual framework and "technical universe" intended to explore and illuminate ideas important in the design of physical configuration -- each game located at the intersection of a particular kind of social organization and a particular configuration of pieces and rules for their manipulation. Habraken and Gross see designing as a social process in which designers manipulate and transform complex configurations while making agreements and rules for doing so.

Louis Bucciarelli also sees designing as a fundamentally social process [4]. However, he is less interested in showing the impact of particular social arrangements on resulting artifacts than he is in exploring the character and consequences of the discourse through which design participants communicate with each other under conditions of sustained uncertainty and ambiguity. He studies design process in situ, through participant observation, and argues on the basis of his fieldwork the crucial importance of multiple representations of objects-in-the-making, how naming things and working out divisions of labor are part and parcel of designing, and how design discourse across multiple representations can both facilitate and inhibit convergence of the description of an artifact.

William Porter and Donald Schon are both concerned with the experienced phenomena of designing, "design phenomenology." Their papers grow out of their collaborative research.

Porter, in his work [5] introduces the idea of underlying "logics" of design, logics implicit in the dialogue an architect conducts between his appreciation of a site and the images, prototypes, and principles he brings to that site. In his two thought experiments, Porter illustrates a way of reconstructing plausible logics of designing on the basis of observed design products -- drawing, in one case, a completed building in the other. In this process, he illustrates both the systems of implications built into a architect's appreciation of a site, and the "types" -- exemplars and archetypes -- from which an architect may draw his response to those implications.

Schon focuses on types [6] -- functional building types, references, spatial _gestalts_ and experimental archetypes -- and their roles in skilled designing. Through analysis of the protocols of seven practiced designers, he argues that design reasoning employs rules derived from, and contextually adjusted and criticized on the basis of, underlying types. He claims, moreover, that types may "loosen up" to yield "worlds" of objects and relations that practiced designers inhabit when designing.

In all this work, one can detect the evocative presence of the computer. Sometimes, as in the work on the Constraint Manager and on Expert Systems, the computer occupies center stage. In the other efforts, though mainly concerned with other issues, we nevertheless draw lessons from our studies for two principal questions: What aspects of designing can be made computable? What are the properties of a computer assistant useful to designers? Porter, for example, argues on the basis of his reconstructions that a computer programmed to "bookkeep designing" must do more than keep a record of spatial position, dimension and shape. Schon claims that useful computer assistance must be able to capture and manipulate the "perceptual geometry," the spatial _gestalts_, with which designers work. Bucciarelli emphasizes that a useful computer assistant must be able to retain traces of multiple representations of "the same objects," while leaving room for appropriate levels of ambiguity and vagueness.

Often (but not always) through the lens of the computer, we concern ourselves with a few central issues: the representation of design knowledge and designers' objects, the distinctive character of design reasoning, and the nature of design discourse.

Representation. The Constraint Manager represents design knowledge in terms of variables, constraints made up of variables, and relationships among constraints. Porter and Schon point to underlying design types and worlds as holding environments for design knowledge, and stress the need for representations that capture the "fullness" (of associated ideas and images), the context dependency, and the transformability of types. Ervin sees Urban Planners using diagrams, simplified images meant to explain, rather than portray, and explores how they contain "useful uncertainty," so valuable in the conceptual stages of designing. Bucciarelli too sees the sketch as representative, not necessarily of spatial form, but of paradigmatic knowledge shared by participants within specialized "object worlds."

What are the characteristics of the objects represented in design knowledge? How are the units of design thinking demarcated, at what levels of aggregation do they operate, how are they "chunked"? How should we take account of the visual objects with which designers think, especially in the early stages of a design process? How should we incorporate multiple representations of "the same objects" and shifts in the sets of objects with which designers think?

Reasoning. Schon observes that designers' patterns of inference are "entirely conventional." What is distinctive about design reasoning is not the most general formal logic by which designers move from premises to conclusions but the substantive systems of implications, or "if/thens" -- illustrated in Porter's descriptions of compositional and archetypal logics -- that seem to occupy an intermediate status between the broadest principles of formal logic and the personal, idiosyncratic styles of thinking characteristic of individual designers.

How should we take account of the distinction between designers' "local reasoning," within some established frame of reference, and shifts from one frame of reference to another? What are the implications of context dependency and shifts for the development of computer tools?

Discourse. Bucciarelli and Porter both stress, albeit in different ways, the "layering" of meanings that is central to discourse among professional participants in design processes: their varied interests, strategies of representation, and languages. These authors begin with the complexity of actual design discourse. In contrast, Habraken and Gross, et.al. present tools for constructing controlled environments in which aspects of social organization can be separated out so as to explore their differential impacts on designing.

This distinction illustrates a more general difference between those researchers who prefer a more naturalistic, observational method of study, yielding descriptions of complex phenomena, and those who strive to build formal models and environments in which complex phenomena can be unpacked and subjected to more nearly controlled experiment. It has been interesting, however, to note the degree to which these two approaches have begun to inform each other -- the developers of the Constraint Manager looking to natural studies as sources of input for their programs; proponents of protocol analysis, design reconstruction and participant/observation looking to controlled computer environments as settings in which to unravel the ambiguous results of their empirical observations.

We are convinced that if research on design is to lead to legitimate theory and the development of improved methods for designing, such a reapproachment is necessary. As design is interdisciplinary, synthetic, so too must its study be. Only in this way can we come to understand the fruitful conjunction of the creative and the instrumental in the design work of the engineer and the architect.

References

[1] Gross, Mark, Ervin, S., Anderson, J., and Fleisher, A., "Constraints: Knowledge Representation in Design." Design Studies, Vol. 9, No. 3, July 1988.

[2] Purcell, Patrick, "Computer Environments for Design and Designers," Ibid.

[3] Habraken, John and Gross, M., "Concept Design Games," Ibid.

[4] Bucciarelli, Louis L., "An Ethnographic Perspective on Engineering Design," Ibid.

[5] Porter, William, "Notes on the Inner Logic of Designing: Two Thought Experiment," Ibid.

[6] Schon, Donald A., "Designing: Rules Types, and Worlds," Ibid.

Use of Systemic Theory to Represent the Conceptual Mechanical Design Process

Manjula B. Waldron, Kenneth J. Waldron, Dean H. Owen
The Ohio State University
Columbus, Ohio 43210

Abstract:

A theoretical framework for representing the mechanical design process based on the use of systemic theory, is presented. The systemic approach is taken since the design process is context dependent and is initially functional in nature. The function-to-form transformation is part of the process. Hence, initially it is the semantic function which is important since the syntactic structure follows from it. Through our observations of the design process, we have found that expert designers often make choices from a set of design options in their experiential repertoire which fit the given problem context. It was observed that the designers identify two types of constraints, namely <u>primary</u> (binding or non-negotiable) and <u>secondary</u> (plastic or negotiable), which they use to set the context and to direct their design. We have taken the systemic functional analogy and defined a set of metafunctions which we believe to be important in describing the conceptual design. The systemic theory representation allows for the use of the problem solving methodology from Artificial Intelligence to generate a system network of the design choices which the designer makes. Identification of the meta-functions allow for the design to be mapped at three levels.

INTRODUCTION

In the literature design has been viewed as a process which operates on an input to produce an output. While studying expert mechanical designers engaged in the design of a robot manipulator arm, the authors observed that the activity of designing consisted of gathering knowledge, maintaining and matching it, distinguishing and mapping it. The information provided to the designers or used by the designer was for the purpose of constraining the design. If the given information was adequate, the designer constructed functional specifications which could be mapped by the design process onto a possible design product. These specifications contained, implicitly or explicitly, the properties of the user, the environment in which the product was to be used, and the conditions or constraints under which this would occur. Usually, the designers considered several possibly viable design options which were available to them to fulfill the same functional specifications. This took the form of using prior designs from their repertoire based on personal knowledge and experience. The design configuration which was ultimately selected and finalized depended on a satisficing procedure among critical functions identified for that design (Chieng & Hoeltzel, 1987; Waldron & Waldron, 1988b).

Recently, researchers have begun to focus on the functional aspects of design (Rinderle, 1987). The focus has quickly shifted into building expert system shells using AI methodology (Lai & Wilson, 1987), rather than formalizing the theory. Yoshikawa (1986) proposes using topology to provide a basis for a theory in which the design process can be modelled. The development by Rosen, Erdman and Riley (1987) of a relationship between function and structure, using the network concept to examine design representations in dwell mechanisms, comes closest to providing a framework for design. Nevertheless, this work is also focused on developing computer tools rather than on developing a theoretical framework. In this paper we attempt to use the systemic theory (Halliday, 1981) to attempt a formal representation of functions in the design process. Patten (1988), combined this theory with problem solving techniques from Artificial Intelligence to generate natural language text. It is our belief that the representation we propose will allow for the development of better CAD tools in the future.

In the course of a time study of complex mechanical design the authors observed that the designers, in interpreting initial design goals, which were often vague, used certain decision making strategies. The designers used their previous experience and knowledge to identify ways in which the problem could be configured or the context set so that certain alternatives were available to them to proceed when embarking on conceptual design. In setting the context, their reasoning focused on identifying the constraints which must be met in order for the design to proceed. If these constraints could not be met by an existing design in their knowledge base, then this was identified as the critical focus of the initial conceptualization. For instance, the constraints on the power demands were initially poorly understood from the specifications in the leg design of the walking machine (Waldron and Waldron, 1988a). This then formed the major focus for the initial design and set the context for the design to proceed.

Based on this work in mechanical design and the work of others in architecture and manufacturing on use of constraints (Gross 1986; Navinchandra and Marks, 1987; Fox, 1982) we proceeded to study context and how constraints might be specified in conceptual mechanical design and how functions fit these specifications. A further subject of study was ways in which these might be represented so that the current computational techniques from AI might apply. This then forms the major thesis of the present paper.

BACKGROUND

It is generally recognized that the skills of expert designers depend on heuristics and creative problem solving techniques, which they develop through their experience in designing (Simon, 1981). Successful educational programs in mechanical design, such as that at Stanford, are based heavily on this premise. These programs are strongly project oriented and subscribe to the philosophy that the only way to teach design is by designing (Roth, 1987). Through repeated exposure, the relationships between function and form may take on a preconscious path, so that the designer may have no conscious awareness of these relationships at the time of conceptual design. That is, the designer may rapidly access compiled knowledge without consciously reasoning. Hence, during the collection of a design protocol, when asking the designer to speak while designing (Ericsson & Simon, 1984). Such relationships may not be verbalized although actively used Through design practice, designers learn to identify relevant principles of previously encountered design situations. The previously developed solutions to these situations become possible designs which form the starting points for solutions of new design problems. This approach contrasts with that of a less experienced designer who starts with the components of the system to be designed, and thus may or may not be able to efficiently synthesize a solution to the problem at hand. If one reflects further, this philosophy is very akin to acquiring many skills, which are learned by first hand experience, rather than by reading or verbal instruction about them. Examples are riding a bicycle, swimming, cooking, painting, etc. Likewise in design, hands-on experience and practice are necessary to develop a knowledge base which can be easily accessed to form a basis for initial choices to start solutions of new problems.

The representation of function is a serious problem in the development of effective conceptual CAD tools. Machine drawings, which represent form, have intrinsically very little information about function. Experienced readers who have seen similar equipment in the past may be able to infer function from a machine drawing, but for novel designs this is not effective. A wide variety of means of explaining function have been employed in the past (Earle, 1983). These range from the use of verbal explanations, often keyed to labels on a drawing of the machine, through various forms of schematic drawings simplified to emphasize which aspects of the system are important to function, to graphic animated models. Most of these techniques rely, to a greater or lesser extent, on inferences drawn by an experienced reader. The only attempt to make fully rigorous and complete descriptions is that of the patent office (Kiverson, 1977). Even so, the system for making patent descriptions is very far from a consistent formal representation. Certainly little information, if any, is obtained from it regarding the design process to which the design belongs.

In the context of CAD systems the problem of representing function has two aspects. Firstly, a consistent and complete formal representation is necessary to allow communication with the computer software. Secondly, it is necessary to have a means of representation which is easily understood by human observers with varying degrees of experience. Likewise, it is necessary for the human designer to enter functional structure easily into CAD machines. Clearly, lengthy and often nonspecific verbal descriptions do not fit this criterion. In the context of machine communication, it is also necessary to provides a means for easily attaching the functional description of machine drawings which describe form (Serrano and Gossard, 1986).

CONTEXT AND CONSTRAINTS IN CONCEPTUAL MECHANICAL DESIGN.

Use of constraints has been recognized as a viable way in which to manage configuration (Frayman and Mittal, 1987), problem solving (Brown, 1987) and design (Ervin and Gross, 1987). They provide a powerful means of generating computational models as distinct from global optimization and rule based decision making. Little work has been done to study constraints and constraint management in conceptual mechanical design.

In order to explore identification of context and constraints in mechanical engineering design, an expert mechanical designer, who had recently completed a power train design for a walking-machine leg, was interviewed. His retrospective account was videotaped and his depositions were recorded, and supportive documents in the form of drawings, models and actual design results were solicited. The interviewer dialogued with the designer to ask if, at the given points in decision making, the designer had alternative choices or constraints; what knowledge and relationships assisted the designer in arriving at a particular decision; and what alternatives were disregarded by the designer and why. The interviewer took the data and interpreted it to identify the types of decisions which were made and when and what types of knowledge were used. Many alternatives which were available to the designer were not initially voiced, but after much discussion verbalized, often with a caveat "but I spent no time on it because I know it would not have worked because of the following reasons...."

The power train transmission design for a walking machine leg was initially constrained by the final leg geometry which was fixed. The designer asked the question where the power had to be transmitted to? That is, the location of the power delivery had to be precise, since the constraining design variable of location was fixed by the previously designed leg geometry. Hence, the function, transmitting power to the legs was viewed as delivering power to a location. Further, the actuator selection dictated to the designer where the pump locations were and how many. The hydrostatic actuator layout required that there be three pump positions to deliver power in each of the three directions, that is, vertical, horizontal, and lateral. Therefore, the initial constraints of the design which had to be met by the designer were to transmit the power to each of these pumps for every leg. The position of the pumps were known exactly by the leg geometry and the actuator design.

The designer also stated that, in distributing mechanical power, the guidelines implicit to him from experience were that the design should be 1) <u>light weight</u> (minimal weight was a global constraint in the overall walking machine design) 2) <u>high strength</u> (adequate strength for the power levels to be delivered). 3) <u>high reliability</u> for the distributed system and 4) the design must deliver power with <u>maximum efficiency</u>. Hence, the overall context of the design was set by the function of the power train which was to distribute a given amount of power to each pump at a specified location, and constraints 1) through 4) which by their very nature plastic and satisficing in nature rather than satisfying. These constraints allowed the designer to prune the selection tree until a satisficing solution could be met so that the design could proceed. These were the functional goals that the designer had set out to meet initially.

In his overall design the designer was set to select his design from the available design choices until these system constraints were met. It is proposed that the constraints in design are of two types. First, the <u>primary</u>, binding or non-negotiable constraints present explicitly in the problem statement which must be met before the design can proceed. These are also called hard constraints by researchers. The location of the pump position is a primary constraint in the above example. Secondly, the subordinate, secondary, plastic, or negotiable constraints which are to be satisficed and are expressed as limits. These are also sometimes referred to as soft constraints by the computer science community. One design may be evaluated to be superior or inferior, relative to the negotiable constraints amongst acceptable solutions. To the designer, a design is not viable if it cannot meet the primary constraints. The constraints together set the context within which the design is conceptualized. It should be noted that the designer had decided early on that power, rather than torque, was the critical factor since he was aware of the constraining relation P = K.T.W. and torque (T) would be inversely related to speed (W) for the same power (P).

The functional specifications in the present example consisted of two primary constraints: namely, the pump position specified in machine coordinates, which must be satisfied to within a tight tolerance, and the peak power requirements. That is, the power which must be delivered to the flywheel. Once the designer had an idea of these constraints, he proceeded to assemble all possible potential distribution systems available to him, based on his prior knowledge which might satisfy these primary constraints. There were three non-overlapping contenders. Namely, flexible shafts powered from a centralized source, a hydraulic system and a series of drive shafts. The designer made sketches on a note pad and found literature on the power capabilities of these. While collecting this information, the designer also brainstormed about hydraulic components and in rough sizing found that these could not provide the required power with high efficiency and low weight. The flexible shafting had a problem with weight. While efficiency was higher, and reliability was reported in the literature to be as good as that of the hydraulic system, the designer had no experience with this system. The third contender was a series of drive shafts which could be designed to satisfice all the secondary constraints. The designer had prior experience with this type of system. While the designer did not list experience as an evaluating criteria, it was clear from dialogues that the familiarity or experience with a design option was a strong evaluating factor in finding, or not finding, a satisficing solution along a secondary constraint. Hence, it would seem that experience would be an important factor in setting context along secondary constraints.

Once it was decided to select drive shafts then the set of primary constraints were generated for the next decision point. The design decisions relating to the series of drive shafts consist of selecting cross-sections and materials of the shafts, choice of coupling elements and specification of positions of all elements. The possible shaft arrangements were a single, central located shaft or three shafts, one through the leg mounting axes on either side of the machine, with the third along the center line of the body of the machine. In both cases the shafts would be parallel to the longitudinal axis of the body. The primary constraint on placement was the functional requirement that the center of the body of the machine be empty for packaging other components and payload. This information was contestably available. Hence, the three shaft arrangement was the only viable alternative. At that point, sagittal symmetry was used to provide the best solution. This decision constrained the arrangement of coupling elements since all three shafts had to be powered from a single prime mover shaft. The available choices of coupling elements were toothed belts, gears, or chains. These had their own added secondary constraints of noise, lubrication, efficiency, and weight, etc. A network representation of this example will be developed in a later section. It is shown as Figure 1.

SYSTEMIC REPRESENTATION OF DESIGN

The basis for the systemic representation, as described in the previous section, is that the expert designer works in terms of functional units which are obtained from a selection of previous designs, or from other sources or, if necessary, created. These functions can be classified into four non-overlapping categories called metafunctions which can then be described in the form of a network. We are providing a theory which can then be related to AI methodology and natural language representation (Patten 1988). This representation is based on two observations. The first is that design is context dependent, as described in the last section and alluded to by Akin, 1986 and is by itself not by self contained. The environment, in which the design will be used and the expected users change the characteristics of the design, and hence, of the process (Waldron, 1988). The second observation is that the design is "functional," i.e., must provide certain functions. We can, therefore, propose the concept of a "system" (hence the term systemic), which is a set of design choices in a specific design context. This systemic description of design (within the context of choice) is "paradigmatic" (function based) rather than "syntagmatic" (form based) (Kress, 1976). It should be noted that systemic design is a descriptive approach and not prescriptive, hence it is different from the "systematic design" (Pahl & Beitz, 1984).

Understanding the conceptual mechanical design process requires an assumption that the design environment contains different kinds of functions which are useful in different phases of the design process. The functional description of the design process addresses what it does. The implementation description describes how it does it. To understand this in more detail, one must represent the relative functions of parts and subparts in such a manner that they represent the total behavior (Sembugamoorthy & Chandrasekaran, 1986).

The system based on the systemic theory, as described above and used in linguistics (Halliday 1981) is a set of mutually exclusive classes (or features) which represent a choice or "potential." This ensures that two choices cannot be considered simultaneously, unless they belong to the same functional class. For example, if the power system is hydraulic it cannot be electric, etc. Each choice must be further qualified within the context. For example, if the flexible shaft system has higher weight it cannot have lower weight, etc.

The system network provides a graphical display of the relationship between classes. The classes which must be present within a context are entry level. The following notation is adapted from (Butler 1985). A system is illustrated by a T intersection representing choices between two or more features. Coupling between classes is represented by the curly brackets { }. Figure 1 shows the use of this notation.

Accordingly, a system network provides a classification scheme for design objects which may be pursued into varying levels of detail. However, the choice of system is functionally based, whereas the classes may be functionally or structurally related. One could therefore group the design functions into abstract categories called metafunctions following Halliday, (1981). The metafunctions in design are identified as:

1. Experiential: This metafunction represents the design world in general. For example, semantics, processes, events, actions and objects and the relationships among them, etc. This is where logic or the propositions of context are defined. The context in design is established by the constraints. As stated before, these are of two kinds, the primary (or non-negotiable) and the secondary or subordinate (negotiable within limits).

2. Communicational: This metafunction represents the expressive roles in design. The designer draws what is being designed in order both to evaluate and satisfy geometric constraints and to communicate with other designers and manufactures. This metafunction is, therefore, a mediary between the conceptual designer and his prior knowledge, or the knowledge of the others in the process. Thus, it can take two forms: self communication and communication with others. Sketches and their verbal descriptions were seen in design protocols to support existence of this metafunction.

3. Textual: This metafunction represents those aspects of the design in which the designer organizes the design to form a cohesive whole. Coordinates of reference, relative dimensions, tolerancing and fit and assembly drawings all ensure the cohesive organization in design.

4. Affordance: This metafunction describes how properties of the design (as a potential device) match properties of the environment described in the specifications. What the specified environment affords the device (including the operator, if necessary, in this unit of analysis) constrains the design in functionally significant ways. For example, the pads on the feet of a walking machine must match the characteristics of the surface over which the machine must walk (hard earth, soft, sand, mud). If the machine must walk on a variety of surfaces, then the foot pads must be modifiable in adaptive ways.

Hence, the same design can be simultaneously described in different functional dimensions depending on what functionality is being considered and in what context. Therefore, the shape arrived at may represent several functions which span different metafunctions.

The structural representation is the explicit representation in the drawings (design output), that is, how are the components laid out, what are they adjacent to, what are their dimensions, as explicitly stated. The functional representation on the other hand involves viewing the drawings as a cohesive whole, which includes implicit information about each component as well . The drawings (including past lists and specifications) implicitly contain the functions they serve in the assembled design and in the information about the process they convey to the person who must manufacture the part (Waldron & Vohnout, 1988).

During the study of expert mechanical designers, at times when they were looking at and talking aloud about assembly drawings, all identified this implicit (functional) information first. This included what the depicted device did and how it could be used, based on their design experience. Naive designers, on the other hand, could only identify the explicit (structural) information (Waldron & Waldron, 1988b).

Each level has its own independent systems of features. In the present work we focus mainly on the structural level, which is arrived at by means of the pragmatics of design or the context of the design. As stated earlier the design is only meaningful in its context. That is, the kind of design situations or constraints demand the kinds of design or the design elements which get selected. In other words, at each level of constraint, whether primary or secondary, certain alternatives exist from which the relevant class is chosen. The selections at the form level are mapped from the functional level which then may have a corresponding graphical representation. Hence, there are three levels, namely the functional, procedural and the graphemic (graphical and/or notes. At the highest level the function is addressed. At the syntatic level the syntax or the logic of the design is selected within the context, and at the graphical level the physical manifestation of the design takes place. Thus, in the systemic theory of design, the constraint (or the context) becomes the key to the paradigmatic or the available choice. As such, the four metafunctions provide the framework for the systemic representations and need to be explored in future research.

EXAMPLES FROM CONCEPTUAL DESIGN:

There is not a great deal of observational data available on the conceptual mechanical design process. The few studies that have been reported which consider functional information from a design methodology viewpoint, report of designers making choices and therefore provide support for the systemic theory representation (Rinderle, 1987; Waldron and Waldron, 1987; Rosen et al., 1987; Ullman, et al., 1987).

In designing the power train for transmitting power to the six legs of the adaptive suspension vehicle at The Ohio State University, as described earlier, the following records were made: the predicted load history determined the power level and the type of actuators to be used in the leg design. These, in turn, fixed the pump positions to which the power was to be delivered. That is, what was to be delivered and where it was to be delivered, were the starting constraints (system). These were the primary constraints for the power train which formed the functional specifications for the design of that mechanical system. The subordinate constraints (or contexts) such as weight, reliability, and efficiency, were expressed as tradeoff limits for optimization criteria.

Figure 1 shows the system network representation for this problem. The experiential metafunction for this design consists of the primary functional constraints (context) namely, PUMP POSITION and POWER, each have three choices (features or functional classes). The pump position has associated with it choices of the flexible shaft and the hydraulic system which are ruled out because of subordinate constraints. It also has the third choice of a series of drive shafts, which satisficed the subordinate constraints. The choice of mechanical power also satisfices the primary and subordinate constraints and the design function moved to the DRIVE SHAFT SYSTEM. There are three entry systems at this point, namely, the COUPLING ELEMENT, PLACEMENT and SHAFT systems. Each system has certain features. For example SHAFTS can be segmented, solid or quill. The first two were rejected on the basis of subordinate constraints, and the quill type was judged to be the best choice. Likewise, the gear belt was found to be the best choice for the COUPLING ELEMENT and the three drive shaft placement was optimal, which then completed the power train design at the conceptual level. The design then proceeded to the detailed level where choices of components such as tapered bushings were made which then translated to functional sub-components such as flanges, tapers, etc. While developing this the designer was pursuing other metafunctional representations, such as placement in the machine environment (affordance), sketches, drawings (communicational) and ensuring that the system subcomponent fitted in a specified manner (textual). Representation of these other metafunctions will in itself be interesting.

It should be noted that other researchers have also observed this selection process in design Ullman et al. (1987) report that their designers considered the outside location then the inside location and then a forked support for the pivot arm. The designers rejected the inside and forked support for safety reasons (subordinate constraints) and proceeded to select the outside constraint. So the following systemic representation would show this process.

	inside	X	poor safety
pivot arm	outside		good safety
	fork support	X	poor safety

These authors also report that their designers each pursued a single conceptual design and they "patched." Our contention is that this is a problem of the protocol methodology. The designer probably did not vocalize the functional choices which he or she has rejected, based on prior experience. Rather, the context of the design allowed the designer to eliminate other choices. If asked, the designers would probably have explained their reasons, as our subject did. The patching process they observed can be described in the systemic theory. Designers are good at knowing what they do not want and at eliminating choices which violate the primary constraints. Thus, the observation of patching could be the design proceeding in the context of the designer's experience. When a designer is at a particular decision point, it represents his best judgment to this point. Back tracking will not serve any purpose unless they scrap the entire design and start obtaining new knowledge. Hence, the further the designer is in the process, the more committed he appears to be. Since the designers in that study were experienced designers, it is highly probable that this is what occurred. The systemic representation allows for this aspect to be investigated in their data.

The same argument holds for the observation that the designers do not keep their designs balanced. The chances are that the expert designers, through their prior experience, were already cognizant of the fact that alternative choices were probably not viable. If this was a part of their compiled knowledge, it would not surface in the talk aloud protocol. In our experience with data collection, we found that in order to obtain the designer's functional knowledge it was important to combine the protocol with discussion to ensure other aspects of design were also verbalized. In the Ullman et al. (1987) study, one also observes the presence of all four metafunctions.

DISCUSSION

The systemic representation allows the same design in the form of engineering drawing to be represented functionally in different ways so as to elaborate on the function of the design. This is important, since the design product specification in the form of blueprints and notes often contains this functional information explicitly as well as implicitly. The systemic representation is based on the paradigmatic axis (choice) and supposes that at each point the designer has a distinct set of options. Hence, concurrent or parallel paths can be represented easily. On the other hand, shape or form representation is on a syntagmatic axis which is a chain where the representation may be sequential.

The advantages, in CAD, of representing design paradigmatically are apparent from the example provided. One could represent the design hierarchically and, as new choices become available they could be added appropriately. The same design specifications can be represented in many ways (different functional specifications) providing for alternative functional representations. Thus, updating the design knowledge flow becomes simplified. The systemic theory allows the design knowledge to be represented in a best first form since at each point there is only one potential choice which is available to the designer. The choice of primary constraint satisfaction allows the search in the synthesis process to be much curtailed. Should new primary or subordinate constraints become available, they can be added to the systemic representation relatively easily. This representation also allows for easy use of AI methodologies of problem solving. Discussions with expert designers show that the systemic representation is closer to the way designers design. Notably their search space is reduced drastically using this technique of organizing context.

The systemic representation of design also accommodates the inclusion of compiled knowledge and of relatively high level representation of previous designs by specifying primary and subordinate constraints along appropriate paths. The primary and secondary constraints concepts are routinely used in the Kepner-Tregoe method in decision theory. The proposed methodology allows for these to be represented appropriately in design. It also allows for modeling the opportunistic (tactical) as well as strategic aspects of the design process. The opportunistic aspect would be linked to previous experience whereas the strategic aspect adds options to the functional network. Multidimensionality of design can be accommodated and linked by this type of representation.

CONCLUSIONS

In this paper, use of a systemic theory for representing conceptual design was presented. It was argued that designers at the time of creating the conceptual design use their prior experiential knowledge. At every step in the design process, based on the context of the design, the designer has several distinct choices of functional classes from which to select.

The systems are selected based on primary and subordinate constraints which set the context of the design. Prior experience of the designer helps to prune the search tree by eliminating alternatives. This idea of design choices within a context lends itself well to representation by the systemic theory. Systemic theory has also been used for written and spoken language representation, where problem solving procedures from AI can be used.

The systemic theory allows for a descriptive or paradigmatic representation of design. It allows the design to be represented as a classification hierarchy which lends itself to easy translation into symbolic computational techniques. In this paper four metafunctions for design are proposed and examples are given to provide a functional representation of conceptual design. Currently, the collected protocols are being studied to identify metafunctions. Based on this model, a systemic grammar for design can be developed. It is believed that these aspects will allow for the development of semiotics in design and lay a foundation for a design language.

REFERENCES

Akin, O. "A Formalism for Problem Restructuring and Resolution in Design." Planning and Design 13:223-232. 1986.

Butler, C.S. (1985) Systemic Linguistics Theory and Applications London, England, Batford Academic.

Chieng, W. H., Hoeltzel, D.A. (1987) "A generic planning model for large scale design optimization with a power transmission case study. Proceedings ASME Computers in Engineering conf. NY 1987 p. 113-124.

Earle, J.M. (1983) Engineering Design Graphics , Addison-Wesley.

Ericsson, K. & Simon, H. (1984) Protocol Analysis , Cambridge.

Fox, M. (1982) "Job shop scheduling: A study of constraint directed reasoning" Carnegie Mellon University PhD dissetation.

Gross, M.D. (1986). "Design as exploring constraints," MIT PhD dissertation.

Halliday, M.A.K. (1981) : Explorations in the Functions of Language. Wheeling, Ill., Whitehall Co.

Kiverson, G. (1977) The Art and Science of Inventing, Van Nostrand and Reinhart New York.

Kress, G.R. (1976) Halliday: System and Function in Language. Oxford, England. Oxford University Press.

Lai, K. & Wilson, W.R.D. (1987) "FDL - A Language for Function Description and Rationalization in Mechanical Design." Proceedings ASME Computers in Engineering conf. 1987 p. 87-94.

Pahl, G., Beitz, W. 1984 (Wallace K.M.ed) Engineering Design, Design Council, London, Springer Verlag.

Patten, T. (1988) Systemic text generation as problem solving. Cambridge, England, Cambridge University Press.

Navinchandra, D. & Marks, D. (1987("Design exploration through constraint relation" in J. Gero (ed) Expert systems in computer aided design. North Holland.

Rinderle, J. R. (1987) "Function and Form relationships: A basis for preliminary design" Proceedings of NSF Workshop in Design Process, M Waldron (ed) August, Oakland. Ohio State University, p. 295-315.

Rosen, D., Erdman, A. & Riley, D. (1987) "A general knowledge based system shell with application to drive mechanism." Proceedings ASME Computers in Engineering conf. Aug. p. 29-36.

Roth, B. (1987) Presentation on "Design education," session NSF Design, Theory, and Methodology workshop, Oakland, CA. Feb.

Sembugamorthy, V., Chandrasekaran, B. (1986) "Functional representation of devices and compilation of diagnostic problem solving" OSU/AI Lab Technical report. June.

Serrano D., Gossard, D.C. (1986) "Combining mathematical models with geometric models in CAE systems." Computers in Engineering Vol. 1, p 277-285.

Simon, H. (1981) The Sciences of the Artificial, Cambridge; Mass, MIT Press.

Ullman. D.G., Stauffer, L., Dietterich, T. G. (1987) "Toward expert CAD" Computers in Mechanical Engineering, Nov/Dec 1987, pp. 56-70.

Waldron, M.B. (1988) "Modeling of the Design Process." to appear in the IFIP 5.2 Intelligent CAD workshop, Yoshikawa and Gossard (ed), North Holland.

Waldron, M.B. and Vohnout, V. (1988) "Formalizing Knowledge in design for CAD/CAM integration." to appear in International Workshop on Engineering Design and Manufacturing Management. Melbourne, Australia November.

Waldron, M.B. and Waldron, K.J. (1988a) "Time Study of the Design of Complex Mechanical Systems" in Design Studies.Vol 9, 2 April p. 95-106.

Waldron, M.B. and Waldron, K.J. (1988b) "Position Paper on Conceptual CAD," in Proceedings of Computers in Engineering Conference San Francisco August p. 203-210.

Yoshikawa, H. (1986) Design Theory collection of papers, University of Tokyo.

Acknowledgments:

This work was partially supported by NSF grant award # DMC-8610111. Help of Mr. Vince Vohnout, President, Adaptive Machine Technology, in obtaining the design example and Dr. T. Patten for introducing us to systemic theory is also acknowledged.

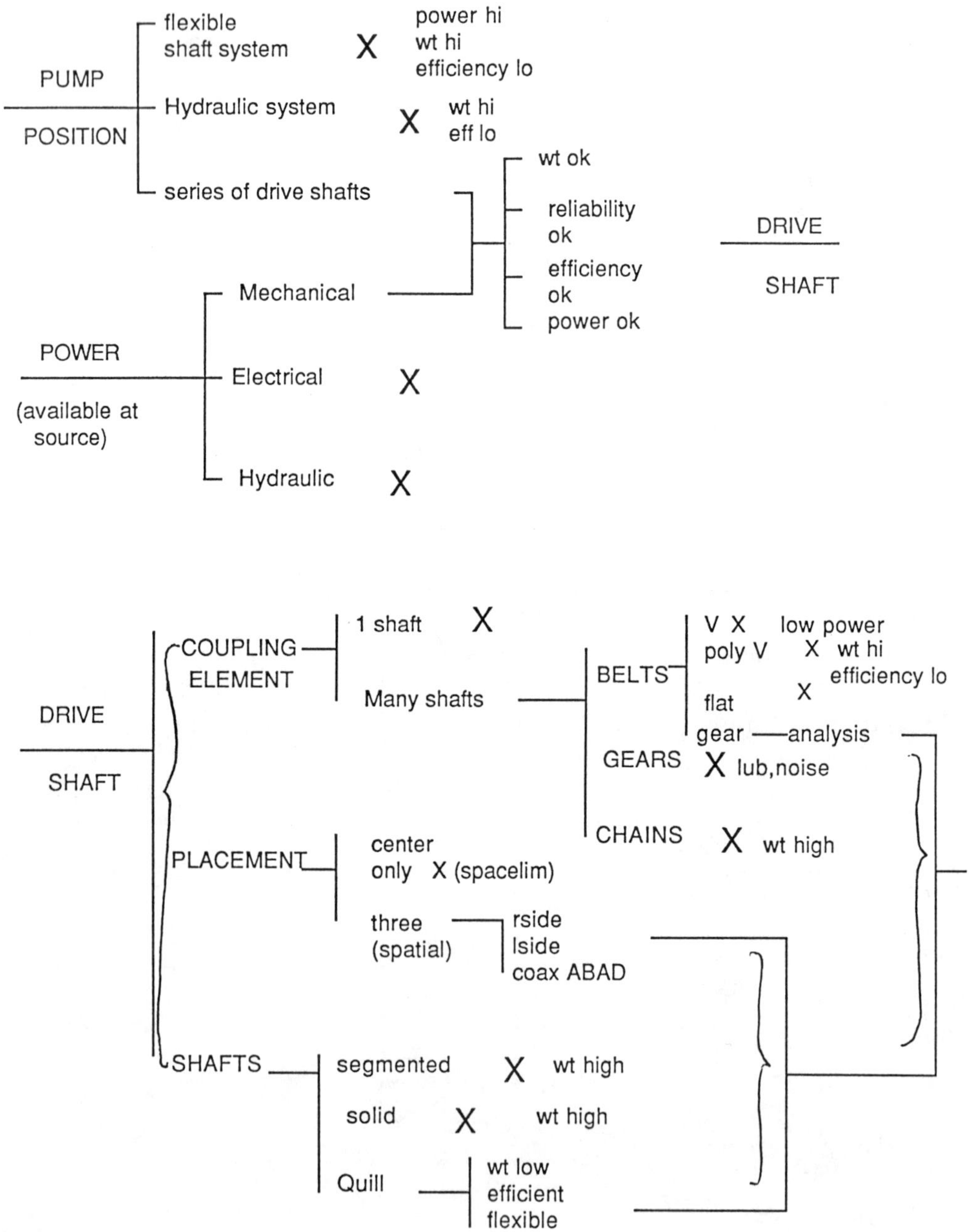

Figure 1 System network of power train

Tools for Expert Designers: Supporting Conceptual Design

Sandra L. Newsome and William R. Spillers
Rensselaer Polytechnic Institute
Troy, NY 12180

Abstract

This paper reviews a few examples of information available from psychology that are pertinent to CAD tools used by expert, creative designers. Well-documented evidence on expertise is used to formulate a set of recommendations that will promote conceptual design when used in a CAD system. The advantages of borrowing information from established disciplines such as psychology to facilitate theory in new domains such as design are discussed.

In order to build tools for engineering design, we must first understand the role of the human designer in the design process. In pursuit of this goal a number of investigators (e.g., Waldron, Waldron, & Owen, 1988; Ullman, Dietterich, & Stauffer, 1988) have undertaken the arduous, time-consuming task of trying to discover exactly what it is that designers do when they design. The primary research method used in these attempts is protocol analysis. This method involves videotaping the behavior of subjects (designers) and a subsequent, lengthy review of the videotape in an attempt to understand the cognitive behavior under study. Although it may be too early to determine how useful this method will be in building design theories, protocol analysis typically requires long hours of labor for small increments in knowledge. Protocol analysis has the additional disadvantage that in this case investigators have started from scratch assuming that there is no existing information that is helpful in the study of the design process.

An alternative approach to building useful design tools is simply to borrow from other domains empirically verified information that appears to be relevant. For example, the psychological literature provides information about a number of cognitive processes such as memory, perception, and problem-solving behavior that relate directly to processes involved in engineering design. Of course, the validity of the borrowed information must be verified in the new domain.

Since research results can be domain specific, borrowing from other domains has the obvious disadvantage that some time will be spent implementing features that do not work. We do not know, for example, that a technique that promotes problem solving in one domain will do so in another. However, two points can be made here: 1) it has proven sound in many cases within science to generalize from one discipline to another and 2) transferring knowledge from one domain to another prevents us from re-inventing the wheel. Although the idea of a design discipline is relatively new, the study of humans is not. Consequently, using our experience from this discipline we should expect to provide some extremely useful tools for designers fairly quickly while serving the long-term goal of building a cognitive theory of design.

The purpose of this paper is to cite a few examples of information available from psychology that could usefully shape current CAD tools used by expert, creative designers. The end-goal, of course, is to provide a computer tool that facilitates (e.g., makes it less time-consuming, less costly in mental effort) the design process. Although there are a number of different ways to improve CAD tools (e.g., Newsome, Spillers, & Vosburgh, 1987), we are particularly interested here in tools that could support conceptual aspects of the design process.

Conceptual Design

The conceptual portion of the engineering design process is particularly difficult to specify because it is not an isolated activity. We would argue that conceptual behavior in design is that portion of the design process that requires complex cognitive information processing including knowledge and information acquired outside of the particular design project at hand. Conceptual activities are also those activities most likely to involve elements of creativity (unique ideas) and consequently represent the most difficult portion of the design process to automate (Spillers & Newsome, in press).

Probably the easiest way to delineate conceptual portions of the design process is to contrast them with routine activities involved in designing. Routine activities are characterized by their predictability or lack of uncertainty. Although routine activities may require some cognitive processing their outcome is pre-determined. In contrast, conceptual activities are characterized by a high degree of ambiguity. In other words, in studying conceptual design we are interested in design activities in which more than one outcome is acceptable.

We often think of conceptual activities as occurring primarily at the beginning of the design process. Given our definition, it is probably the case that there are more conceptual activities at the beginning of a design because there are more degrees of freedom in a totally unspecified design. However, it is possible that certain types of conceptual activities occur at a number of different intervals within the design process.

The Relationship Between Conceptual Design and Expertise

Since we have defined conceptual activities as complex cognitive behavior that often involves knowledge acquired through a variety of experiences, the most relevant area of the cognitive psychological literature appears to be that which addresses the issue of expertise in problem solving. Specifically, we have examined the empirical literature that discusses how expertise changes problem-solving approaches and how expertise impacts on the dialogue between the user and a computer. From this review, we have formulated a set of testable hypotheses about CAD tools that could be implemented within a variety of different design environments.

What Do We Know About Experts?

One well-documented characteristic of experts in computer programming is that they tend to approach a problem-solution in a breadth manner unlike novices who utilize a depth approach (Anderson, 1985). In other words, experts tend to break a problem into a complete list of sub-problems before working through any sub-problem to its lowest level. For example, expert programmers would identify all of the subroutines or procedures prior to writing the steps for any one subroutine or procedure. Novices, on the other hand, tend to identify a single sub-problem and work its solution out completely before identifying any other sub-problem. According to Anderson (1985), the differences are not striking in simple problems but tend to be dramatic when programs require complex planning. Mikaelian (1988) has even shown that expert programmers prefer information outside of their domain of expertise to be organized in a breadth manner versus a depth manner.

Computer programs are only one type of design domain. However, it seems to us that almost all types of design by experts could be characterized by the depth approach. For example, the designer of a large building may initially divide a project into sub-problems concerned with structural elements, heating and ventilation, and wiring. These sub-problems are all identified in the early stages of design without completing the details of any one sub-problem.

A related characteristic of experts in a number of different domains is their tendency to utilize relatively abstract representations or patterns in solving a problem. In contrast, novices tend to concentrate on smaller, more concrete features of a problem. Adelson (1981;1984) demonstrated this phenomenon in computer programmers but it has also been shown for chess (Chase & Simon, 1973), the Oriental game of GO (Reitman, 1976), and electronics (Egan & Shwartz, 1979). In computer programming, this tendency to use abstract representations is exemplified by the expert programmer's emphasis on function or semantics. Novice programmers, on the other hand, tend to be concerned with syntax or surface features of the computer language (Adelson, 1981).

Once again, we believe that expert designers may also initially conceive of design solutions in relatively abstract terms. A CAD system for architects described by Lansdown (1987), for example, uses what he calls prototypes as basic building blocks. These prototypes are elements that have only the most basic features but can be easily altered to create unique elements or arrangements. Lansdown (1987) suggested that the advantage of using these relatively abstract elements was the speed at which it was possible to go subsequently from a vague concept to a finished presentation.

As a final example, consider the performance of expert designers relative to memory and the ability to process information relevant to a problem solution. It has been known for sometime (De Groot, 1965) that chess masters are able to reconstruct the positions of more than twenty pieces after studying a game board for just five seconds while novices are only able to reconstruct four or five pieces after the same amount of study. This difference in experts' memory has also been found for expert computer programmers (Sheiderman, 1976; McKeithen, Reitman, Rueter, & Hirtle, 1981). Moreover, the differences in memory cannot be dismissed as individual differences between experts and novices because memory span appears to be the same for the two groups when tested with scrambled computer programs or randomized groupings of chess pieces. It seems likely that expert designers charged with the primary responsibility for a complex design project (e.g., large office buildings, bridges, etc.) develop similar abilities regarding expanded memory for each of the design components.

These three characteristics of experts (i.e., a breadth problem-solving approach, use of abstract representation, and expanded memory of problem-solution information) are not independent of one another. An expert's breadth approach to problem-solving is based to some extent on his/her ability to effectively utilize relatively abstract representations of the problem. Similarly, the abstract representations of problem information allow the expert to retain in memory larger chunks of information that represent familiar patterns.

Although these characteristics (as well as others) have been empirically documented in certain domains such as chess, geometry, physics, and computer programming (Anderson, 1985), empirical documentation of these characteristics in expert designers is still lacking. Nevertheless, these findings provide a basis for incorporating highly specific elements in a CAD tool for designers which then can be tested to determine their effectiveness relative to more traditional CAD tools. The first question that must be answered in this endeavor is how the knowledge of experts should be used to build better CAD systems.

CAD Dialogues That Support Expertise

The most straightforward method of incorporating our knowledge of expertise in a CAD tool involves changing the dialogue that occurs between the designer and the particular software package being used. With this in mind, we recommend the following modifications in the user interface of CAD packages intended to support conceptual portions of the design process.

In order to encourage and support a breadth approach to design problems, software interfaces should allow the designer to map out rough global plans for different portions of a design prior to determining all the details for any portion of the design. For example, a free-hand sketching recognition system such as the one developed by Fang and Ullman (1988) could be used to allow a designer to draw rough sketches of a plan (e.g., building, wiring, ventilation, etc.) to be stored for later retrieval and modification as details become more specific.

This recommendation is similar to the suggestion of Simes and Sirsky (1985) who argued that as expertise increases the need for immediate closure decreases in a human-computer dialogue. In terms of design problems, we are suggesting that closure on small parts of the design problem too early in the process discourages a breadth approach and locks the designer into ideas about the design that later require extensive patching.

An expert's ability to use abstract representations in thinking about a design solution can be supported in a CAD environment by adopting an approach similar to the one used by Lansdown (1987) in his CAD system. In other words, using the most basic or prototypical representations allows the designer to utilize relatively abstract patterns until he/she is ready to be more specific.

We also recommend that queries made by the computer to aid in design solutions be formulated so that abstract qualities of the design are represented. Lamberti and Newsome (in press) found that abstract queries in a computer-diagnostic expert system were answered by experts more quickly and accurately than queries about specific details. Abstract queries in the study were defined as those questions that asked about relationships among different components of the problem. Concrete queries asked specifics about a single component of a problem. Abstract and concrete queries in a CAD system could have similar forms. Abstract questions would call the users attention to the more global relationships among design elements (e.g., how wiring and ventilation plans would be related) whereas concrete questions would direct the user towards specific details of a single design component (e.g., where ventilation shafts on a specific floor would be located).

Our final recommendation concerns experts' ability to process larger chunks of information because of their expanded memory of problem-solution information. This characteristic of expertise is particularly difficult to support in a CAD environment. Lansdown (1987) complains that most rule-based productions in knowledge-based systems are limited in their design usefulness because they do not capture adequately the "chunks" of design knowledge. According to Lansdown, the interconnectedness of design knowledge creates much larger chunks than most productions can handle.

Although it is difficult to specify correctly the exact size of the piece of information that an expert designer can utilize, we can at least avoid the opposite error of making the piece of information presented to the designer too small. Unfortunately, most current CAD systems err in the direction of presenting too little information at any one time (or too detailed information). We recommend that representations be more global (less detailed) allowing the expert designer to study overall patterns while filling in details from memory.

Summary

We have attempted to show how well-documented evidence on expertise can be used to formulate a set of recommendations that will promote conceptual design when used in a CAD system. It should be noted that we have only presented a few examples of the type of information that can be useful in the study of engineering design. Although our recommendations seem intuitively plausible for the design field, clearly their effectiveness in this domain will have to be determined by comparing CAD systems that make use of these suggestions to systems that do not. However, these recommendations provide specific, testable hypotheses that have the advantage of being based on a systematic body of knowledge that already exists. All this is to say that we should approach the design of CAD tools by using robust empirical research already available in other disciplines.

Preparation of this paper was sponsored by the Design Theory and Methodology Program of the National Science Foundation.

References

Adelson, B. (1981). Problem solving and the development of abstract categories in programming languages. Memory & Cognition, 9(4), 422-433.

Adelson, B. (1984). When novices surpass experts: The difficulty of a task may increase with expertise. Journal of Experimental Psychology: Learning, Memory, and Cognition, 10(3), 483-495.

Anderson, J.R. (1985). Cognitive Psychology and Its Implications. New York: W.H. Freeman and Co.

Chase, W.C. & Simon, H.A. (1977). Problem solving in semantically rich domains. Cognitive Science, 1, 193-215.

de Groot, A.D. (1965). Thought and Choice in Chess. Paris: Mouton.

Egan, D.E. & Schwartz, B.J. (1979). Chunking in recall of symbolic drawings. Memory & Cognition, 7, 149-158.

Fang, R.C. & Ullman, D.G. (1988). Free-hand: A sketching recognition system for conceptual mechanical design. Design Process Research Group Tech. Report #DPRG-88-5, Oregon State University, Corvallis, Or.

Lansdown, J. (1987). The creative aspects of CAD: A possible research. Design Studies, 8(2), 76-81.

Lamberti, D.M. & Newsome, S.L. (in press). Presenting abstract versus concrete information in expert systems: What is the impact on user performance? International Journal of Man-Machine Studies.

McKeithen, K., Reitman, J.S., Reuter, H., & Hirtle, S.C. (1981). Knowledge organization and skill differences in computer programmers. Cognitive Psychology, 13, 307-325.

Mikaelian, D.M. (1988). The effects of programming expertise on the organization of task-relevant information. Unpublished masters thesis, Rensselaer Polytechnic Institute, Troy, NY.

Newsome, S.L., Spillers, W.R. & Vosburgh, A.M. (1987). Quantitative evaluations of human factors savings in CAD workstations. Proceedings of the Third International Conference on Civil and Structural Engineering Computing, Civil-Comp 87, 1, 105-108.

Reitman, J.S. (1976). Skilled perception in GO: Deducing memory structures from inter-response times. Cognitive Psychology, 8, 336-356.

Shneiderman, B. (1976). Exploratory experiments in programmer behavior. International Journal of Computer and Information Services, 5, 123-143.

Simes, D.K. & Sirsky, P.A. (1985). Human factors: An exploration of the psychology of human-computer dialogues. In Hartson, R.H. (Ed.) Advances in Human-Computer Interaction, Norwood, NJ: Ablex, 49-104.

Spillers, W.R. & Newsome, S.L. (in press). Design theory: A model for conceptual design. In this volume.

Ullman, D.G., Dietterich, T.G., & Stauffer, L.A. (1988). A model of the mechanical design process based on empirical data. Design Process Research Group Tech. Report #DPRG-88-1, Oregon State University, Corvallis, Or.

Waldron, M.B., Waldon, K.J. & Owen, D.H. (1988). Use of systemic theory to represent the conceptual mechanical design process. Preprints of Design Theory '88, Troy, NY: RPI.

Modeling Software Design
Within a Problem-Space Architecture

Beth Adelson
Tufts University

Also to appear in: *Applications of Artificial Intelligence to Educational Testing.*
R. Freedle (Ed.) Erlbaum: Hillsdale, NJ.

1 Introduction

In this paper[1] we describe research on modeling software design skills within the Soar problem-solving architecture (Laird, Newell & Rosenbloom, 1987; Steier & Newell, 1987). We focus on an analysis of expert designers of communications systems designing an electronic mail system. The research addresses the issues of:

1. Guiding mental simulations of a design-in-progress using learned schemas:
When designing a complex piece of software impasses will arise. To resolve these impasses skilled designers will run mental simulations of their designs-in-progress(Adelson & Soloway, 1985, 1986; Steier & Kant, 1985; Kant & Newell, 1984). However, these simulations can be done from many points of view. For example, the simulations can depict the system as a set of functions, as a set of data objects or as a set of interacting functions and data. We have found that skilled designers have schemas that result in a systematic and effective ordering of the viewpoints taken during a design session.

2. The interaction of general knowledge of design with domain knowledge about the system being designed:
As stated above, the designers choose viewpoints of the system and then simulate their designs-in-progress from those viewpoints. For example, using general knowledge about designing communications systems one designer chose to simulate the system from the

[1]We are grateful to David Steier for his continuing generous help. Also, we would like to thank Minnie Adelson for carefully chewing over earlier drafts.
This work was supported by grants from the Design, Manufacturing and Engineering Program and the Knowledge and Data Base Systems Program at NSF.

viewpoint of the system's concurrent processes. However, in order to *carry out* the simulation the viewpoint was mapped into domain knowledge and instantiated as a simulation of multiple users concurrently issuing commands such as send and receive.

3. Progressive deepening of problem representations during problem-solving:
   Progressive deepening occurs when a problem-solver retraces steps along a path he previously has taken. The retracing is done, however, in light of new information that has been acquired since the previous trip down the path. The phenomenon was first noticed in studies of chess players but occurs in design as well (Newell & Simon, 1972; de Groot, 1965). It is clear that progressive deepening is needed when a problem is too complex (and novel) to be understood in a single pass. However, an explanation of the phenomenon is needed. This explanation should include an account of the mechanisms that guide the selection and integration of new information with the information gathered in earlier passes.

   In the research presented here we find that progressive deepening takes the form of repeated simulations of the same set of mail commands at increasingly specific levels of detail. The repeated simulations can be used by designers to increase their understanding, both of the system's *mechanism* and *behavior*.

Below we describe currently existing Soar systems which form the theoretical perspective for the research described here. We then discuss our protocol data on software design. This is followed by a description of the Soar system suggested by the data, and a discussion of the issues relevant to developing an architecture that can support a unified theory of cognition.

2 Soar: A General Architecture for Cognition

Ultimately, the Soar architecture is intended to embody a *unified theory of cognition*; capable of accounting for the entire range of cognitive problems or "tasks". Additionally, it is expected to be able to do so by relying on the mechanisms of recursive sub-goaling and chunking.

Currently, Soar can solve a wide range of standard AI problems. It can solve most of the "toy" problems such as eight puzzle and tower of Hanoi (Laird & Newell, 1983). Toy problems are ones which require goal-oriented action, without requiring knowledge about the problem domain. It can also solve knowledge-intensive, expert system tasks, such as those solved by the VAX configuration system, R1; and the medical diagnosis system, Neomycin. The weak methods such as generate and test, means-ends analysis and a variety of search strategies have been implemented. Additionally, Soar exhibits learning phenomena such as learning with practice, transfer across tasks and generalization. Figure 1 provides a list of the tasks now performed by Soar.

In terms of accounting for tasks that are central for a theory of cognition, research is now being conducted to look at learning by analogy; and reasoning with mental models (Golding,1988; Polk, 1988; Steier & Kant, 1985). The research presented here is intended to be part of this effort to extend the range of Soar's performance to complex cognitive tasks.

2.1 The Nature of Problem-Solving in Soar

As to the nature of the theory, Soar grows out of three decades of work by Newell and his colleagues (Newell, Shaw, & Simon, 1960; Newell & Simon, 1972; Card, Moran & Newell, 1980; Forgy &

Small, knowledge-lean tasks (typical AI toy tasks):
Blocks world, eight puzzle, eight queens, labeling line drawings (constraint satisfaction), magic squares, missionaries and cannibals, monkey and bananas, picnic problem, robot location-finding, three wizards problem, tic-tac-toe, Tower of Hanoi, water-jug task

Small routine tasks:
Expression unification, root finding, sequence extrapolation, syllogisms, Wason verification task

Knowledge-intensive expert-system tasks:
R1-Soar: 3300 rule industrial expert system(25% coverage)
Neomycin: Revision of Mycin(initial version)
Designer: Designs algorithms(initial version)

Miscellaneous AI tasks:
Dypar-Soar: Natural language parsing program(small demo)
Version-spaces: Concept formation(small demo)
Resolution theorem-prover(small demo)

Weak methods
Generate and test, AND/OR search, hill climbing (simple and steepest-ascent), means-ends analysis, operator subgoaling, hypothesize and match, breadth-first search, depth-first search, heuristic search, best-first search, A*, progressive deepening (simple and modified), B* (progressive deepening), minimax (simple and depth- bounded), alpha-beta, iterative deepening, B*

Multiple organizations and task representations:
Eight puzzle, picnic problem, R1-Soar

Learning:
Learns on all tasks it performs by a uniform method(chunking). Detailed studies on eight puzzle, R1-Soar, tic-tac-toe, Korf macro- operators
Types of learning:
Improvement with practice, within-task transfer, across-task transfer, strategy acquisition, operator implementation, macro-operators, explanation-based generalization

Figure 1: Summary of Tasks Performed by Soar. (From Laird, Newell & Rosenbloom, 1987.)

McDermott, 1977; Newell, 1980). In this body of research problem solving is characterized as movement through successive *states of knowledge* in a *problem space* in order to achieve a *goal.* The problem-solver starts out in an initial state which contains an incomplete representation of the problem solution and a description of what would constitute a sufficient solution. The description of the solution could be, for example, the desired behavior for the mailer, whereas the solution itself would be a pseudocode specification of the mechanism producing the behavior.

The problem solver's relevant knowledge is then brought to bear and the initial representation of the problem is transformed in a way that brings it closer to the goal state representation; the problem solution. Relevant knowledge may consist of specific information about the problem domain as well as general problem-solving strategies.

2.2 Elements of the Architecture

To model problem-solving as it is framed above we need to be able to provide accounts of:

1. The representation of the current problem solution at varying stages of completion.
2. The representation of whatever is known about the desired problem solution.
3. Knowledge about how to assess and transform the partial solution with regard to the desired solution.

The above are realized using the following elements of the Soar architecture:

- Production Memory (PM). This encodes long-term knowledge that can be applied during problem-solving. (Point 3 above.) This can include factual knowledge about the problem being solved, strategic knowledge about how to proceed in problems like the current one and operational knowledge about specific problem-solving moves to be made in a given situation. It is the use of this operational knowledge that transforms the problem solution from an early insufficient version to a desired or goal state version. The productions that contain this type of operational knowledge place operators into working memory. When these operators are applied (Section 2.3), they transform the solution-in-progress.
- Working Memory (WM). This holds the representations of the current and desired problem solution described in points 1 and 2 above. WM also holds long term knowledge that has been identified as relevant.
- The Decision Cycle. This brings the appropriate knowledge in production memory to bear given the state of things in working memory. The difference between the starting and goal states is reduced through the decision cycle. The decision cycle is made up of two phases:
    1. An *elaboration* phase that causes already known information in production memory to be added to working memory. Information in production memory is added to working memory if it is relevant to what presently is in working memory. Elaboration is acheived by a matching process: The antecedents of all productions in PM are matched against the contents of WM; all productions that do match "fire" causing the objects described in the productions' consequents to be placed in WM.
    2. A *decision* phase that makes problem-solving decisions based on the information in working memory. The decision process begins once the elaboration process has added all that it currently can to working memory.

2	8	3
1	6	4
7	b	5

1	2	3
8	b	4
7	6	5

Figure 2: A Starting State (top) and A Goal State (bottom)for the Eight Puzzle. (The "b" represents the blank cell.)

2.3 The Architecture in Action: An Example of Problem-Solving in Soar

Below we present a description of Soar solving the eight puzzle. It illustrates how the elements of the architecture function in order to move the problem-solver through the problem space towards the goal state.

The eight puzzle is a typical AI toy problem, calling for goal-oriented, but knowledge-lean behavior. The problem-solver is presented with a board containing nine squares and eight numbered tiles. The goal is to transform the tiles' starting configuration into the goal state configuration (Figure 2). The eight puzzle is chosen here *not* as a representative cognitive task, but because its simplicity allows us to focus on the Soar architecture. Issues of Soar systems as cognitive models are dealt with in Section 4.

As problem-solving begins, Soar begins its problem solving by explicitly setting the goal of solving the eight puzzle. In the first decision cycle, "Eight-Puzzle" is proposed and selected as the current problem space, so that only productions relevant to that problem space will fire.

As we can see in line 0D of the trace of the program (Figure 3), Soar begins its problem solving by setting a top-level goal of solving the problem at hand. In the first decision cycle, "Solve-Eight-Puzzle" is proposed and selected as the name of the current problem. Additionally, "Eight-Puzzle" is proposed and selected as the current problem space, so that only productions relevant to that problem space will fire.

As we will see below, problems can be divided into sub-problems, each having its own problem space with operators appropriate to the sub-problem. During the second decision cycle, The initial and goal board configurations are placed in WM.

The third decision cycle begins, as usual with an elaboration phase; based on the initial configu-

```
Cycle Action
0D    Solve the problem is the current goal

1E    Propose Solve-Eight-Puzzle as the name of the current goal
1E    Propose Eight-Puzzle as the problem space
1D    Select Solve-Eight-Puzzle as the name of the current goal
                           Select Eight-Puzzle as problem space

2E    Place description of goal state board  configuration in WM
2E    Place description of current board  configuration in WM
2D    Select S1 as the current state (with the initial and
      goal states in WM)

3.1E  Propose 01 (down)
3.1E  Propose 02 (right)
3.1E  Propose 03 (left)
3.2E  Mark down as best (it moves 6 into its goal position)
3D    Select 01 (down) as operator

4.1E  Propose S2 (swap 6 and blank)
4.2E  Copy the unmodified parts of the board
4D    Select S2 as state

5E    Propose 04 (down)
5E    Propose 05 (right)
5E    Propose 06 (left)
5E    Propose 07 (up)
5E    Mark 7 as worst (it undoes the last move)
5D    Tie impasse, create subgoal
```

Reading from left to right:
The left most column (labeled "Cycle") indicates the decision cycle (D and E stand for decison and elaboration respectively). The next column indicates the action taken in that cycle.

Figure 3: Trace of 5 Decision Cycles For 8 Puzzle. (From Laird, Newell & Rosenbloom, 1987.)

2	8	3
1	6	4
7	b	5

(starting state)

2	8	3
1	b	4
7	6	5

(new state)

Figure 5: Board State After Applying **down** Operator to Starting State

into play.

2.3.1 Sub-Goaling to Resolve Impasses

When an impasse is reached the Soar architecture sets up a sub-goal to resolve the impasse[3]. For a tie impasse in the eight puzzle the sub-goal is to "select" an operator from the set of possible ones. This is achieved by moving into the *selection problem space.* A further sub-goal results in which the candidate operators are actually tried out (by moving back into 8 puzzle space) and the state that will result from each one is evaluated. For this example (Figure 5, bottom), **down** is found to be best since it will lead to a follow up move that will allow the 2 to be moved into place. The other two candidates are found to lead to moves that will move tiles out of their desired spots.

Three points are important here: 1. The detection of the impasse and the setting up of the appropriate type of sub-goal is *not* done by the task specific eight puzzle productions; it is done by the architecture. 2. Once a sub-goal is established it is pursued and resolved in the *same* way as a higher level goal. A problem space is selected; a current and goal state are defined; and operators are then applied to the current state in order to transform it into the goal state. 3.

[3]Here we see a tie impasse, however, no-change, conflict and rejection impasses occur as well (Laird, Newell & Rosenbloom, 1986). In fact, "no-change" impasses often arise in complex problem solving domains such as design. A no-change impasse can occur when a problem-solving action is decided on, but how to take the action is not known. (The impasse is called no-change because an action has been decided upon but does not lead to a change in the problem-solver's state). For example, when trying to solve the problem of how to spend lottery winnings a decision may be made to take the action of buying a new car, but how to go about choosing one may not be known. A no-change impasse has now occurred. The impasse can be resolved by setting a sub-goal to gather information relevant to making the choice.

Production: Instantiate-Operator:
If:
the current problem space is 8 puzzle & the current board has a tile in a cell adjacent to the blank's cell
Then:
place a new operator in WM that will move the tile into the blank's cell
& place in WM the information that the operator is acceptable

Production: A-Means-Ends-Analysis-Operator
If:
the current problem space is 8 puzzle
& an operator will move a tile into its goal state position
Then:
place in WM the information that the operator is the best choice

Production: Avoid-Undo:
If:
the current problem space is 8 puzzle
Then:
place in WM the information that the operator to move the tile that was last moved is a worst choice.

Figure 4: Productions creating candidate operators and their acceptability values

ration, "`6 down`", "`5 left`" and "`7 right`" are all proposed as "operators" (moves). Information is also placed in WM indicating that the `down` operator is preferable since it will result in moving the 6 into its desired position.

This information about operators and their preferability is placed in WM as a result of productions firing based on the contents of WM. Figure 4 shows the productions that fired during the third decision cycle. Instantiate-Operator fired three times, proposing `down`, `left` and `right` as acceptable operators (line 3.1E in Figure 3). A-Means-Ends-Analysis-Operator fired next (line 3.2E) marking `down` as the best operator.

When all of the productions that can fire have fired, elaboration is said to have reached "quiescence"; the decision process begins and `down` is selected as the best operator in the current situation or "context"[2]. This ends the third decision cycle. The fourth decision cycle ends when the `down` operator is "applied"; the productions relevant to the operator fire (lines 4.1E and 4.2E), a new board configuration results (Figure 5) and a new "current state" is decided upon (line 4D).

Looking at the elaboration phase of the fifth decision cycle in Figure 3 we see that `down`, `right`, `left` and `up` are all placed in WM as acceptable candidate operators. Additionally, `up` is marked as unacceptable by the Avoid-Undo production (Figure 4). We also see that after the elaboration phase ends the decision process reaches a "tie impasse" (line 5D); WM contains four candidate operators and all of the information provided in the elaboration phase says *only* that three are acceptable. The information in production memory about this problem space is incomplete; it cannot resolve the tie. However, here is where the notion of sub-goals and problem spaces comes

[2]In Soar the current situation is termed the current context and is specified by the value of the current goal, problem-space, state and operator.

This sub-goaling can occur to an arbitrary depth. These three points lead to some of the appeal that Soar has as a theory. By being able to detect impasses and set up appropriate sub-goals, the architecture, the part of Soar which is specified in advance and remains constant across tasks, does a good deal of the problem-solving. Additionally, the ability to solve problems in this uniform way (by recursive sub-goaling) allows Soar to provide a parsimonious account of complex problem solving.

Turning to software design we will see that organizing the problem-solving into problem spaces continues to be ueful. We will also look at the way in which the problem spaces are related and how information from one problem space can further problem-solving in another.

3 Modeling Software Design within Soar

3.1 Method

Below we present our data on software design.
Subjects.
Three expert software designers served as subjects[4]. Each of the experts had worked for at least eight years in commercial settings designing a wide variety of software.

Procedure.
We presented each of the designers with the following design task to work on.

Task:
Design an electronic mail system around the following commands:

> `READ`, `REPLY`, `SEND`, `DELETE`, `SAVE`, `EDIT`, and `LIST-HEADERS`. The goal is to get to the level of *pseudocode* that could be used by professional programmers to produce a running program. The mail system will run on a very large, fast machine so hardware considerations are not an issue.

This task[5] had several important properties: (1) It was non-trivial, it required close to two hours of the subject's time. (2) It was novel, none of the designers had designed a solution to the problem previously. These two properties meant that we would have the opportunity to see not only "routine cognitive skill", but problem-solving as well. (3) The problem we chose was similar to the types of problems which the subjects had to deal with professionally. As a result we would be able to see them using the general design skills that they had acquired over time.

3.2 Analysis of the Protocol Data

Generally protocol data can be seen as a series of episodes, with each episode reflecting the single, current focus of the subject's attention.

---

[4]The protocols from all three subjects were analyzed In Appendix 1 we present a long section of the protocol of S1. S1's protocol is representative of the protocols of all three subjects and is focussed on in this chapter.

[5]We have also studied novices designing the same system as well as this group of experts designing an interrupt handler and a data base system (Adelson & Soloway, 1985).

3.2.1 Pairs of Episodes

One striking aspect of the protocol discussed here is that the episodes formed related pairs. The first episode in a pair appears to take place in a general design space and the second episode appears to take place in a space containing knowledge about mail systems. The first four episodes from S1 are representative of this phenomenon.

Episode 1:

> S: "....I'm going to start working here, functions of an electronic mail."
> {writes 'Functions' and 'Data' in two separate columns}

Episode 2:

> "We must be able to:
> Prepare,
> Send,
> Receive...
> {writes prepare, send, receive under the heading 'Functions'}
> ...the system must be able to store them,
> the system must be able to handle abnormalities throughout it."

In episode 1 S1 decides to view the system as a set of functions. In episode 2 he goes on to enumerate what those functions would be.

In episode 3 S1 decides to view the system as a set of data objects and in episode 4 he elaborates the features of those objects.

Episode 3:

> "Let's stop there and take a fresh cut at the data."

Episode 4:

> "there must be some data of the users,
> e.g the destinations,
> the addresses;
> also data of update formats;
> data.....,
> there must be the message store itself.
> The message store would come in several forms:
> the store for messages sent but not received,
> the store of messages received by some individual,
> I think of unix, of the dead letter concept
> - the store of messages that we can't do anything with
> (what) if we want to store them.
> "

In Table 1 we present a summary of episode pairs for the first 10 episodes of S1's protocol. The first episode in each pair establishes the goal of viewing the sytem from a particular *perspective*:
View 1. View the system as a set of functions: Episode 1.
View 2. View the system as a set of data objects: Episode 3.
View 3. View the system as a set of concurrent functions: Episode 5.
View 4. View the system as a state machine for states of user: Episode 7.
View 5. View the system as a state machine for messages: Episode 9.

The second episode instantiates the view as a simulation. Both the content and the ordering of these episode pairs suggest that the designer is using an experience-based schema to direct the design session. It appears that the designer starts off with an incomplete representation or model of the design and then, to attain a finished design, simulates that model from each of these perspectives in order.

3.2.2 Schemas

The design process seems driven by an experience-based schema for two reasons: First, successive episodes do not appear to arise from the context that immediately precedes them; in episodes 3 and 4 the system is viewed as a set of data objects and then in episodes 5 and 6 as a set of concurrent processes. Second, the particular views chosen, such as dealing with concurrency issues, would be ones to develop given these designers' experience with communications systems.

The structure of the schema is also interesting; taken in order the five views comprise a set that would be effective in uncovering most of the aspects of the system that need refinement. The first view looks at the commands of the mailer, the second view at the messages themselves. Once the commands have been specified it becomes possible to look at the interactions produced when they are used in sequence, as they will be by the system's users. This is uncovered by the fourth view which looks at the system as a state machine from the user's perspective. It also becomes possible to look at their concurrent functioning (view 3). The fifth view looks at the interaction of commands and messages; it therefore, is dependent upon having specified the commands and messages in the first and second views.

The issue arises as to when during problem-solving will this schema be used. It is likely the prolem-solver will re-use previous solutions or good analogies whenever possible (Carbonell, 1983, 1986; Burstein & Adelson, 1987; Burstein & Adelson, this volume; Kolodner, 1984; Adelson & Soloway, 1985). However, in the absence of right, or "almost-right" (Sussman, 1975) solutions it then becomes likely that the problem-solver will turn to higher level, experience based schemas.

The pairing of the episodes is explained by the schema that was used. The schema suggested views of the system that would lead to refining the design. The first episode in a pair sets up a view and the second episode simulates it.

3.2.3 Interaction of Domain and General Knowledge

In designing a large software system the designers employ three types of knowledge; knowledge of *design*, knowledge for representing *systems as pieces of pseudocode* and knowledge of how mail systems *behave*. From our perspective, these bodies of knowledge can be seen as three *problem spaces*; a *design* space, a *pseudocode* space and a *mail* space (see Section 4). In simulating views of the system there has to be a mapping between the the designer's high-level problem space for

Episode 1. View the system as a set of functions.
Episode 2. Simulate the behavior of the system's functions. The commands prepare, send receive and store must be included in specifying the mailer's functionality. Discover that error recovery must be handled gracefully throughout the system.

Episode 3. View the system as a set of data objects.
Episode 4. Elaborate the features of the data objects . In the mailer messages are the data objects. Messages have destinations of senders and receivers. Additionally, messages are grouped together in stores. The stores can have various functions. For example the mail system needs a store for messages that the user has received but not yet read, as well as a store for messages that have been read but not yet saved or deleted.

Episode 5. View the system as a set of concurrent functions.
Episode 6. Simulate the behavior of a system in which there are concurrent senders and receivers. Discover that the design needs to specify when users should be notified that new mail has arrived (as it arrives, only at logon, etc). Also discover that, since mail is both being sent and received, more than one type of processing must be handled and therefore a "dispatch demon" is needed to handle the flow of messages.

Episode 7. View the system as a state machine for the states of a user.
Episode 8. Simulate the behavior of the system as a scenario in which the user logs on and issues a sequence of mail commands. He is notified that he has mail, he lists the headers, he reads a message, he makes some disposition of the message and then is able to begin again (listing headers, etc.). The designer discovers that the post-conditions of the commands need to be enumerated both to refine the command definitions and to understand the potential interactions between commands. For example, if `READ` includes an implicit and immediate `DELETE` it will prevent the user from being able to save or forward the message.

Episode 9. View the system as a state machine for the states of messages.
Episode 10. Simulate the behavior of the system in terms of the states of the messages. A message is created, sent, received, read and disposed of.
These actions described at the level of files and locations within files are sufficient to generate pseudocode.

Table 1: Description of Episodes 1 through 10.

design, in which the schema resides and the *domain* space where knowledge about the *behavior* of mailers resides; it is in this domain space that the view is instantiated, run, and evaluated. For example, in episode 7 the designer chooses to view the system as a state machine in which the user goes through a sequence of state transitions. This gives rise to episode 8, in which the designer instantiates this state machine by constructing a simulation in which the user logs on, is notified that he has mail, lists the headers, reads a message, makes some disposition of the message and then begins again.

In order to model this piece of problem-solving, what needs to be be accounted for?
The goal to look at the system as a state machine in which a user goes through state transitions needs to be represented. A mapping is then needed in order to instantiate this goal in design space as an executable "reception scenario" in mail space. This means that the user's state transitions need to be put into correspondence with the issuing of commands such as `READ` or `SAVE`. Additonally, data objects must be understood to correspond with messages.

In the domain space there needs to be enough knowledge about the behavior of the mail commands to propose and simulate candidate versions of them. The candidate versions then have to be evaluated by comparing their behavior to some representation of the ideal behavior. The candidates can then be modified in accord with the results of the evaluation. The ability to modify a representation of a command in pseudocode space based on the results of a simulation in mail space implies a mapping between mail space and pseudocode space. All this of reasoning is done by the designer in Episode 10. The designer simulates `SAVE` and then realizes that in order to save a message its number must be specified when the command is issued.

> "...We can save it somewhere – it goes into some state...
> When we save them off in a file like that are we going to save them as mail objects, or are we going to just append them on so they're not distinguishable. I now need to start making some decisions on that.
>
> "Looking at the (commands) and what they operate on, we're going to assume in many cases that there are some *store* of messages.
> We're operating on individual messages themselves.
> So I've got to have some way of identifying where it is within that store...
> ...Let's just number them (the messages).
> Save is going to take some mail number from his mail store, and move it over to his saved mail."

3.2.4 Progressive Deepening

As mentioned above, progressive deepening is the retracing of steps along a previously taken problem-solving path. The retracing is done because the first trip down the path was not sufficient; and new, relevant information has been acquired. Both deGroot (1965) and Newell and Simon (1972) found progressive deepening in chess and Newell and Simon suggest that progressive deepening may be one of the hallmarks of complex problem-solving. Below we present an example of progressive deepening from our data. This is followed by an explanation of why progressive deepening would occur within Soar.

In the following sequence of episodes we see the progressive deepening of the definition of the store command.

In episode 2 the commands are just written down by the designer as a list of functions to be specified.

> "So I'm going to start working here, functions of an electronic mail, we must be able to prepare, send, receive...
> the system must be able to store messages,
> the system must be able to handle abnormalities throughout it."

In episode 8 the designer simulates these same commands using each one's output as input to the next and discovering that the side-effects of each have to be elaborated. For example, below he decides that a user should be able to list all the message headers without being committed to then reading them.

> "Let's look at a reception scenario ...
> He'll look at all of them in the store...
> (I am) making a conscious decision that he is going to have the option of looking at all the mail items before he jumps in and starts reading them, so he's got a survey capability here.
> That tells me that we've got to start making more design functionality decisions as we go through here...
> (Looking at) this store (of messages) brings him to a new state where he knows what mail he has
> (If, as a result of listing the headers) he reads it, he's got to make a disposition of it, what he's (going) to do with it. Whatever disposition we decide - destroy, store, most of those dispositions are going to get him back to his initial state."

In episode 10 the designer finally simulates the commands at a level sufficient to generate pseudocode. Here he uses language that refers to reading from and writing to files:

> "Each user of the system will have his own mail store,
> I can save a message into an identified store of messages.
> So I've got to have some way of identifying where it is within that store.
> Let's number the messages in the store.
> So when I store, I want to identify the mail number and the mail store..."

Simulation *and* progressive deepening arise naturally within a Soar architecture. There are two reasons why simulation occurs within Soar. The first reason has to do with seeing complex problem solving as occuring in a *set* of related problem spaces. Problem-solvers have different *types* of representations of the problem solution in different problem spaces. Additionally, the information contained in one type of representation may contribute to the development of another. In design, simulations arise because detailed information about the behavior of a mail command, obtained from a simulation in mail space, can help in developing the representation of the command in pseudocode space.

The second reason for simulation has to do with comparing current and goal states. In designing the mail system, the designers *represent* the goal state in terms of the desired *behavior* of the mail system. However, the problem solution is a pseudocode description of the mailer. In order to compare this pseudocode representation to the goal they need to *simulate* the pseudocode.

The simulations done by the designers need to go through progressive deepening; at the beginning of the design session the designers' representation of the mailer is in terms of the high level *behavior*

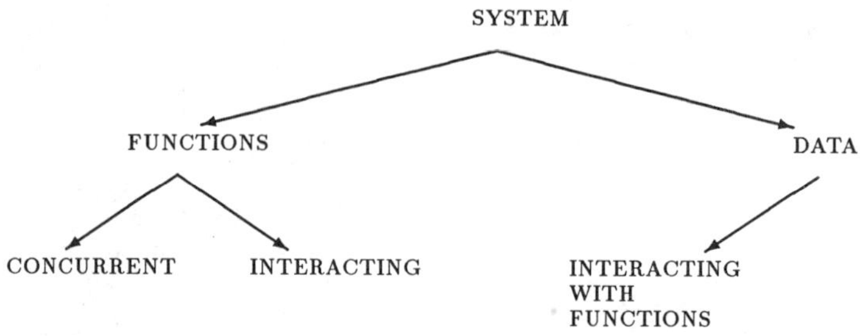

Figure 6: Hierarchy of Design Space Operators

of the system. This is the representation that would be likely for a person who had used, but not actually designed such a system. The representation of the behavior needs to be refined to a degree that allows that behavior to be expressed as pseudocode. But in a task that has a complex solution there are many aspects to the refinement. The use of repeated simulations, from different perspectives, allows the designer to attend to different aspects of the refinement in a systematic way. This allows the designer to bring his understanding to the required degree of specificity without overloading working memory.

4 Sketch of the Mail Designer-Soar System

Recall that in the context of Soar, problem solving is characterized as movement through successive *states of knowledge* in order to achieve a *goal*. The states of knowledge contain representations of the problem at various points in the problem-solving process. Additionally, the different aspects of the problem are regarded as different problem spaces in which different, appropriate, kinds of knowledge are brought to bear.

In order to model the design process within the Soar framework we need to provide accounts of:

1. The initial and goal state representations which form the system's input and output.
2. The problem spaces with their appropriate operators.

As to point 1: In the initial state there is a representation of the desired behavior of the mailer. The goal state would be a pseudocode description of the mailer's commands.

On point 2: The problem solving consists of trying first to apply existing knowledge relevant to a *pseudocode* problem space in order to represent the design as pieces of pseudocode. If existing knowledge is not sufficient to directly represent the design in terms of pseudocode the behavior of the mailer is simulated in *mail* space. The simulations are repeated, from varying perspective and in increasing detail until the designer understands the system's behavior at a level that allows it to be expressed as pseudocode. The set of perspectives used in the simulations in pseudocode and mail space are generated by the strategic knowledge in the *design* problem space. Table 2 lists the operators that apply in the "design", the "pseudocode" and the "mail" problem spaces.

The design space operators form a hierarchical tree of perspectives from which to view the system (Figure 6). The perspectives direct simulations of the design-in-progress both in pseudocode and

Design Space Operators:
THE OPERATORS CAUSING DIFFERING VIEWS OF THE SYSTEM TO BE TAKEN.

1. Design the system as a set of functions.
2. Design the system as a set of data objects.
3. Design the system as sets of concurrent functions.
4. Design a state machine of the states a user goes through.
(Resulting in a focus of attention on interactions between functions.)
5. Design a state machine of the states a message goes through.
(Resulting in a focus of attention on interactions between function and data.)

Pseudocode & Mail Space Operators:
THE OPERATORS TO GENERATE, RUN AND EVALUATE CANDIDATE VERSIONS OF THE FUNCTIONS' MECHANISMS IN PSEUDOCODE SPACE AND THE FUNCTIONS' BEHAVIOR IN MAIL SPACE.

1. Prepare.
2. Send.
3. Receive.
4. List-Headers.
5. Read.
6. Store.
7. Delete.

Table 2: Operators in the Design and Mail Spaces

mail space. The tree is traversed in a breadth-first order, this leads to the systematic ordering of the perspectives taken during the design process.

In "pseudocode space" the system is represented as pieces of pseudocode. These pieces are repeatedly executed and refined until the design of the mailer is complete.

5 Conclusions

We have described modeling software design within a Soar framework. Using this framework we are able to provide accounts for the following:

1. The role of schemas in bringing general knowledge to bear on knowledge about a domain: In the data presented the designer uses a high-level schema in order to create an ordered set of pairs of episodes in which a variety of inter-dependent aspects of the mailer are considered and refined.

2. The role of simulation:
Simulation supports the design process in two ways. It allows the comparison of current and goal states when the current state is represented as a mechanism and the goal state is represented as behavior. Additionally, simulation supports the development of a representation of the mechanism of the system being designed when the system's behaviors are simulated at a level of detail that allows the behaviors to be expressed as mechanisms.

3. The role of progressive deepening:
Because a set of simulations from a variety of perspectives are needed to complete the design we find that the same set of commands is simulated repeatedly at increasing levels of refinement.

We are optimistic that a Soar framework will continue to support detailed accounts of the mechanisms that underlie cognitve problem-solving skills.

REFERENCES

Adelson, B. and Soloway, E. The role of domain expertise in software design. *IEEE:TSE*, November, 1985.

Adelson, B. and Soloway, E. A model of software design. *International Journal of Intelligent Systems*, Fall, 1986.

Burstein, M. Concept Formation by Incremental Analogical Reasoning and Debugging. In Machine Learning. Los Altos, CA: Morgan-Kaufmann. 1985.

Burstein, M. and Adelson, B. Analogical learning:
Mapping and integrating partial mental models. In *Proceedings of the Cognitive Science Society, 1987.*

Card, S., Moran, T. and Newell, A. Computer text editing. *Cognitive Psychology, 12, 1.* (1980) 32-74.

Carbonell, Jaime G. Learning by Analogy: Formulating and Generalizing Plans from Past Experience. In Michalski, R.S., Carbonell, J.G. and Mitchell, T.M. (editor), "Machine Learning: An Artificial Intelligence Approach", Pgs. 137-162. Tioga Publishing Company, Palo Alto, California, 1983.

Carbonell, Jaime G. Deprivational Analogy: A Theory of Reconstructive Problem Solving and ExpertiseAcquisition. In Michalski, R.S., Carbonell, J.G. and Mitchell, T.M. (editor) "Machine Learning: Volume II", Pgs. 371-392. Morgan Kaufman Publishers, Inc., Los Altos, CA, 1986

de Groot, A.D. Thought and Choice in Chess. Paris: Mouton & Cie. 1965.

Forgy, C. and McDermott, J. OPS, a domain-independent production system language. Proceedings Fifth International Joint Computer Conference, MIT AI Lab, Cambridge, MA. 1977.

Golding, A. Learning to Pronounce names by taking advice. Thesis proposal, Stanford U. 1988.

Kant, E. and Newell, A. Problem solving techniques for the design of algorithms. *Information Processing and Management.* 1984. pp. 97-118.

Kolodner, J. *Retreival and Organizational strategies in Conceptual Memory: A Computer Model.* Hillsdale, NJ: Erlbaum. 1984.

Laird, J. Newell A. and Rosenbloom, P. Soar: An architecture for General Intelligence. CMU CS Tech Report. 1986;

Newell, A. The Problem Space as a Fundamental Category. In Attention and Performance VIII, R. Nickerson, Ed. Erlbaum, Hillsdale, NJ, 1980.

Newell A., Shaw, J. and Simon, H. A. Peport on a general problem-solving program for a computer. Proceedings of the International Conference on Information Processing, UNESCO, Paris, 1960.

Newell A. and Simon, H. A. Human Problem Solving.
Prentice-Hall, 1972.

Polk, T. Fourth Annual Soar Workshop. University of Michigan, Ann Arbor. January, 1988.

Schank, R., and Abelson, R. Scripts, Plans, Goals and Understanding. Hillsdale, NJ: Erlbaum, 1977.

Steier, D. M. Proceedings of the 1987 IJCAI Conference.

Steier, D. M. and Kant, E. *IEEE:TSE*, November, 1985.

Sussman, G. *A Computer Model of Skill Acquisition.* NY, NY: American Elsevier Publishing Corp. 1975.

Steier, D. Integrating Knowledge into an Algorithm Designer. Thesis proposal, Carnegie-Mellon U. 1986.

Appendix 1: Episodes in the Protocol

EPISODE 1
L1: (writes 'functions') The view I'm going to take as I go through this system,

L2: (writes 'data') keep looking at the data and the information component behind us,

EPISODE 2

L3: so I'm going to start working here, functions of an electronic mail,
L4: we must be able to prepare,
L5: send,
L6: receive
L6a: (writes prepare, send, receive)
L7: in receiving a number of choices and decisions have to be made
L7a: as to the disposition of it
L8: we'll get into that later;
L9: the system must be able to store them,
L10: the system must be able to handle abnormalities throughout it.

EPISODE 3
L11: O.K. Let's stop there and take a fresh cut at the functions.
L12: And take a first cut at the data and what would be behind it.
L13: E: Can you say why you stopped at that point?
L14: I feel I've gone as deep as I need for the moment.
L15: I recognize I'm going to go deeper into each of these,
L16: but I'm just taking a top layer of it first of all.
L17: {I'm looking} over here and see if from the data side I can recognize other views that I haven't seen through this approach
L18: and we'll be sure they interact between the two of these approaches, each time going down more and more in detail.

EPISODE 4
L19: (begins writing under 'data') For data,
L20: there must be some data of the users,
L21: {e.g} the destinations,
L22: the addresses;
L23: {also} data of update formats;
L24: data.....,
L25: there must be the message store itself.
L26: The message store would come in several forms:
L27: the store for messages sent but not received,

L28: the store of messages received by some individual,
L29: I think of unix of the dead letter concept
L30: - the store of messages that we can't do anything with
L31: {what} if we want to store them.

EPISODE 5
L32: Throughout the system, it is going to be handling concurrency,
L33: so that is the initial event I need to address.

EPISODE 6
L34: For the moment, let us break it down.
L35: Concurrency among the senders and receivers.
L36: There must be some function other than senders and
receivers, which is periodically monitoring what is
happening to our message store;
L37: I think of unix.
L38: We have to make strategy decisions on how we deliver messages
to the receiver of it;
L39: whether we deliver when the receiver logs on or when the
message is sent,
L40: or whether we deliver through having some continuously running
program,
L41: some demon type of program that comes in periodically, looks around
and says, 'Can I do anything with the message?'
L42: So there must be some sort of dispatching philosophy.
L43: (writes 'dispatching philosophy')

EPISODE 7
L44: What's coming to mind is that here we've got,
L45: in fact we can look at the states for our messages,
L46: the states of this whole system,
L47: and take another view as well as the function-data view.
L48: Think of it as a big state machine,
L49: what is happening to it so that we can see that for a while...
L50: (paper change, comments about papers)
L51: States:
L52: What are the things we are dealing with?
L52a: Objects, (writes 'Objects) {in} terms of events.
L53: Objects: we've got users,
L54: we've got message mail.
L55: Events, (writes 'events')
L56: new mail ready,
L57: receivers log on.
L58: O.K. Let me start looking at the states of a user first
of all. How in the world is this going to be seen from
the view of the user?

EPISODE 8

L59: He starts out, start him out with some state of, let's call it an initial state first of all.

(begins drawing a state diagram)

L60: Let's look at a reception scenario and
L61: through log on he becomes capable of receiving messages,
L62: receiving mail
L63: --either there is mail or there is not mail.
L64: If there is mail,
L65: and this may be because he gets special messages on,
L66: or because some time later, after he's logged on, he is informed that there is mail and is willing to receive it.
L67: There must be another case where there is mail but not willing.
L68: There is mail and willing to receive it.
L69: And he starts.
L70: He'll look at all of them in the store,
L71: {I am} making a conscious decision that he is going to have the option of looking at all the mail items before he jumps in and starts reading them, so he's got a survey capability here.
L72: That tells me that we've got to start making more design functionality decisions as we go through here and,
L73: in effect, build up the user's specification over the functional specification of the system as seen by the user.
L74: For the purposes here, this will go into user manuals
L75: - Through the lack of any user input we're going to give it this (survey) capability
L76: That is our decision here.
L77: (Looking at) this store {of messages} brings him to a new state where he knows what mail he has
L78: He can choose at this state to exit and get out to some other place we don't know yet where it is.
L79: Going to read mail,
L80: read monitoring.
L81: When he reads it, it's got to make a disposition of it what he's got to do with it.
L82: Most of those are going to get him back to his initial state.
L83: Whatever disposition we decide - destroy, hold, reply, whatever else will get him back there.
L84: I think it will be very obvious now that I've got to decide what those functions are from the users point of view before I can continue much farther with this design.
L85: So let me note that. (makes a note)
L86: At this stage, I need the functional specifications.
L87: I'll come back and do that in a little while, time permitting,
L88: and carry on making some other assumptions here.
L89: (Pointing to and looking at each successive command in the drawing of the test case being simulated in the last 5

episodes):
L90: Store...
L91: Read, OK, That's good enough for the user view of it.

EPISODE 9
L92: Let's look at mail messages.

EPISODE 10
L93: Any one mail message starts out conceptually as nonexistent and then we create it, prepare it.
L94: So there must be a function of editing mail
L94a: and deciding whether we are going to send it.
L95: Let's hide all of that for the moment and say we prepare and send.
L96: When we say send, there must be some function of routing it.

L97: Prepare and send -
L98: So then that mail message will normally go into a case where it's waiting to be received,
L99: but the receiver may be available immediately, in which case, it would be presented to him.

L100: Let's split apart those two actions into two different states, so it's waiting to be received
L101: Then there's some command that says, "Read it," -- becomes read.
L102: Then there's some command that says, "Delete"
L103: and goes back here -{to read}
L104: we can save it somewhere -- it goes into some state.
L105: I'm thinking, what is the form of these message.
L106: When we save them off in a file like that are we going to save them as mail objects, or are we going to just append them onto they're not distinguishable. I now need to start making some decisions on that.
L107: Let me look back at my primitives.
(looks back to previous sheet)

L108: Read, Reply, Send, Delete - I'm going to make an assumption that these primitives are operating on mail themselves - read a mail message as well as to read a line of something. Reply to a mail message, delete a mail message, store, the data view of this system, works in units of mail messages.

L109: O.K. I don't see here any primitives that will help me in preparing a message. There's an edit primitive which edits an existing message?.......
L123: Looking at the primitives and what they operate on.
(begins looking back at primitives)
L124: We're going to assume in many cases that there are some

store of messages
L125: We're operating on and individual messages themselves.
L126: For instance "list headers",
I can give that an argument that says, list headers
in this particular store of messages, and it will list
them out.
L127: I can send an individual message and the primitive
will have enough intelligence to take the addresses
within it and dispatch it off to the correct recipient.
L128: I can save a message into an identified store of messages.
L129: I can read a message from a store of messages.
L130: So I've got to have some way of identifying where it is
within that store.
L131: So if I make a mental model of this thing called
message store, mail store, which contains mail
messages. Let's just number them. The first one is
always one, etc. So when I read, I want to identify
the mail number and the mail store.
L132: Each user of the system will have his own mail store,
there will be many of them, and the primitive will be able to
operate on the appropriate one.
L133: Reply, and what this does is it presents the mail message
to user.
L134: Delete again deletes a particular mail number from the mail
store.
L135: This will ... Decision to be made. When we delete, it will be
based on a user request, so we are in an interactive session
with the user. He's going to have this as his mental model of
what's happening with his mail messages. If we delete one of
them in the middle, we've got to decide if we change the order
of the messages or not.
L136: My decision is that this delete primitive will make no change
in the ordering of messages. The effect of delete takes place
at some later time after the session is over.
L137: Reply, takes from the user and replies to an unidentified mail
number in the mail store. So the reply primitive uses
this mail message to get the address of the sender of it and
replies back to him.
L138: Whereas Send, more general, send is just a new message, and
out of that new message the primitive is able to abstract
the destination of addresses.
L139: Save, save, save, brings us to our two main stores that we had
already made up there...save state.
Now we've got our model elaborated on to reveal that there is an
addition to each user having its mail store, he's got his saved
mail.
L140: does he get rid of saved mail? An issue to be decided.
(marks it)

L141: We may need to come back and give him the capability to re-read saved mail and decide what to do with it.
That grows slightly large.

L142: Save is going to take some mail, number from his mail store, and move it over to his saved mail.

L143: Edit. What's going through my mind here is that in giving me these seven primitives is what's helping me and hindering me because you're giving me a constraint that I've got to work with and not letting me construct that I would like to have, and I now find myself working around the seven primitives, trying to build up a system that will work in the presence of these seven primitives.

L144: This says edit. I've been assuming that a lot of the editing takes place hidden underneath it. I don't know why I would ever want to edit something that is already sitting inside in one of my mail stores. The editor normally takes place at the moment of preparation of the message which I have identified before, but I don't have primitives to deal with and that phase - under it. Edit, edit, edit, I'm at a loss as to how to interpret edit right now. Let's leave it aside for the moment.

L145: List headers is straight forward. List headers by how mail is stored.

L146: The function that's missing for this one (save) would be some form of pointing to the saved mail instead of the mail store. We would normally need to save, I'm aware of that.

L147: There might be some copy function, copying saved mail over to mail store, that is nothing new. We could make some other definitions of these primitives and play on the bahavior as to what happens in our model between mail stored and saved mail.

L148: For instance, it could be that when I log on to a session with the mail system, I list out the headers and then I go through reading. When I read one, it is automatically taken out of mail store and then my decision is whether I delete it or save it. Deleting just discards it; saves it would automatically move it over to saved mail, as opposed to this primitive of where we don't move things from the mail store automatically.

L149: So we get more decisions that have to be made to relate to the functionality as seen by the user. I think I've gone as far as I want to go with that one for right now.

PART 3

COMPUTATIONAL MODELS OF DESIGN PROCESSES

COMPUTATIONAL MODELS OF DESIGN PROCESSES

Gale E. Nevill, Jr.
Department of Aerospace Engineering,
Mechanics and Engineering Science
University of Florida
Gainesville, Florida 32611

ABSTRACT

This paper attempts three major tasks. The first is to give an overview of computational models of design processes. This provides a context for the second major task, to provide a descriptive review of the research on computational models of design being sponsored by the NSF Design Theory and Methodology Program. The third is to provide a critique of the Program and of the work being sponsored.

SCOPE

The focus of the paper is on models which describe what machines do autonomously. This may be part of an interactive system. Emphasis is on models which have evolved from VLSI design in EE, simple device design in ME, process synthesis in ChE, building design in CE, planning in AI and, to a limited extent, architecture and software engineering. The paper focuses on knowledge based design of physical systems and artifacts, with emphasis on preliminary (or rough) design and detail design phases.

The paper says little about problem description and clarification, representation, analysis in support of design or optimization approaches (covered elsewhere) or about problems with uncertain knowledge or specifications, how to match models to domains, human/machine cooperation and communication, explanation, getting CAD tools to work together, knowledge acquisition, machine learning, programming or hardware. The paper does not attempt to give a comprehensive review of the literature but does provide a number of general references plus cite specific sources that illustrate particular points.

1. INTRODUCTION

Computational models of design play two major roles. First, they are a necessary part of the development of more effective CAD tools in various fields. All CAD tools are based on some sort of model, either implicit or explicit. The effectiveness of the tools is strongly influenced by the power of their model and by its match with the domain, as well as the knowledge and computational resources available. The explicitness of the model strongly affects the users ability to understand and to improve the model.

The second major role of computational models is to support research into design theory and methodology. They serve as a readily controlled and understood test bed for exploring and validating ideas. Perhaps more important, efforts to develop autonomous design tools force us to make explicit aspects of design that are masked by the enormous flexibility and capability of human designers. Development of autonomous tools thus leads to an enhanced level of understanding and precision regarding basic design principles.

In general, design process models have evolved in response to the unique domain characteristics, needs and traditions of specific fields. This has led to significant differences as well as the expected commonalities. Domain differences and the differences in nomenclature make it particularly difficult for researchers in one field to follow and utilize developments in another. For this reason models tend to develop in relative isolation. Therefore some attempt has been made here to discuss models from all of the fields mentioned earlier. The principal domain factors influencing model development as well as applicability are discussed in Section 2.2 below.

A particular warning is in order regarding nomenclature. Different researchers in closely related fields as well as in different fields may use the same word to mean different things as well as different words for the same concept. For example, use of the term Goal [Mostow 85] for the concept which I call Subproblem caused initial confusion; in various places the terms Constraint, Interaction, Goal, Requirement and Decision have the same meaning (I think). Authors [Steinberg 87a] sometimes point out possible confusion such as the use of Refinement to mean structural decomposition rather than going from a more abstract level to a more specific one as in MOLGEN [Stefik 81a], but usually it is up to the reader to figure out such problems. Thus the warning; take nothing for granted and regard every new author and every term with a healthy suspicion. Also, undoubtedly there have been omissions and misunderstandings in the preparation of this paper. Please call them to my attention.

The rest of this paper is organized as follows. First, a brief overview of computational models of design is given. This includes sections dealing with the nature of design problems, the distinguishing characteristics of domains, design knowledge, general model issues, specific models, control of the design process and current research issues. Second, the research of each Program grantee in this area is described and evaluated. Third, a critical evaluation of the work supported in this area in the context of the overall state of the field is given. This is followed by a listing of references cited.

2. AN OVERVIEW OF COMPUTATIONAL MODELS

There are a number of sources which are particularly recommended for additional general insights, details, and the results of domain specific experience. An Electrical Engineering view is given in [Tong 87a, 88], [Mostow 85a, 85b], [Steinberg 88] and [Mitchell 85, 87]. A Chemical Engineering view is presented in [Nishida 81], [Westerberg 81, 84, 85] and [Lien 87]. Mechanical views are found in [Brown 86], [Mittal 86], [Dixon 86a] and [Libardi 88]. A Civil Engineering perspective is given in [Fenves 87]. General ideas for search are given in [Pearl 84]. [Barr 81] and the various AAAI and IJCAI proceedings are good starting points for the AI literature.

2.1 The Nature of Design Problems Considered

Desirable attributes of computational models are dependent on the nature of the problems considered. This paper focuses on models for problems with reasonably clear, complete and stable initial descriptions. Thus it is oriented toward problems arising in engineering more than in art, architecture or economics for example.

Further, the focus is strongly on models which essentially map function into structure, not models intended to discover domain properties, invent new ways to exploit basic domain properties or to design specifications for systems. The focus is also on models intended for actual implementation.

Here design is considered to be a process which maps an explicit set of requirements into a description of a physically realizable artifact which would satisfy these requirements plus implicit requirements imposed by the domain/environment. Typically, problems are specified explicitly in terms of:

- Function: allowed and required behaviors of the artifact (transfer energy, separate component streams,...),
- Performance: desired use of limited resources (space, money,...), desired values of artifact functional parameters (speed, stress,...),
- Context: requirements of the operating environment (legal, reliability, maintainability,...),
- Domain Resources: available components, fabrication capabilities.

In addition, problems usually have implicit (and sometimes explicit) specifications in terms of:

- Design Process: resources available (machines, software, models, time,...), performance (finish by next Tuesday),
- Physical Laws: domain-specific limitations (speed of light, transition temperature,...),
- Design Criteria: such as style, aesthetics, modularity imposed by expectation, tradition.

The design task is thus to create a solution artifact description meeting the requirements of the problem. The computational model should be configured so as to facilitate the creation of "good" solutions with the expenditure of "reasonable" quantities of design process resources.

2.2 Distinguishing Characteristics of Domains

A number of characteristics of design domains (for example, circuits, buildings, mechanisms) strongly affect the suitability of computational models. This has influenced both the development and use of computational models in various domains. Several of the most important of these characteristics are described below (see [Steinberg 88] for additional ideas).

The first characteristic is the natural separability or modularity of the domain, which might facilitate partitioning into subproblems. In general, separability requires that there are clusters of decisions with richer interactions within clusters than between clusters. These clusters might be based on function, form or level of abstraction for example. The next concern is the nature of the interactions between subproblems. Do they interact through the requirement that the output of one be compatible with the input of another? Are influence loops present? Do they compete for the use of limited resources? Does the extraneous heat generated in one cause malfunctioning in another? Are the interactions weak and easily dealt with or strong and difficult?

Domains differ widely with regard to the number, size, and complexity of natural subproblems and their associated solution modules. For instance, there are commonly a modest number (perhaps dozens) of physical units involved in chemical process design but they may require thousands of simultaneous equations to describe each. In contrast, VLSI design involves very large numbers of basic components which each require a modest number of equations for characterization [Westerberg 81a].

The next characteristic is the amount of a priori knowledge regarding subproblem (and thus solution) structure and interaction. If little is known, the task is one of exploration and invention. In this case structure and interactions must be created or discovered and managed as the design progresses. As more becomes known, tasks move through intermediate types, involving the selection and organization of known major components, to "routine" design in which only parameter values remain to be determined.

Another important characteristic is the nature of the problem specifications. Principal factors include:

Can they be precisely described, quantified, measured and calculated?

Are they complete and correct?

What is the nature of the behaviors specified (functional inputs and outputs, pointwise requirements, limitations on linearly additive resources such as weight)?

The nature of the solution space is also important. Is the space discrete or continuous? Is the solution space densely populated with feasible solutions so that the task is largely one of optimization? Is the space sparsely populated so that the task is to find any solution which satisfies the constraints? What is the level of detail of the solution sought? Is the problem one of system design or of basic component design? Usually, the lower the level, the more idiosyncratic the design model is to the domain.

The final characteristic, and an extremely important one, is the nature of the knowledge of the domain available. This is discussed in the following section.

2.3 Knowledge in Design

Knowledge plays a critical role in all design and the quantity and nature of available knowledge are key factors in the development and use of computational models. In knowledge based design, emphasis is on the explicit representation and use of knowledge. Knowledge can be organized and classified in many ways. These include:

By source (heuristic versus theoretical),

By subject (domain, design process, control),

By purpose (generate subproblems, select strategy, estimate value,...).

A fundamental concern of knowledge based design is to acquire, organize, represent and effectively utilize a variety of types of knowledge. Knowledge is acquired from human experts, from theoretical domain understanding, from past problem solving experience and from intermediate steps in the current design task. It may be organized to match the structure of the design process model chosen, to match the expected solution structure, and in various other ways including combinations of the categories suggested above. Knowledge is represented in terms of features, production rules, objects, arrays, and so fourth. The use of design knowledge is the subject of much of the remainder of this paper.

Before continuing, two important points are noted. One is that the effective combination of different types of design knowledge is difficult and that there is considerable current attention being devoted to this problem. The second is that throughout knowledge based design there is the basic dilemma of how to get machines to use human knowledge effectively without imposing human limitations on them.

2.4 Some Important General Issues

In all but the simplest design tasks, a final design is developed through a series of steps, thus design is ordinarily incremental. Further, when the partial design resulting after each step is evaluated to provide information which guides subsequent steps, the process is called iterative. Iteration thus plays a fundamental role in design. The general nature of iteration is discussed here. More specific aspects are introduced in Section 2.5.

Iteration involves a basic cycle which may be divided into the following phases:

- Evaluation: analysis of the current situation (partial design state, rate of progress, ...) to determine important characteristics, patterns,
- Generation: suggestion of candidate next steps,
- Prediction: determination of implications of the candidate steps,
- Selection: choice of the actual next step,
- Implementation: commitment to the step selected,
- Notification: communication of the implications of the step implemented throughout the design system.

Executing the iterative cycle requires a number of capabilities. First, there is the need to analyze partial design states and design process status. Features significant to the current situation at various levels of abstraction must be chosen and represented. Mathematical and symbolic models are sought; often little is available since most analytical tools are intended for application to complete designs. New techniques must be developed, a particularly difficult task when both discrete and continuous variables are present.

The next phase involves generation of a "good" set of candidate steps. This rarely means all possible legal steps; usually we seek a limited set of steps which (we hope) contain only steps which effectively lead toward the problem solution. Suggestions may be based on heuristic knowledge regarding how to map function into structure or may involve recognition of partial design state patterns which indicate applicability of specific transformations or previously used design plans. The nature of possible steps ranges widely and includes knowledge seeking, design process control and domain commitment types. Steps may involve suggestion of new topology, revision of existing parameter values or retraction of prior commitments.

The prediction phase provides the basis for choosing among the candidate steps. This phase analyzes the partial design states and design process status which would result from each candidate step and attempts to predict how much closer to the solution each would be than the current situation. Analysis may utilize mathematical models but ordinarily must rely primarily on heuristic methods. Predictions should include the effects of the proposed step on other subproblems and on expected future design states. Since there is rarely any adequate model to guide this process, it is usually difficult.

The selection phase is relatively straightforward if the prediction process is effective. It should entail application of explicit decision criteria. This may be rather complex if there are multiple, competing, non-commensurate goals.

Implementation involves changing the current situation description to reflect the selected design step. Each step involves some sort of commitment but not necessarily one in the problem solution domain. Implementation includes determining the effects of the changes imposed throughout the entire design problem solving system. These may include new goals, constraints or performance limitations for other subproblems.

Finally, it is necessary to communicate these effects appropriately throughout the design problem solver. This involves decisions regarding where and to what extent new information should be communicated and how it should be stored.

Designers often find it valuable to explore design problems using different points of view, usually oriented to different functions or requirements [Libardi 88]. For example, in structural design, the problem might be viewed from a stability or from a stress point of view. Computational models are expected to benefit from a like capability. It may also be desirable to utilize multiple representations at the same level of abstraction. Of course, these extensions require additional capabilities for evaluation, reasoning, decision making, testing for consistency and so on. We also note the desirability of allowing the designer to maintain inconsistent models under some circumstances.

2.5 Specific Design Process Models

The following sections describe the principal approaches being taken in engineering and related fields. There is some overlap between the models described, and it is common for components of several of the models described to be combined in a single application.

2.5.1 Top down refinement

The fundamental idea of top down refinement is to start with the initial problem specifications in highly abstract form and to refine them by adding detail and by decomposition to the point at which primitive operators are available to carry out all the subtasks defined. Most commonly this process is executed using iterative steps. The success of the approach depends on many factors, particularly on the identification of suitable levels of abstraction and problem decompositions and on the management of conflicts which arise during the design process. Abstraction and decomposition are discussed in this section; conflicts are discussed in Sections 2.5.2 and 2.5.3.

Here a higher level of abstraction means use of a representation in which more important details are included and less important details are omitted (problem linearization is a simple example). Many levels of abstraction may be used, for example MOLGEN [Stefik 81a] uses six. Each level is commonly oriented toward one dominant concern or viewpoint. This process generally leads to abstract design problems which are simpler and easier to solve than the corresponding complete detailed problems [Nevill 87, 88]. In general, problem specifications at one level of abstraction guide specification at lower levels of abstraction (higher levels of detail). The intent is that higher level specifications constrain lower level efforts to more promising directions but do not prune viable alternatives. One meaning of refinement is thus to move down through various abstraction levels, adding detail.

Problem specifications may also be partitioned at the same level of abstraction. This action will be called decomposition. Successful decomposition is most commonly based on natural separability or modularity as described earlier. If decomposition continues with each subunit in turn being decomposed, this is called hierarchical decomposition. Another meaning of refinement is thus to decompose at the same level of detail. Combinations of abstraction and decomposition refinements are common.

This "divide and conquer" approach is powerful because it limits the number of design decisions to be considered at any one time. It does create several complex problems, however. If good abstractions and decompositions are not known a priori, they must be created dynamically. Issues of granularity, completeness, transmission of performance goals, constraints and limited resource budgets between levels, and generation and selection of good partitions must be considered. Top down refinement is discussed in more detail in [Tong 87a] and [Steinberg 87a, 87b, 88]. [Libardi 88] includes further discussion of several of the issues above.

2.5.2 Avoiding conflicts

Problem specifications can rarely be decomposed into completely independent sets of subproblem specifications. When these subproblems interact, conflicts will inevitably occur (note that subproblem interaction is equivalent to interaction between subproblem solutions). Interactions may arise in a number of ways, both within and between levels of abstraction. The input required for one subproblem may depend on the output of another. The subproblems may compete for limited resources. Interface compatibility may be required. Some emission of one may be dangerous to another. In general this necessitates consideration of the requirements imposed on each subproblem by others.

A first step is to seek partitions with a minimum of interaction [Alexander 64]. When important interactions do occur, conflicts may often be avoided by a process called constraint propagation [Stefik 81a, 81b]. Interactions must first be identified. Then they are attached to affected subproblems as constraints (requirements). A common strategy, termed least commitment, is to delay refinement of a subproblem until all of the anticipated constraints have been fully specified. This is not always possible, however, for there may be loops of subproblem dependencies. In this case it is usually necessary to make reasonable initial guesses regarding constraints and to continue iteratively. The use of constraint propagation, though effective, has proven to be computationally very expensive [Steinberg

87a, 87b]. There is a clear need for techniques to make constraint propagation more efficient.

2.5.3 Resolving conflicts

Conflicts cannot always be avoided. They may arise due to interaction loops, due to faulty prior decisions, refinements and commitments, or be due to additional details not considered at higher abstraction levels. They may also be present when existing designs are being reused or when an imperfect trial design is chosen as a starting point. Two basic approaches are used to try to resolve conflicts, backtracking and modification.

Backtracking involves returning to an earlier refinement state and then attempting to continue along different and more successful lines. Chronological backtracking retraces steps in the reverse order in which they were initially taken and supports systematic exploration of alternatives. Dependency directed backtracking seeks to evaluate the conflict to determine what combination of decisions, assumptions, etc. is responsible for the difficulty and to return directly to the responsible location. In either case information regarding the nature of the conflict and means for avoiding it is needed.

Modification generally means making changes at the same level of detail. The term patching is applied to one step fixes. Extension refers to fixes dealing with incompleteness. Iterative redesign refers to multiple step efforts and typically applies a hill-climbing approach [Dixon 86b], [Orelup 87]. In any of these approaches there is the need to determine the nature of the conflict, what is causing it and to suggest effective design steps to improve the situation [Mitchell 83].

There is of course no guarantee that all conflicts can be resolved. As a last resort it may be necessary to seek a change in initial specifications or even admit failure.

2.5.4 Bottom up composition

Whereas top down refinement starts from problem specifications and works toward basic solution components, bottom up composition does the opposite. The fundamental concept is to explore what can be constructed using available components. In simple problems this may involve exhaustive enumeration of possibilities. More complex problems usually require strong guidance to avoid the well known exponential explosion of possibilities. Bottom up composition seems to be most useful in cases where solutions are specified largely in terms of thematic consistency and when the goal is to find "interesting", innovative solutions.

The concept of generative spatial grammars has been developed in architecture [Stiny 80, 85] and more recently in engineering [Fenves 87b], [Fitzhorn 88b]. The approach is to develop generative rules which ensure that only correct (usually meaning physically realizable and consistent with some theme) combinations of elements will be created. These rules are then applied repeatedly to generate design solutions.

In other areas heuristic rules are used in a similar manner to create promising designs but without any guarantee of correctness. This seems to be particularly appropriate when innovations are sought [Dyer 86], [Coyne 87], [Murthy 87]. Composition is also used in systems which adopt a means-ends approach and seek to identify measures of the progress of candidate steps toward the goal.

Composition is also appropriate for creating macro-components and thus dynamically raising the basic component level. There has been little use of multiple levels of abstraction in compositional approaches although the idea seems to have considerable merit.

2.5.5 Search

The modeling of problem solving as search in a space of possible solutions has received considerable attention in AI research over the years. Design is rarely modeled purely as search; however, ideas associated with a search model are useful in support of other models. The value of early pruning to reduce search space size and the use of branch and bound techniques are evident in a number of contexts. The use of concepts of heuristic guidance and best first exploration as exemplified by A* and related algorithms are also widely applicable. Unfortunately, the difficulties encountered in creating functions for estimating design step values are commonly severe.

2.5.6 Global optimization

When design problem specifications, solution components and domain and design process knowledge can be accurately represented mathematically, modeling design as global optimization becomes attractive. In this case, extensive knowledge from optimization theory plus an array of readily available and efficient computational tools may be applied to design problems. This approach has probably been most successful in Chemical Engineering Process Design where techniques from mixed integer linear and nonlinear programming have been successfully used [Duran 86], [Kocis 88], and techniques such as tearing can be applied to facilitate the solution of very large sets of nonlinear equations [Westerberg 81]. The key consideration is the existence and ease of obtaining good mathematical models for all of the relations and knowledge involved.

2.5.7 Other approaches

Numerous other models of design have been proposed and used for various purposes. For instance, the foundations of design theory are being explored modeling design as the enumeration of a string in a formal design language with a Turing machine as an enumerator [Fitzhorn 88a]. Design has been modeled as distributed problem solving involving cooperation among independent modules [Meunier 88], and as a process of negotiation between stakeholders. A transformational model has been explored which converts a specification into an implementation using a series of correctness-preserving transformations from one complete description to another [Mostow 85]. This model is similar to the abstract refinement model but has a looser control structure. Design has also been modeled from the perspective of automating decision support [Mistree 86] and using a pure expert system perspective in which the process used by the expert is directly captured [Kowalski 85].

2.5.8 Problem simplification using domain knowledge

Unique characteristics of a particular domain sometimes allow for dramatic simplification by allowing assumptions which severely restrict available alternatives, by providing bounds which facilitate pruning, by suggesting partitions or by identifying particularly important subproblems which deserve priority attention. Examples include a priori specification of requirements on interfaces between modules in circuit design [Rubin 87], the use of pinch points to suggest desirable decompositions in heat exchanger design [Westerberg 85], use of Burmester theory in mechanism design [Kota 86], and consideration of "bottlenecks" to direct focus in circuit design [Tong 87a]. While not a computational model per se, these simplifications often have a powerful influence on the model required. In many cases, for instance, the simplification allows straightforward exhaustive enumeration of alternatives.

2.6 Control of the Design Process

Control of the design process basically involves two questions which must be answered repeatedly: What decision should be made next? and Which choice should be selected for that decision [Tong 88]? In routine design the solution structure, order of decisions and available alternatives may be known in advance. In problems where the solution structure is known a priori, compatible rigid control can be effective. When the solution structure is unknown but the general nature of the domain is well understood, limited flexibility can be added through establishing appropriate strategies such as early commitment or least commitment. Flexibility can also be obtained by choosing among a set of strategies. At the extreme of uncertainty regarding domain structure, knowledge can be clustered into specialists and an opportunistic, blackboard approach taken [Hayes-Roth 78, 85], [Nii 86a, 86b].

At its most general, effective control involves a meta design problem, using meta knowledge to direct the design process. Precompiled knowledge plus information acquired during the design process is used to guide the process toward convergence on a satisfactory design as rapidly as possible. The goal directed planning approach [Tong 87b, 88] is based on top down refinement of a design plan guided by information obtained from the results of a rough design simulating an abstract design plan. It seeks to make mostly good choices the first time and to identify and improve poor choices.

2.7 Current Research Issues

The following seem to be particularly important research issues in the area of computational models of design at this time:

explicit representation of different types of performance goals and resource limitations, use of non-scalar goals and resources, strategies for dealing with multiple goals and resources, budgeting of resources to subproblems, adapting problem solving process to goal and resource configuration, maintaining multiple viewpoints,

dynamic generation of information during the design process, determining which and how much information to seek, evaluation of partial design states, dealing with imperfect information, learning and reusing previously developed designs,

dynamic creation and use of abstractions of subproblem descriptions and design operators, geometric abstractions and abstract geometric reasoning, treatment of dependencies between abstraction levels, abstract analysis,

automatic identification of decompositions, prediction of interactions involving subproblems not yet identified, controlling constraint propagation,

automatic compilation of domain knowledge into efficient models and processes, explicit use of domain theory in knowledge based systems, identification of bottlenecks, combining algorithmic with heuristic methods, combining models,

goal directed constructive grammars, constraining generators to produce primarily good moves, maintaining multiple designs, search for novel alternatives.

Additional recent ideas are presented in [Mostow 85], [Libardi 88] and [Tong 88].

3. DTM PROGRAM GRANTEE RESEARCH

The preceding section gave an overview of computational models. This section reviews the work reported by DTM Program grantees in this context. The review of each project will be divided into two sections, the first giving an overview of project activities and the second an evaluation.

3.1 A Domain-Independent Model of Knowledge-based Design

JACK MOSTOW, LOU STEINBERG, NOSHIR LANGRANA & CHRIS TONG

3.1.1 Overview of project activities

This project involves the integrated exploration of knowledge based, top down refinement with constraint propagation for (1) interactive NMOS digital circuit design (VEXED, BOGART), (2) controlling the design process (DONTE), (3) mechanical rotary power transmission design (MEET, DPMED, SPIKE, Floor-planner), and (4) domain independent approaches (EVEXED) and knowledge compilation.

Testing and use of VEXED, both by the research team and by a number of undergraduate student teams, has led to several significant conclusions relating to needs for improvement. First, constraint propagation is a very computationally expensive process and must be speeded up considerably if larger problems are to be tractable. Second, explanation to the user of the current state of the design and of predicted ramifications of user-proposed actions is both difficult and important. Third, the basic top down refinement plus constraint propagation model needs to be augmented with additional capabilities, in particular, for backtracking and redoing operations, for identifying and replaying previous action sequences and for modifying and rearranging modules and allowing sharing of functionality between modules. Retraction and replacement of constraints for these operations is being investigated. Fourth, users need more help with the control decisions required by VEXED. The BOGART system which allows for replaying previous design decisions supports this need. VEXED is currently limited to designing circuits with approximately 20 modules.

DONTE is a program for intelligent control of the design process. It uses a top down refinement approach called goal directed planning to plan the design process so as to adapt to the specific problem. Generation of information during the design process is emphasized. The DONTE development addresses many complex and important issues. This report notes particularly several issues and ideas relevant to scaling up VEXED-like systems. One idea involves dealing with repeated structures in terms of virtual copies of a prototype plus distinguished copies for first and last elements. Another idea being explored is the dynamic chunking of decisions into macro-decisions based on interdependencies revealed by constraint propagation. The problem is thus automatically reformulated in terms of higher order features. Another idea is the use of solutions to abstract problems (termed rough design) to generate information to guide the design process. DONTE also attempts to identify and focus on critical design areas (bottlenecks) by budgeting limited resources and observing the ease with which budgets are met. Extension is underway to deal with multiple resource limitations and with multiple performance goals.

The project is also exploring design of simple mechanical systems, focusing on rotary power transmissions. The mechanical work began with the adaptation of the VEXED shell to create MEET. Characteristics of the mechanical domain, particularly the required greater (compared to the circuit design domain) level of detail of basic domain components, led to the development of other models. These include DPMED which uses a general iterative hill-climbing approach for parameter value selection and can handle multiple design criteria, SPIKE, which employs a bottom up heuristic guided search for selecting gear sequences, and Floor-planner, which uses a search model to fit gears into available space. One conclusion of this work is that design model suitability is quite sensitive to domain characteristics. Current work is directed toward expanding capabilities to enable solution of larger, more realistic, problems and the integration of multiple subsystems embodying different models of design.

The fourth project area involves the generalization of VEXED and MEET into a domain independent shell called EVEXED. This effort led to the conclusion that in use of EVEXED there was a high price paid for generality and that the development of specialized design algorithms that do not fit the EVEXED model was important. This in turn led to a project focus on exploring the possibility of automatic knowledge compilation (a process for transforming knowledge about a specialized class of design tasks into an efficient procedure for that class). This work is pursuing two related directions, the rational reconstruction of existing specialized algorithms, and the automatic transformation of design knowledge into an algorithm. The work uses a house floorplanning task as a testbed for learning-oriented explorations of methods for optimizing the generate and test algorithm. Approaches include using constraints to guarantee generation of only acceptable alternatives, use of a generate, test and patch approach and the generation of abstract or rough solutions followed by refinement.

3.1.2 Comments

This project involves the highly productive cooperation of a number of researchers from several disciplines. The work is at the forefront of research in several important areas. These include: (1) development of top down refinement models with constraint propagation for large scale, realistic problems, (2) generation and use of knowledge for efficient control of the design process, (3) adapting models to task and domain requirements, and (4) knowledge compilation to produce efficient specialized procedures. There appears to be substantial and fruitful interaction between the researchers and a clearly conceived plan of research. At the same time the group has demonstrated the flexibility to learn from its experience and to redirect its effort into more promising areas. This project is an example of the synergisms made possible by assembling and supporting a competent multidisciplinary group.

3.2 Hybrid Computer Aided Design Methodology: Coupling Numerical Analysis and Reasoning Aids

TAI H. KWON, P.W. WEEKS & H.H. LEE

3.2.1 Overview of project activities

This project investigates the development of an interactive CAD system which integrates pre- and post-processing and numerical analysis of complete designs with intelligent evaluation of designs based on analysis results, and suggestions for design steps to improve the current design. The model used is essentially an iterative redesign one very similar to that developed by Dixon and associates [Dixon 86b].

The Prototype-I system was developed for application to polymer injection molding cooling system design. The user acts as top level controller. Analysis to determine temperature distributions and heat flow is based on modeling as steady state heat conduction and carried out using Objective-C. Symbolic evaluation and step suggestion uses PROLOG and heuristic redesign rules. The Evaluation module is currently capable of checking the validity of user input, of recommending tolerances for evaluation criteria based on estimates of parameter significance, and for checking if design performance meets criteria. The Decision Making module identifies the worst design criteria violation based on fixed prioritization and uses heuristic rules to suggest the best redesign step. Possible steps include add, move, delete or increase size of cooling liquid channel and lower coolant temperature. Difficulties regarding reasoning about geometry are noted for further study.

Exploration of both strongly coupled methods, in which analysis and heuristic reasoning use common languages, data structures, etc., and weakly coupled systems, where the analysis and reasoning communicate through datafiles, is planned. Initial system development sought strong coupling through use of PROLOG and Objective-C. Experience illustrated the shortcomings of PROLOG for design (as discussed in depth in [Dietterich 87]) and a switch to KEE and SUN COMMONLISP is in progress.

3.2.2 Comments

The work reported is sound, interesting, ambitious, and is directed generally toward worthwhile goals but to date breaks little new ground. The project represents the application of a well known model to a new domain. Results thus far confirm similar conclusions reached by others. In order to generate new basic knowledge, more limited goals and more specific identification of basic issues to be addressed seem desirable. If the comparison of strong and weak coupling methods is to be the primary focus of the work, then methods for quantitative evaluation of performance need to be established. I am concerned here that the result of this work will be another interesting system which does not clearly identify significant new basic knowledge or insights.

3.3 Research on Expert Systems in Automated Manufacturing, Knowledge Acquisition and Design Methodologies

ALICE M. AGOGINO

No report describing this project was provided. These comments are based on an examination of several papers [Choy 86], [Agogino 87], [Cagan 88] and the questionnaire response which was provided.

3.3.1 Overview of project activities

One area of activity has been the development of symbolic monotonicity analysis. This work begins with the classical model of global optimization [Wilde 78] and extends the concept of monotonicity analysis to allow symbolic rather than numerical application. SYMON applies monotonicity rules to the equations representing the problem in parametric form and identifies all active constraint sets which yield possible optimal solutions. When global monotonicities can be found, this can result in a significant reduction in the dimensionality of the optimization problem. Maintaining the results in parametric form also means that the analysis need not be repeated for similar problems with different parameter values. The SYMFUNE program extends this work by providing symbolic functional evaluation which often allows further reduction in the problem to a smaller set of parametric cases presented symbolically in functional form. It also sometimes provides insights into problem structure (such as mathematically defined regions in which there are no feasible solutions) that are usually undetectable with numerical optimization.

An approach to innovative design based on first principles (expressed mathematically) is also being explored. This work begins with a initial design configuration provided by a heuristic expert system (not described) and uses monotonicity concepts to innovate new geometries. This involves defining new primitives, such as by dividing integrals (expressed as summations) into multiple discrete regions. New geometries are then generated iteratively. Examples given are limited to very simple cases such as design of the cross section of a cantilever beam.

3.3.2 Comments

The extension of monotonicity concepts to allow symbolic analysis is certainly worthwhile and offers promise of expansion of the role of optimization tools in design. Further extension to relations defined heuristically would be natural and very valuable, but undoubtedly difficult. The basic idea of incorporating first principles explicitly into the design process is very attractive. However, the applications thus far have been so simple that it is hard to judge the potential of the approach for substantially more complex problems. Extension to more complex problems is encouraged.

3.4 Engineering Design Synthesis: A Domain Independent Representation

MARY LOU MAHER

3.4.1 Overview of project activities

This project explores application of a knowledge-based, top down, hierarchical decomposition model to preliminary structural synthesis of buildings. Knowledge is classified as either planning knowledge, which guides goal (subproblem) ordering, or design knowledge, which indicates how to find a solution for a given goal. EDESYN acquires knowledge through four user-created files: (1) a preconditions file containing top level problem specifications, (2) a constraints file which indicates infeasible combinations of goals for various conditions, (3) a decomposition file which provides a complete set of possible goal structures plus all means for satisfying the goals, and (4) a plan file which contains rules for selecting and ordering further decompositions.

The knowledge is represented by four classes of frames which in turn may contain rules. The four classes are: (1) precondition-frames, which store requirements and specifications, (2) goal-frames, which contain information about how to satisfy goals (through selection from a set of alternatives, direct calculation or further decomposition), about constraints identifying infeasible solutions, and about which decomposition plan is appropriate, (3) component-frames, which contain design alternatives for goals which are not to be further decomposed, and (4) plan-frames, which guide the decomposition of goals. Plan frames are associated with goal frames that require decomposition.

Synthesis is implemented as a constraint directed search for feasible solutions in the space defined by the given hierarchy of possible decompositions. At present a depth first, exhaustive search is conducted with goals being decomposed until they reach the component level. Planning is limited to ordering of the set of subgoals resulting from a decomposition. All feasible solutions are sought and stored. No complete examples illustrating performance were described.

A number of directions for further development have been identified. These are: (1) partial design evaluation to support pruning, (2) graphical visualization for the users benefit, (3) interactive knowledge acquisition, and (4) improved constraint handling.

3.4.2 Comments

This work is a relatively straightforward application of well known concepts to the area of structural synthesis. Its state of development is quite limited, at least as compared to VEXED and DONTE for example. It uses exhaustive search, it maintains all solutions found, it uses a fixed decomposition structure, and it does not deal with interactions between subproblems at the same level. The principal value of the work thus far is in its representation and use of the specific features of the building structure domain. There appears to be a need to more clearly identify specific basic issues, to make more use of the ideas currently being explored by others, and to direct the research so that careful exploration of selected issues is emphasized.

3.5 Methodology of Intelligent Computer-aided Design and Its Application to DC/AC Electronics

KWA-SUR TAM

3.5.1 Overview of project activities

The long term goal of this project is to use basic domain principles plus expert knowledge for the autonomous generation of promising designs. Initially a generate and test approach has been implemented for the autonomous creation of designs of voltage multiplier circuits. This program, called VMD, starts with the desired voltage multiplication factor n, an integer greater than 1, plus three prioritized selection (performance) criteria. Specialized domain knowledge (and limitation to capacitors and diodes as components) is then used to implement a unified approach for the generation of n-fold voltage multipliers using n capacitors and n diodes. A pair of special capacitor trees is generated and then used together with domain specific constraints to generate all possible legal voltage multiplier circuits, with attention to avoiding generation of identical circuits.

The designs generated are then evaluated one at a time to identify the best. Performance measures are first calculated. Designs are compared based on the highest ranking criteria first and lower ranked criteria are used only in case of ties. Two somewhat redundant representations are maintained to facilitate computation.

Future directions for the project are indicated to include: (1) focus on exploiting basic domain principles, (2) matching representation to needs of the design process, (3) enhancing the creativity of the generation process, (4) making the generation more efficient, and (5) using knowledge obtained during the design process to guide design generation.

3.5.2 Comments

This project is seeking to address a number of broad and important issues, with emphasis on the use of basic domain principles to guide design and exploitation of the capabilities of computers for tireless search. Preliminary results are interesting but are based on a very simple exhaustive generation and test model. It does not seem feasible to do significant research on all of the issues mentioned. Considerable focusing of attention with a clarification of which goals are to be emphasized appears desirable. There is certainly much to be learned regarding automatic compilation of domain knowledge, for example.

3.6 Iterative Redesign and Respecification: Research on Computational Models of Design Processes

JOHN R. DIXON

3.6.1 Overview of project activities

This project focuses on design problems in which the design variables are known and their values are to be determined (parametric design) and for which an initial trial design is readily obtainable. Emphasis is on determination of general concepts and methods which are domain independent. Concepts are tested by application to simple mechanical design tasks. The _iterative redesign_ approach explores problems which are manageable without decomposition. The _iterative respecification_ model is applied to problems in which a decomposition is given a priori.

Iterative Redesign

The basic architecture of the iterative redesign model, incorporated in Dominic (I and II), is a cycle which involves evaluation of the current design with regard to performance parameters and constraints, the testing of results for acceptability, and revision of parameter values (redesign) based on use of dependencies (expressed numerically) between design variables and performance parameters. It is a hill-climbing approach in which one performance parameter at a time is selected and its value modified. Dominic does not seek an optimal solution but rather a satisfactory one.

Dominic I solved problems in two different domains but demonstrated some important deficiencies, particularly a need for improved control. This led to development of Dominic II, which is capable of dealing with discrete variables and modifies its strategy based on problem solving progress. New values for discrete variables are suggested as if they were continuous and then made discrete. Dominic II monitors its problem solving performance by recording, for a number of iterative loops, (1) the design variable modified and how much, (2) the target performance parameter and the consequent satisfaction change, (3) active constraints, (4) the current design, and (5) the overall satisfaction level resulting. This record is then analyzed to determine if the design has improved its level of satisfaction and if specific unproductive efforts are present.

Six specific unproductive design efforts have been identified: Creeping, Cycling, Floundering, Confined, Blocked, and Constraint-bound. Dominic II has also incorporated a library of six distinct redesign strategies: Vanilla (the original one), Aggressive, Semi-aggressive, Re-order-performance-parameters, Constraint-adapter, and Big-jump. A set of heuristic rules, relating strategy choice to strategy history (also recorded) and unproductive behaviors, is then applied to determine which strategy to use next.

Dominic II has been applied to problems in five different domains and has proven clearly superior to Dominic I. There remain, however, additional ways to improve performance. Four issues to be addressed are: (1) more effective handling of system level variables, (2) using dependency analysis to generate information, (3) utilizing previously solved problems, and (4) more effective use of dependency information to guide design.

Iterative Respecification

This project effort extends the concept of iterative improvement of designs to enable dealing with more complex problems which benefit from decomposition. The iterative respecification model is designed for parametric design of mechanical systems in which a fixed prior hierarchical decomposition of the problem is assumed. This leads to an assumed structure for the problem solving process in which control is implemented by manager nodes in the hierarchy. Communication within the model is limited to parent-child paths.

The basic idea of iterative respecification is that managers solve their assigned problem by determining values for design variables at their own level and by iterative specification of requirements to subordinate subproblems. Appropriate subproblem design modules solve each specified subproblem and report the results back to the manager. The manager then analyzes the results reported, determines design quality, and creates new subproblem specifications using additional information regarding dependencies. This might be called iterative backtracking. The system begins in a rough design mode which ignores extremum objectives. Once an acceptable design has been developed, the system activates the extremum objectives and continues in optimization mode. One of the interesting aspects of this model is its use of dependencies indicating relationships between (A) system level design variables and subproblem specification parameters and (B) system level satisficing and extremum objectives.

The model has been implemented and tested by application to some twenty cases involving a platform, an electric motor driven pump and a heat exchanger. The problem was decomposed into a simple two level hierarchy and a fixed subproblem solution ordering was assumed. Acceptable designs were obtained in all cases in which they existed.

Testing revealed a number of weaknesses. Explorations are planned in the areas of simultaneous adjustment of two or more subproblem specification parameters, more explicit representation of domain knowledge related to respecification, richer forms of communication, representation and use of dependencies, reuse of designs and multiple extremum objectives.

3.6.2 Comments

Iterative Redesign: This project started with a simple but significant problem class and has carefully explored the application of the iterative redesign model. The work has created a solid foundation of definitions, concepts, experience with implementation, and understanding, and has systematically extended that foundation.

Issues have been clarified and explored and results clearly communicated in the literature. Integration with the study of representation, particularly the features work, has been valuable. The work has clarified the unique aspects of mechanical engineering design and has contributed to the understanding of matching model to domain. It has also contributed to the use of information acquired during design problem solving to control the design process. This has been one of the important design research efforts of recent years.

Iterative Respecification: This work seeks to extend the basic idea of iterative improvement of designs to deal with more complex problems which may be (and need to be) hierarchically decomposed. Consistent with their basic "build up from a solid foundation" philosophy, the project has started with problems with a priori decompositions. Hopefully this will provide a basis for future development of dynamic decomposition models. The approach combines ideas from top down refinement, backtracking and iterative redesign; it explores the use of iteration between levels of a hierarchy. Since the concept of iteration is central to much of design theory, the work has considerable potential value.

3.7 An Investigation of the Interdisciplinary Aspects of Electrical Engineering Design

SHELDON S. L. CHANG

3.7.1 Overview of Project activities

The stated objectives of this project are (1) to advance engineering design methodologies as a coherent science, (2) to develop a framework for accumulating human experience in electrical engineering design, (3) to investigate the possibility of enhancing a designer or a team of designers' innovativeness with interactive computer-aided design, and (4) to gain insight and better understanding of the generic properties of design methodologies, design problems, and matching between the two. A multiple feedback loop model of design, involving a hierarchy of stages, has been chosen. The stages are identified as I. conceptual design and systems selection, II. system design and component selection, III. modular detail selection, and IV. detail design. A design methodology matrix has been formulated with the vertical axis containing the design stages and the horizontal axis containing various electrical product domains (motors, filters, ...). The approach is to study vertical integration of design through the stages, and to investigate horizontal comparisons with the eventual goal of extending these comparisons to other domains. Emphasis on knowledge-based systems for conceptual design and on interactive design to enhance proficiency and innovativeness is planned.

Investigation of a conceptual framework for vertically integrated design focuses on the increased detail of lower stages, on the importance of prevailing design standards, on the dependencies associated with the multiple feedback paths and on optimization at each stage. Important questions to be addressed are identified as: (1) the possibility of developing a domain independent model which will be optimum for all domains, (2) can the model and the vertically integrated process be used to guide standardization, (3) is the model sufficiently general that methodologies developed will be transferrable across disciplines, (4) can vertical integration enhance the efficiency and innovativeness of designers, and (5) does vertical integration reduce cost and time delay in manufacturing.

Cross disciplinary comparisons are planned with concentration on formulation of early stage design methodology. Use of fuzzy set theory plus knowledge based systems methods is anticipated.

Three specific areas of project activity are mentioned. The first area involves investigations of the feedback process. Here focus is on the desirability and existence of "dichotomy" (separability, independence) between stages and on the possibility of reducing the need for feedback adjustments. The second area involves interactive design processes and plans to focus on the computer presenting the designer with alternative choices and computing consequences of candidate design decisions. Emphasis will be on enhancing innovativeness. Third, plans are mentioned to make use of a nearly-in-operation CAD-CAE laboratory.

Results to date are indicated to be: (1) a description of design automation for the component parts industry, (2) a comparison showing the potential advantages of a vertically integrated automated approach over an unautomated approach to a sales to engineering to production cycle, and (3) an exploration of the usefulness of fuzzy sets and heuristic classification to design. No computer implementation or acquisition of data from designers is indicated.

3.7.2 Comments

The effort described in this report is directed toward a number of worthwhile goals. However, I find several aspects of the report disturbing. First, I am concerned with the breadth of the work being attempted. Work is proposed in the areas of knowledge acquisition, knowledge-based systems, computer-human interaction, design process integration, innovation, optimization, dealing with uncertainty, cross disciplinary knowledge transfer, changing industry standards, automated backtracking, and problem decomposition. I believe that a more limited scope and focus on a limited number of clearly defined issues is desirable.

Second, the report suggests effort in several areas being addressed by other researchers without reference to their work. For example, efforts directed toward determining "Under what condition is it possible to make the optimum design decisions completely in the hierarchical order without feedback adjustments?" is closely related to work in top down refinement with constraint propagation such as is described in Section 3.1 above. Extensive work in planning and coordinating the use of multiple design tools [Parker 87], [Bushnell 86] also seems relevant. Taking more advantage of the work of others could help reduce the scope of the project.

Third, it is not clear what implementation is underway or planned. I am concerned that the project will develop too much formal structure without subjecting the ideas to the harsh reality test of implementation. I believe that the best way to do research in this general area is through a series of tightly coupled cycles of conception, implementation and evaluation.

3.8 Toward a Formal Theory of Design
PATRICK FITZHORN

3.8.1 Overview of project activities

This work is very different from the other work described. It is highly and purely theoretical, with no implementation and thus far little attention to implementation. The work views design primarily from the perspective of enumeration of alternatives. Its focus is on two principal areas: (1) representation of design artifacts as multi-dimensional sentences generated from formal grammars, and (2) describing the design process as an enumerator of these sentences. It attempts to generate a domain independent theory of design, capable of spanning all design areas.

In this work a language is defined as a set of finite length strings over some finite alphabet of characters. Strings of characters are called sentences. Sentences are generated by grammars which recursively apply productions building from a starting set of characters. The work builds on results from explorations of phrase structure grammars, graph grammars and shape grammars. It assumes that any designed artifact can be equivalently specified as a sentence derived from a grammar.

Results of applying graph grammar concepts to the topology of solid object descriptions have been interesting. Grammars have been shown to be syntactically complete (capable of modeling any solid object) and syntactically closed (generate only descriptions of physical realizable objects). Extensions to geometry are now being explored using attributed graph grammars.

Shape grammar concepts are being applied to the exploration of design features. This work advances the view that the feature set of a domain is the set of all right-hand sides of the defining productions of a grammar. This perspective allows exploration of the completeness and equivalence of feature sets and of the decidability of such questions. Applications of multiple feature sets for integrating design and manufacturing process planning are suggested.

The purely generational view of design has led to the definition of design as "a Turing machine whose (i) enumerated sentences describe the artifacts of design, and whose (ii) state transition function delta describes the rules of design in some domain." This view allows for conclusions relevant to the Turing machine, the most general computational automaton known, to be extended to design. The investigator recognizes concerns with the "sterility" of this work and states that research using this new formal paradigm for viewing design is being carried out with great concern about the nature of effective computational procedures.

3.8.2 Comments

Evaluation of this work involves examination of several very basic issues. First, how basic and how general is appropriate? Although this work contributes insights regarding what is possible, what is not possible, what is complete and what is closed, the results appear far from immediately useful. The attempt at complete generality (spanning all design areas) is accompanied by the danger that results will be useful to none.

A second major concern involves the purely bottom up, constructive view of design which ignores completely goals, interactions, requirements, resource limitations (particularly issues relating to efficiency), and evaluation of design performance. In problem domains where consistency with a theme is the dominant concern, this formally structured compositional approach seems appropriate. In other problem areas, however, designers have concluded that design should be viewed as constraint satisfaction, goal directed search, resource budgeting, decision making or negotiation. It is hard to see at present how a single paradigm (view) can be effective in all domains. Here it is important, however, not to impose human limitations on the models of design to be implemented by machines. There seems to be no end in sight to the rapid increase in the capabilities of computer hardware. Thus surely models of design are needed which exploit machine capabilities far beyond what we have today, even if those models would be most inappropriate for human use. Of course, it is quite helpful if humans can understand and guide the use of these models.

Another concern is with issues of selecting good grammars and good feature sets. Thus far the focus is on possibilities, not desirabilities. Emphasis thus far is on exhaustive enumeration of possibilities. There is a related need to test and evaluate the designs generated and (perhaps) to guide the process. Here again the efficiency issue arises. If our machines become sufficiently fast then perhaps efficiency will become unimportant. Although history has proven it dangerous to bet against technological development, it is doubtful if cost, time and other limited resources can ever be neglected. It is rather more likely that problems will increase in difficulty to match the improvements in capability.

Overall, this is considered to be interesting, provocative and potentially useful work but near the boundary of too abstract for this Program.

4. CRITIQUE OF DTM PROGRAM RESEARCH

The research in the area of computational models of design processes supported by the DTM Program will be evaluated according to several criteria: quantity of output, technical quality of output, focus on key issues, balance, scope, completeness, coordination, efficiency of process, immediate practical applicability and enhancement of foundation of the field. The quantity of output of the projects is generally quite high. Extensive publication in respected and readily available locations is occurring. Related to this is the quality of the work reported. This is generally quite high, with the publications resulting being cited frequently and several considered to be very important. There is no question that the work is dealing with many of the key issues of the field. In some cases, however, there is little focus and thus a danger that results will be interesting but not sufficiently precise to be useful to others.

The work supported seems to be reasonably well balanced and to provide adequate scope along several appropriate dimensions, such as application domain and location in the commonly accepted design phase hierarchy. The work supported can certainly be faulted for lack of completeness; however, the priorities seem sound and the primary limitation is insufficient funds.

There seems to be little coordination among the various projects, with some appearing to be unaware of relevant results from others. Efficiency is hard to judge but I suspect that more explicit planning of process and goals would be beneficial. The work is not as yet having significant impact on practical applications but that is as it should be, I believe. Finally, and probably most important, this work is contributing substantially to creating a foundation for the DTM field.

Based on these observations, I have two principal recommendations. The first is that the Program put more emphasis on grantees clearly identifying a limited set of specific issues to be addressed by their project and resist the temptation to be broad and fuzzy. The second is that more cooperation and communication among grantees be encouraged. This would facilitate the synergistic exchange of ideas and experience and help build the design community. These nearly-annual grantee conferences help, but more is needed.

Overall, I feel that the DTM Program is spending its money wisely and obtaining truly significant results.

ACKNOWLEDGEMENTS

A number of people shared ideas with me and sent copies of papers. I would like to express my thanks to Lynn Conway, Steve Director, Jack Dixon, Jim Douglas, Arthur Erdman, Steve Fenves, Patrick Fitzhorn, Ignacio Grossmann, Mary Lou Maher, Sanjay Mittal, Jack Mostow, Rudy Motard, Alice Parker, George Stiny, Chris Tong and Art Westerberg for their help.

REFERENCES

[Agogino 87] Agogino, A. M. and Almgren, A. S., "Symbolic Computation in Computer-aided Optimal Design", Expert Systems in Computer-Aided Design, Gero, J. S., Ed., pp.267-284, North-Holland, 1987

[Alexander 64] Alexander, C., Notes on the Synthesis of Form, Harvard University Press, 1964

[Barr 81] Barr, A. and Fiegenbaum, E. A., The Handbook of Artificial Intelligence, William Kaufmann, 1981

[Brown 86] Brown, D. C. and Chandrasekaran, B., "Knowledge and Control for a Mechanical Design Expert System", IEEE Computer, July, 1986

[Bushnell 86] Bushnell, M. L. and Director, S. W., "VLSI CAD Tool Integration using the ULYSSES Environment", Proceedings ACM/IEEE 23rd Design Automation Conference, pp. 55-61, 1986

[Cagan 88a] Cagan, J. and Agogino, A. M., "1stPrince: Innovative Design from First Principles", Submitted to AAAI-88

[Choy 86] Choy, J. K. and Agogino, A. M., "SYMON: Automated Symbolic Monotonicity Analysis for Qualitative Design Optimization", Proceedings of the ASME Computers in Engineering Conference, Vol. 1, pp. 207-212, July, 1986

[Coyne 87] Coyne, R. D., Rosenman, M. A., Radford, A. D, and Gero, J. S., "Innovation and Creativity in Knowledge-Based CAD", Expert Systems in Computer-Aided Design, Gero, J. S., Ed., North-Holland, 1987

[Dietterich 87] Dietterich, T. G. and Ullman, D. G., "FORLOG: A Logic-Based Architecture for Design", Expert Systems in Computer-Aided Design, Gero, J. S., Ed., pp.1-17, North-Holland, 1987

[Dixon 86a] Dixon, J. R., "Artificial Intelligence and Design: A Mechanical Engineering View", Proceedings AAAI-86, Vol. 2, Philadelphia, PA, August, 1986

[Dixon 86b] Dixon, J. R., Howe, A., Cohen, P. R., and Simmons, M. K., "Dominic I: Progress Towards Domain Independence in Design by Iterative Redesign", Proceedings of the ASME Computers in Engineering Conference, Vol. 1, July, 1986

[Duran 86] Duran, M. A. and Grossman, I. E., "A Mixed-Integer Nonlinear Programming Algorithm for Process Systems Synthesis", AIChE Journal, Vol. 32, No. 4, pp. 592-606, April, 1986

[Dyer 86] Dyer, M. G., Flowers, M., and Hodges, J., "Edison: An Engineering Design Invention System Operating Naively", International Journal of Artificial Intelligence in Engineering, Vol. 1, No. 1, pp. 36-44, 1986

[Fenves 87a] Fenves, S., "Role of Artificial Intelligence and Knowledge-Base Expert System Methods in Civil Engineering", EDRC-12-17-87, Engineering Design Research Center, Carnegie-Mellon University

[Fenves 87b] Fenves, S. J. and Baker, N. C., "Spacial and Functional Representation Language for Structural Design", Expert Systems in Computer-Aided Design, Gero, J. S., Ed., North-Holland, 1987, also EDRC-12-18-87, Engineering Design Research Center, Carnegie-Mellon University

[Fitzhorn 88a] Fitzhorn, P., "A Computational Theory of Design", Design Computing, Vol. 3, No. 1, 1988

[Fitzhorn 88b] Fitzhorn, P., "Toward a Formal Theory of Design", Preprints, Design Theory '88, Section 3.9, Troy, New York, June 1988

[Hayes-Roth 78] Hayes-Roth, B. and Hayes-Roth, F., "Cognitive Processes in Planning", Rand Report R-2366-ONR, December, 1978

[Hayes-Roth 85] Hayes-Roth, B., "A Blackboard Architecture for Control", Artificial Intelligence, Vol. 26, No. 3, pp. 251-321, July, 1985

[Kocis 88] Kocis, G. R. and Grossmann, I. E., "Solution of Mixed-Integer Nonlinear Programming Problems in Engineering Design with DICOPT", Preprints, Design Theory '88, Section 5.6, Troy, New York, June 1988

[Kota 86] Kota, S., Erdman, A. G., and Riley, D. R., "Development of Knowledge Base for Designing Linkage-Type Dwell Mechanisms: Part 1-Theory", ASME Design Engineering Technical Conference, Paper 86-DET-47, Columbus, Ohio, October, 1986

[Kowalski 85] Kowalski, T. J., Knowledge Representation, Learning, and Expert Systems: An Artificial Intelligence Approach to VLSI Design, Kluwer Academic Press, Hingham, Mass, 1985

[Libardi 88] Libardi, E. C. Jr., Dixon, J. R., and Simmons, M. K., "Computer Environments for the Design of Mechanical Assemblies: A Research Review", to be published in Engineering with Computers

[Meunier 88] Meunier, K. L. and Dixon, J. R., "Iterative Respecification: a Computational Model for Hierarchical Mechanical System Design", Proc. ASME Computers in Engineering Conference, San Francisco, August, 1988

[Mistree 86] Mistree, F. and Muster, D., "The Decision Support Problem Technique for Design", Proceedings, ASEE Annual Conference, pp. 117-126, Cincinnati, June, 1986

[Mitchell 83] Mitchell, T., Steinberg, L., Kedar-Cabelli, S., Kelly, V., Shulman, J., and Weinrich, T., "An Intelligent Aid for Circuit Redesign", AAAI-83, pp. 274-278, 1983

[Mitchell 85] Mitchell, T. M., Steinberg, L. I., and Shulman, J. S., "A Knowledge-Based Approach to Design", IEEE Transactions on Pattern Analysis and Machine Intelligence, Vol. 7, No. 5, pp.502-510, September, 1985

[Mitchell 87] Mitchell, T. and Mostow, J., "Artificial Intelligence and Design", Tutorial No: TP 2, AAAI 87

[Mittal 86] Mittal, S. and Araya, A., "A Knowledge-Based Framework for Design", Proceedings AAAI-86, Vol. 2, Philadelphia, PA, August, 1986

[Mostow 85a] Mostow, J., "Toward Better Models of the Design Process", AI Magazine, Vol. 6, No. 1, pp. 44-57, Spring, 1985

[Mostow 85b] Mostow, J., Editor, Special Issue on Artificial Intelligence and Software Engineering, IEEE Transactions on Software Engineering, Vol. SE-11, No. 11, November, 1985

[Murthy 87] Murthy, S. S. and Addanki, S., "PROMPT: An Innovative Design Tool", Expert Systems in Computer-Aided Design, Gero, J. S., Ed., North-Holland, 1987

[Nevill 87] Nevill, G. E., Jr. and Paul, G. H., Jr., "Knowledge-Based Spatial Reasoning for Designing Structural Configurations", Proceedings of the ASME Computers in Engineering Conference, Vol. 1, pp. 155-160, August, 1987

[Nevill 88] Nevill, G. E., Jr., Jackson, L. A., and Clinton, J. H., "Automated Hierarchical Planning for Structural Design", To be published, Proceedings of the ASME Computers in Engineering Conference, August, 1988

[Nii 86a] Nii, H. P., "Blackboard Systems: The Blackboard Model of Problem Solving and the Evolution of Blackboard Architectures", AI Magazine, Vol. 7, No. 2. pp. 38-53, Summer 1986

[Nii 86b] Nii, H. P., "Blackboard Systems: Blackboard Application Systems, Blackboard Systems from a Knowledge Engineering Perspective", AI Magazine, Vol. 7, No. 3. pp. 82-106, Conference 1986

[Orelup 87] Orelup, M. F., Dixon, J. R., and Simmons, M. K., "Dominic II: More Progress Towards Domain Independent Design by Iterative Redesign", Preprint of paper presented at ASME WAM, December, 1987

[Parker 87] Parker, A. C. and Hayati, S., "Automating the VLSI Design Process Using Expert Systems and Silicon Compilation", Proc. IEEE, Vol. 75, No. 6, pp. 777-785, June, 1987

[Pearl 84] Pearl, J., Heuristics: Intelligent Search Strategies for Computer Problem Solving, Addison-Wesley, 1984

[Rubin 87] Rubin, S. M., Computer Aids for VLSI Design, Addison-Wesley, 1987

[Stefik 81a] Stefik, M. J., "Planning with Constraints (MOLGEN: Part 1)", Artificial Intelligence, Vol. 16, pp. 111-139, 1981

[Stefik 81b] Stefik, M. J., "Planning and Meta-Planning (MOLGEN: Part 2)", Artificial Intelligence, Vol. 16, pp. 141-169, 1981

[Steinberg 87a] Steinberg, L. I., "Design = Top Down Refinement Plus Constraint Propagation Plus What?", AI/VLSI Project Working Paper No. 67, Computer Science Department, Rutgers University, also, IEEE Systems Man and Cybernetics Conference, Fairfax, Virginia, October 1987

[Steinberg 87b] Steinberg, L. I., "Design as Refinement Plus Constraint Propagation: The VEXED Experience", Proc. AAAI 87, Vol. 2, pp. 830-835

[Steinberg 88] Steinberg, L. and Tong, C., "Artificial Intelligence and Design", AI/VLSI Project Working Paper No. 95, Computer Science Department, Rutgers University, also preliminary AAAI88 Tutorial Course Description

[Stiny 80] Stiny, G., "An Introduction to Shape and Shape Grammars", Environment and Planning B, Vol. 7, pp. 343-351, 1980

[Stiny 85] Stiny, G., "Computing with Form and Meaning in Architecture", Journal of Architectural Education, pp. 7-19, Fall 1985

[Tong 87a] Tong, C., "Toward an Engineering Science of Knowledge-based Design", Artificial Intelligence in Engineering, Vol.2, No. 3, pp.133-166, 1987

[Tong 87b] Tong, C., "Goal Directed Planning of the Design Process", Proceedings, Third IEEE Conference on Artificial Intelligence Applications, February, 1987

[Tong 88] Tong, C., "Knowledge-Based Circuit Design", PhD Dissertation, Department of Computer Science, Stanford University, 1988

[Westerberg 81] Westerberg, A. W., "Computer-Aided Design Tools in Chemical Engineering Process Design", Proceedings, IEEE, Vol. 69, No. 10, October 1981

[Westerberg 84] Westerberg, A. W. and Chien, H. H., Eds., Proc. Second Intl. Conference on Foundations of Computer-Aided Process Design (FOCAPD-83), CACHE 1984

[Westerberg 85] Westerberg, A. W., "Process Synthesis - with an Emphasis on Energy Efficient Design", Third Hougen Lecture, Department of Chemical Engineering, University of Wisconsin, March 1985

[Wilde 78] Wilde, D. J., Globally Optimal Design, Wiley, New York, 1978

A Fuzzy Set Algorithm for Engineering Design with Applications to the Component Parts Industry

Sheldon S.L. Chang

Department of Electrical Engineering
State University of New York at Stony Brook
Stony Brook, N.Y. 11794

Summary

A fuzzy set algorithm is formulated for System Selection and Design Specification. To meet the requirements of a given application (Design Goal) the SSDS selects one among a finite number N of alternative systems and specifies the design parameters of the selected system optimally. Optimality is in the sense of highest membership function of the finally designed system in meeting the Design Goal. An application of the SSDS to electrical motor design is discussed.

Using SSDS as an essential element, design automation in the component parts industry from purchase inquiry to finished product delivery is described. The computerized process simulates human decision makers, eliminates bottlenecks, and frees human decision makers from doing routine work.

Introduction

A few years after L.A. Zadeh's publication of his original work on fuzzy set analysis [1], its potential for application to engineering design has been recognized [2, 3, 4]. In a recent paper, Wood and Antonsson gave an excellent review of the applications of fuzzy set theory to engineering design, in addition to reporting their own work [5]. The present paper does not deal with the general or philosophical aspects of the two associated entities: fuzziness and engineering design. Rather, it points to Zadeh's mathematical analysis [1] as an accurate representation of the mental process of an expert engineer in an early stage of the design process, that of system (product) selection and design specification (SSDS), for at least one indus-

This work was supported by the National Science Foundation under Grant DMC - 8702465.

try: the component parts industry. Based on this hypothesis, an algorithm is workout for computerized SSDS, which is an important but missing link in a design automation process.

Design automation of the component parts industry is significant for two reasons:

(i) The component parts industry, counting both the suppliers and supplied is a substantial part of manufacturing. Purchasing specialized component parts in a manufacturing process is a proven way of assuring high quality, low cost, and sometimes least time consumed. The practice is theoretically sound as it heightens competition. Design automation is significant because of the sheer size of the industry.

(ii) The price to pay for purchasing component parts is a somewhat fragmented product design process. The product designer does not have the knowledge nor the freedom in trading between features of the purchased component. This knowledge is provided to him through design automation of the component's supplier. Working with a supplier's design automation terminal in his location, the product designer can design a component which best suits his requirements.

Thus design automation in the component parts industry allows the industry to have both advantages: strong competitiveness in the use of components, and an integrated final product design. However fuzzy set analysis is not the only solution to SSDS. A well-recognized solution is a knowledge based heuristic selection system [6]. However, knowledge acquisition, representation, and software design of such a systems is by no means an easy task. A SSDS modeled on fuzzy set analysis is at least a workable alternative.

Fuzzy Set Representation of Expert Reasoning

In an engineering design process, the first step is to select among a finite number of design alternatives one alternative which best suits the design goal or application. This is usually done by an "expert" or experienced engineer. Our hypothesis is that fuzzy set analysis accurately represents an expert engineer's mental decision process. Let us examine this hypothesis here:

There is a very significant difference between fuzzy set analysis and probability theory. In probability theory, let A and B be two independent events with probabilities of occurence 0.8 and 0.5 respectively. The probability that both A and B occurring is $0.8 \times 0.5 = 0.4$. Both probability values enter into this computation. In fuzzy set theory, let the memberships of A and B be 0.8 and 0.5 respectively. The membership of $A \cap B$ is then 0.5. Its value is independent of the higher membership value of 0.8. Our question is then: "Which method of computation is more representative of an expert engineer's mental process?"

There are a number of features which constitute a design goal. Some of the design alternatives do not meet certain feature requirements, and these alternatives are eliminated. Some feature requirements are well-met by all remaining alternatives, and they are not considered. In making his selection of the design alternatives, an expert engineer considers only a number of critical features. This behavior of ignoring completely the well-met features is best represented by Zadeh's intersection rule.

One may add the following considerations to the above observation:

1. A probability computation would not be correct if one does not consider the correlations between features. The knowledge on correlations between features is usually missing or incomplete.

2. A time-consuming or effort-consuming computation at the preliminary design stage is meaningless, because completed designs on a few alternatives can be made and compared.

3. At this stage of SSDS development, we seek only a workable representation. The nature of an optimum SSDS will be better understood and simulated with its usage.

With the above understanding one possible formulation is as follows: Let D_i , i=1, 2, ...n be the product types or design alternatives, and λ_j , j=1, 2, ... be the performance features. Let $\lambda = \left\{\lambda_j\right\}$. Each design alternative D_i has a membership function $\mu_i(\lambda)$ on the feature space. The design goal can be expressed as a fuzzy set $\mu_G(\lambda)$ on the feature space. Then, using Zadeh's intersection rule, μ_{iG} expresses the suitability of design alternative D_i for the specified design goal:

$$\mu_{iG} = \max_{\lambda} \mu_i(\lambda) \Lambda \mu_G(\lambda) \tag{1}$$

The design alternative D_a with highest μ_{aG} is then selected. Highest μ_{aG} means that D_a is least objectionable when every feature or factor is taken into consideration.

Quite often the performance features λ can be grouped into independent subsets S_k , where each subset S_k may represent one or more related features.
The meaning of independence can be expressed mathematically as:

$$\mu_i(\lambda) = \prod_k \mu_{ik}(\lambda_k) \tag{2}$$

$$\mu_G(\lambda) = \prod_k \mu_{Gk}(\lambda_k) \tag{3}$$

where λ_k represents features in the subset S_k , and the product $\prod_k$ is in Zadeh's sense of fuzzy set intersection. Equation (1) becomes

$$\begin{aligned} \mu_{iG} &= \max_{\lambda} \left[\prod_k \mu_{ik}(\lambda_k)\right] \Lambda \left[\prod_k \mu_{Gk}(\lambda_k)\right] \\ &= \max_{\lambda} \prod_k \left[\mu_{ik}(\lambda_k) \, \Lambda \, \mu_{Gk}(\lambda_k)\right] \\ &= \min_k \max_{\lambda_k} \left[\mu_{ik}(\lambda_k) \, \Lambda \, \mu_{Gk}(\lambda_k)\right] \end{aligned} \tag{4}$$

The membership functions $\mu_{ik}(\lambda_k)$ represent stored knowledge which is generally derived from two sources:

(i) design experience from experts, which are the chief engineer and experienced engineers in this case, and

(ii) accumulated knowledge from previous designs, and large scale explorative designs.

Both membership functions $\mu_G(\lambda)$ and $\mu_{ik}(\lambda)$ are selected from a number of prototype functions:

1. Binary function B(λ)

$$B(\lambda) = \begin{cases} 1 & \text{if } \lambda \in \text{A} \\ 0 & \text{if } \lambda \notin \text{A} \end{cases} \tag{5}$$

where A is the "acceptable" class.

2. Feature functions $F_1(x), F_2(x)$.

The feature x is a function of λ.

$$F_1(x) = \begin{cases} 0 & \text{if } x \leq L \\ (x-L)/(U-L) & \text{if } L < x < U \\ 1 & \text{if } x \geq U \end{cases} \tag{6}$$

$$F_2(x) = \begin{cases} 1 & \text{if } x \leq L \\ (U-x)/(U-L) & \text{if } L < x < U \\ 0 & \text{if } x \geq U \end{cases} \tag{7}$$

Equation (4) determines not only the product type, but also the ranges of performances features λ_k. The cost factor can be treated in the same way as just another design feature, or it can be treated separately. In the latter case, there can be a few alternatives with different costs and different memberships μ_{iG}.

The Critical Feature Group

In (4), the membership functions of certain feature groups are binary. These features are used first to reject certain design alternatives and narrow down the field of choice. Feature groups yielding low intersection membership values for some design alternatives serve the same function. On the other hand a feature group yielding high intersection membership values for all remaining design alternatives become irrelevant. The feature group which is pertinent to the final selection process is called the critical feature group.

Theorem 1: If for a critical feature group the following conditions are satisfied:

(a) the goal membership function is independently specified as a monotonic function of each component λ_{kj} in the feature group λ_k :

$$\mu_{Gk} = \prod_{j \in k} \mu_{Gj}(\lambda_{kj})$$

where $\mu_{Gj}(\lambda_{kj})$ is monotonic in λ_{kj} for each j.

(b) the design choice membership function $\mu_{ik}(\lambda_k)$ is monotonous in every component λ_{kj} of λ_k , such that

$$\frac{\partial \mu_{ik}}{\partial \lambda_{kj}} \cdot \frac{\partial \mu_{Gj}}{\partial \lambda_{kj}} < 0 \tag{8}$$

in the active domain i.e., in the domain of $0 < \mu_{Gj}(\lambda_{kj}) < 1$ for every j, and

(c) the functions μ_{Gj} and μ_{ik} are continuous and differentiable, with continuous first derivatives; then the maximizing solution to the equation

$$\mu_{iGk} = \max_{\lambda_k} [\mu_{ik}(\lambda_k) \wedge \mu_{Gk}(\lambda_k)]$$

satisfies the following condition:

$$\mu_{iGk} = \mu_{ik}(\lambda_k) = \mu_{Gj}(\lambda_{kj}) \tag{9}$$

for all j.

Proof: If for any feature x, $\mu_{Gx}(\lambda_{kx})$ is larger than μ_{iGk} by a finite amount, a contradiction to μ_{iGk} being maximum will be shown. Let $\in$ be a very small positive number:

$$\in \ll \mu_{Gx}(\lambda_{kx}) - \mu_{iGk}$$

Let the constants a_j and b_j be defined by:

$$a_j = \frac{\partial \mu_{ik}}{\partial \lambda_{kj}}$$

$$b_j = \frac{\partial \mu_{Gj}}{\partial \lambda_{kj}}$$

where the partial derivatives are evaluated at the present values of λ_k . Let a new set of values μ_{kj} be defined by:

$$\lambda'_{kj} = \lambda_{kj} + \frac{\in}{b_j} \quad \text{for all } j \neq x$$

$$\lambda'_{kx} = \lambda_{kx} + \frac{\in}{a_x}[1 - \sum_j{}' \frac{a_j}{b_j}]$$

where $\sum'$ means that the term with $j = x$ is not included in the summation. Then

$$\mu_{ik}(\lambda'_j) = \mu_{ik}(\lambda_k) + \in + 0(\in)$$

$$\mu_{Gj}(\lambda'_{kj}) = \mu_{Gj}(\lambda_{kj}) + \in + 0(\in) \ , \ j \neq x$$

where $0(\in)$ means of the order $\in^2$ or smaller. With $\in$ sufficiently small

$$\mu_{ik}(\lambda'_k) > \mu_{ik}(\lambda_k)$$

$$\mu_{Gk}(\lambda'_k) > \mu_{Gk}(\lambda_k)$$

and the contradiction is proved. Similar proof can be given if $\mu_{ik}(\lambda_k) > \mu_{iGk}$.

The significance of Theorem 1 is that it reduces the peak-seeking problem to a Newton's approximation problem. Let f_{kj} denote the inverse function of μ_{Gj} . Let u denote μ_{iGk} . Then from (9)

$$\lambda_{kj} = f_{kj}(u)$$

and solution of the equation

$$u = \mu_{ik}(f_k(u))$$

yields the value of u. Condition (8) is an important property of the critical feature group. For otherwise λ_{kj} would take on an extreme value and is no longer relevant. Theorem 1 can be called the critical feature theorem.

An Example Problem: Design an electric motor for the garage door opener

The design goals are formulated from the design problem: A man's short interval mechanical power is estimated at 1/3 H.P. To be on the safe side, a motor output power above 1/2 H.P. is preferrable. Most house power outlets are rated at 120 volts, 20 amperes, and 60 hertz. It is preferred that the motor is to draw current well within this range. The starting torque is preferrably large to overcome static friction. However, since chain drive is used, there is some slack during starting, and a starting torque equal to full load running torque is acceptable. With some obvious additions, the following set of design goals is formulated:

Design Goals

(a) Power: 120 volts, 60 hertz, 20 amperes, single phase.
(b) Horsepower: approximately 1/2 H.P., preferrably more.
(c) Running current: approximately 7 amperes, preferrably less.
(d) Starting current: approximately 15 amperes, preferrably less.
(e) Starting torque: 100% of full-load torque or more.
(f) Efficiency: The higher the better, no less than 50%.
(g) Cost: Lowest possible.
(h) Life expectancy: Longest possible.
(i) Radio noise: None.

With the exception of items (a) and (i), all the goal statements are not precise, and to some degree interlinked. The motor manufacturer produces the following types of fractional horsepower motors:

1. hysteresis motors
2. reluctance motors
3. stepping motors
4. permanent capacitor motors
5. capacitor start motors
6. capacitor start, capacitor run motors
7. resistive split phase motors
8. shaded pole motors
9. d.c. motors
10. universal motors.

TABLE 1 Fraction Horsepower Motor Characteristics

Motor Type	Power Source	Radio Noise	H.P.	Speed	Eff.	I_{run}	I_{start}	Starting Torque	Cost	Weight	Life
1	AC	no	*L*	syn	M	M	M	M	*H*	H	M
2	AC	no	A	syn	M	M	M	M	*H,M*	H	M
3	AC	no	A	syn	M	M	M	M	*H*	H	M
4	AC	no	A	syn-	H	L	L,M	M	M	M	H
5	AC	no	A	syn-	M	M	L	H	M	M	M
6	AC	no	A	syn-	H	L	L	H	H,M	M	M
7	AC	no	A	syn-	M	M	H	H,M	M	M	M
8	AC	no	L,M	syn-	*L*	*H*	*H*	L	L	H	H
9	*DC*	yes	A	var	H	L	H	H	M	M	M,L
10	A	yes	A	var	M	M	M	H	M	L	*L*

Note: A = all, syn- = below synchronous speed, H = high, M = medium, L = low.

The pertinent motor characteristics are summarized in Table 1. The features which are responsible for eliminating various motor types are underlined. They produce a maximum possible μ_{iG} which is either zero or very low in the binary operations. The motor types which remains are 4, 5, 6, and 7. They belong to the class of single phase induction motors. The feature groups which produce high values of $\mu_{iG}(\lambda_k)$ for all these types are insignificant. The feature group which is most significant in the refinement process consists of two features:

λ_b = motor horsepower rating

λ_c = motor running current

Referring to (5), the stored membership function can be expressed as $F_1(x)$ with L = -0.05, U = 0.05 and

$$x = C_1 \log \lambda_b + C_2 - \frac{746\,\lambda_b}{120\,\lambda_c} \tag{10}$$

The values of C_1 and C_2 for different motor types are listed in Table 2:

Table 2

Motor type	i	C_1	C_2
permanent capacitor	4	0.20	0.72
capacitor start	5	0.32	0.68
capacitor start, capacitor run	6	0.20	0.74
resistive split phase	7	0.32	0.67

The derivation of (10) will now be explained. Let E_a denote the apparent efficiency:

$$E_a = \frac{\text{Output Power}}{\text{Input Voltampere}} = \frac{746\,\lambda_b}{120\,\lambda_c} \tag{11}$$

Let $[E_a]$ denote the ensemble average of E_a . An expression which represents accumulated experience is

$$[E_a] = C_2 + C_1 \log \lambda_b \tag{12}$$

Equation (12) means that $[E_a]$ is a linear function of log λ_b . Its value is reduced as the motor horsepower is reduced.

Equation (10) can be rewritten as:

$$[E_a] - E_a = x \tag{13}$$

The membership function

$$\mu_{ik}(\lambda_b\,, \lambda_c) = F_1(x) \tag{14}$$

means that

1. $E_a > [E_a] + 0.05$ is almost impossible to achieve and $\mu = 0$

2. $E_a < [E_a] - 0.05$ is easily within range and $\mu = 1$.

The design goal membership function can be expressed as

$$\mu_{Gk}(\lambda) = F_1(\lambda_b) \wedge F_2(\lambda_c) \tag{15}$$

For F_1 : $L_1 = 0.25$, $U_1 = 0.75$

For F_2 : $L_2 = 5$, $U_2 = 9$

Maximization Computations

The following conditions are readily checked:

$$\frac{\partial \mu_{Gk}}{\partial \lambda_b} > 0$$

$$\frac{\partial \mu_{Gk}}{\partial \lambda_c} < 0$$

$$\frac{\partial \mu_{ik}}{\partial \lambda_b} < 0$$

$$\frac{\partial \mu_{ik}}{\partial \lambda_c} > 0$$

Thus the condition (8) is satisfied. From (9), μ_{Gk} and μ_{ik} are identical. Using (5) and (b), λ_b and λ_c can be expressed in terms of a parameter u, where u $= \mu_{ik} = \mu_{Gk}$:

$$\lambda_b = 0.25{+}0.50u \tag{16}$$

$$\lambda_c = 9{-}4u \tag{17}$$

Substituting (16) and (17) into (10) gives

$$-0.05 + 0.1u = C_2 + C_1 \log\ (0.25 + 0.50u) - \frac{746\,(0.25 + 0.50u)}{120\,(9 - 4u)} \tag{18}$$

Equation (18) can be solved by using Newton's approximation. Results of the computation are listed in Table 3.

Table 3

Motor type	i	μ_{iG}	λ_b	λ_c
permanent capacitor	4	.757	.6285	5.972
capacitor start	5	.68	.59	6.28
capacitor start, capacitor run	6	.778	.639	5.888
resistive split phase	7	.67	.585	6.32

The capacitor start capacitor run motor gives a higher membership function for the H.P.-running current feature group, but its cost is considerably higher than that of the permanent capacitor motor. The remaining two motor types have lower membership functions at approximately the same cost. A permanent capacitor motor is selected, the design specifications are roundoff to be

λ_b = 0.625 H.P.

λ_c = 6 amperes

The above results are illustrated in Fig. 1, which shows the constant μ contours on the H.P. versus motor current (λ_b versus λ_c) plane. The solid lines are constant μ_G contours, and the broken lines are constant μ_4 contours. The optimum point is marked with a star. At the optimum point $\mu_{Gb} = \mu_{Gc} = \mu_4 =$ 0.75.

In the following sections, an application of SSDS to the component parts industry will be discussed. Since the design alternatives in this industry are usually product types, we refer to SSDS as PSDS, which is abbreviated from Product Selection and Design Specification.

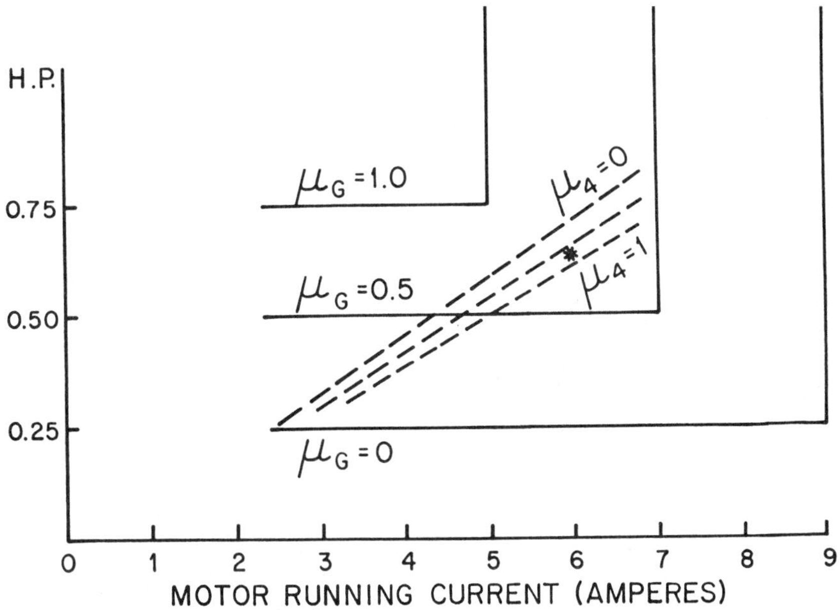

Fig.1 MEMBERSHIP FUNCTION IN FEATURE SPACE

Design Automation for the Component Parts Industry

The component parts industry is an important element in manufacturing. Very frequently some components in a manufactured product are purchased from different companies, e.g. the electrical motor in a typewriter, washing machine, or machine tool, switches, filters, and IC's in electronic circuits, and etc. The manufacturer usually has more than one source of supply to insure the component's quality, price, and punctuality in delivery. The selling and purchasing processes of these components are an active and integral part of manufacturing.

The design automation process to be described applies to a generic class of the component parts industry in which the components to be supplied are not shelf items but engineered specifically for the application. Electrical motors and filters are well known examples. The appliance manufacturers are generally not satisfied by the characteristics and external dimensions of a standard motor. Motor manufacturers (the component suppliers) are willing to design specific motors for such applications because of the large production volume involved.

A generic property then is the necessity of going through engineering during the bidding and sales phase. Since there are a large number of sales offices distributed throughout the world and far fewer engineering offices, the above mentioned need represents both a drain on the engineering manpower and a substantial delay in responding to an inquiry. Vertically integrated design automation (VIDA), eliminates these problems in addition to doing what it normally does: facilitating the design and manufacturing processes.

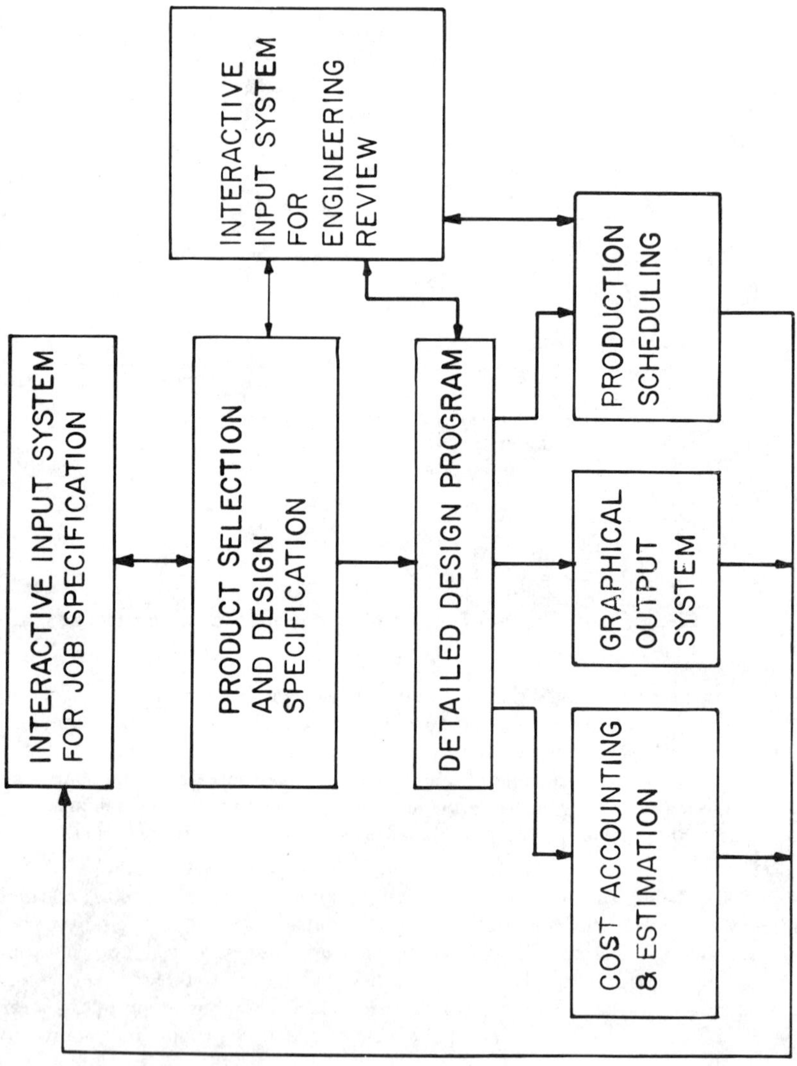

Fig. 2 VIDA System

Knowledge Based VIDA System

Figure 2 illustrates the major software components in a VIDA system. It is organized as a parallel to human operations. A detailed description of the major software components is given below:

a. Interactive Input System (IIS)

Customer's job specifications can be entered in two different ways:

(i) An input routine is provided by the IIS, and a salesperson may type in the job specifications accordingly.

(ii) The IIS prompts the salesperson with a set of nested questions. A salesperson may select either method (i) or (ii) as listed above. IIS acknowledges when the input data is sufficient for the PSDS to proceed. If the salesperson selects (i) but his input data is insufficient, the IIS automatically goes into (ii).

b. Product Selection and Design Specification System (PSDS) (Chief Engineer, experts)"

The PSDS is a knowledge-based system and is a most important part of the VIDA. It will be discussed in some detail. The PSDS can be implemented in two ways:

(i) Heuristic Selection

The heuristic selection algorithm for realizing PSDS is illustrated in Fig. 3A. Customer's specifications are used to generate a set of qualitative or binary features through a process of abstraction. A product class is then selected by heuristically matching the qualitative features to the product classes. Knowledge based rules or decision trees are used in the matching process. Since a substantial amount of information in the customer's specifications is not used in feature abstraction, the unused information is then used for refinement of the selection process: from product class to product type, with design specifications further defining the product.

(ii) Fuzzy Set Representation

There is a certain parallelism between heuristic classification (HC) and fuzzy set analysis (FSA) as illustrated in Fig. 3B. Abstraction of qualitative features in HC is similar to formulation of the goal membership function $\mu_G(\lambda)$ from customer's job specifications in FSA. Matching the qualitative features with similar features in the solution space is parallel to the maximizing operation of equation (1). The membership functions of certain groups of features λ are binary or almost binary in nature. Using these features first, a product class is selected. The final maximizing operation is performed over the selected product class to arrive at the product type and design specifications. The last operation in FSA is parallel to the refinement process in HC.

c. Detailed Design Program (Design Engineer)

The detailed design programs are easiest to computerize. Most engineering offices have clearly written design procedures for their products, including all pertinent equations, constants, and curves. Once the product type and design specifications are made available, detailed design is a matter of routine.

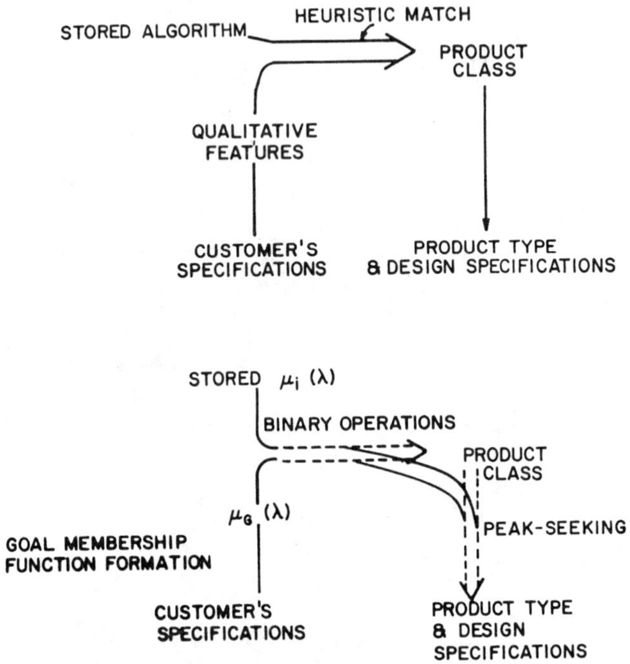

Fig. 3 TWO WAYS OF REALIZING SSDS

d. Cost Analysis (Cost-accounting)

Cost analysis of the detailed product includes (i) labor, (ii) materials, (iii) production tooling, and (iv) overhead. Equations with cost factors which may vary from time to time constitute this program. Quite often, a few alternatives, representing production tools with different degrees of automation are available. The costs for all alternatives are computed, and the lowest is selected.

e. Production Scheduling and Estimation of Delivery Date (production)

The methods used for estimation of a delivery date may have varying degrees of sophistication:

(i) Estimated waiting time plus production time

The estimated waiting time is a function of the current state of production, including all scheduled work in the production queue. It is a value posted by the Production Department periodically. The production time of a designed product depends on the product and its quantity only. It can be readily calculated once a design is made.

(ii) Estimation of Delivery Date from Scheduled Production Routing

Production scheduling can be done heuristically or by mathematical programming. Once a production routing is determined by either method, the job completion date is then calculated step-by-step according to the routing selected.

In general, method (i) is used in the bidding - sales cycle. Once a sale is confirmed, its production is then worked in with all other scheduled work already in the production line. An accurate estimate of the delivery date using (ii) then can be made.

f. Graphical Output Program (GOP) (draftsman)

An important component in the GOP is a Parts Library. Most sales drawings can be done by calling items in the Parts Library with appropriate transformations.

g. Interactive Input System for Engineering Review (IISER)

The IISER allows an engineer to override computer decisions where alternatives are allowed, or to enter certain design constants where "default values" (nominal values) are normally used. The effects of such alterations, in the form of DDP computed results, will be fed-back to the engineer.

A Comparison of Two Systems

The flow of information in an unautomated sales to engineering to production cycle is as shown in Fig. 4: Responding to a customer's inquiry, a sales representative obtains the pertinent specifications and relays the information to engineering, which makes a preliminary design of the product, has it costed and tentatively scheduled. An unscaled drawing of the product, sometimes with calculated performance curves, price, and estimated delivery date are then sent back to the customer through sales. If a purchase order follows, engineering makes the final design and production scheduling. Since one engineering office serves many widely distributed sales offices, it becomes a bottleneck in the sales phase.

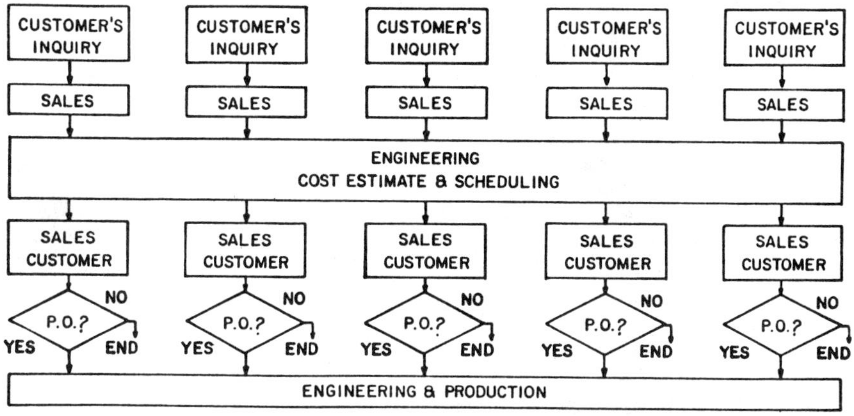

Fig.4 SALES-ENGINEERING-PRODUCTION CYCLE WITHOUT AUTOMATION

There are three essential disadvantages:

1. Since most purchase inquiries do not result in a purchase order, much of the engineering work in the preliminary design phase is wasted.

2. There is substantial delay in the preliminary design phase. The purchase specifications usually go to a chief engineer or another experienced engineer, who, based on his broad experience makes a number of preliminary choices, so that a less experienced engineer can follow through with detailed design work. Usually each engineer has a number of jobs lined up, and the job has to wait in its queue.

3. The tentative production schedule is very uncertain when there is a large number of jobs in the purchase inquiry and response phase. One has to be conservative in estimating a delivery date.

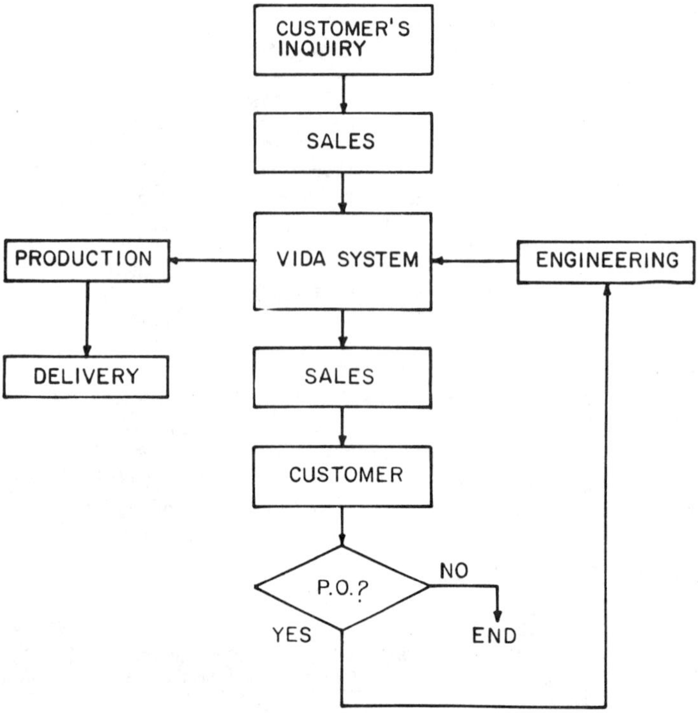

Fig. 5 SALES-ENGINEERING-PRODUCTION CYCLE WITH VIDA

Figure 5 shows a vertically integrated design automation system (VIDA), from sales to production. After obtaining specifics in a customer's inquiry, the sales representative inputs such information into a VIDA terminal interactively. The VIDA program asks clarification if the specifications are incomplete. With a complete set of specifications, the product selection and design specification (PSDS) program makes the preliminary choices and passes this information to the Detailed Design Program (DDP) which makes the detailed design to meet the specifications. The completed design is then entered into a cost analysis and scheduling program which yields price and delivery information. Only successful bids are reviewed by engineering and modified interactively.

The VIDA system eliminates unnecessary use of engineering manpower, and speeds up the sales and bidding operations. Preliminary design by a computerized VIDA can be made within minutes.

Another distinct advantage of VIDA is that its preliminary design results can be conveyed directly to a graphical output program (GOP). Professional grade drawings of the component, and characteristic curves representing its performance are then obtained at very little cost. They give a favorable impression in the bidding process.

Conclusion

Fuzzy set analysis is used to represent expert reasoning in the first phase of an engineering design, that of system selection and design specification (SSDS). Protype membership functions are introduced to represent both the design goal and design alternatives in feature space. Goal membership functions are mathematical representations of the goal specifications. Design alternative membership functions represent accumulated previous design experience or distilled knowledge from a large number of explorative designs. If certain feature values are easily obtained for a design alternative, its membership at such feature values is 1. If certain feature values are difficult or impossible to obtain for a design alternative, its membership at such feature values is then zero. The design alternative having highest intersection with the design goal is selected. The parallelism between the above analysis and an expert engineer's reasoning process is discussed and demonstrated with an example on electrical motor design. A critical features theorem is proved. Using the theorem, the problem of peak-seeking for highest membership function is reduced to a Newton's approximation problem.

An immediate application of the SSDS is vertically integrated design automation of the component parts industry (VIDA). VIDA enables a component parts manufacturer to respond quickly to purchase inquiries, to improve its chance of a successful bidding, and to save engineering manpower for research and development. Its engineering expertise is documented and accumulated automatically.

References

1. Zadeh, L.A., Fuzzy Sets, *Information and Control*, Vol. 8, pp. 338-353, 1965.

2. Becker, J.M., *A Structural Design Process Philosophy and Methodology*, Ph.D. Thesis, University of California, Berkeley, 1973.

3. Tsukamoto, Y. and Terano, T., Failure diagnosis by using fuzzy logic, in *Proceedings IEEE Conference Sensory and Control*, pp. 1390-1395, 1974.

4. Blockley, D.I., The role of fuzzy sets in civil engineering, *Fuzzy Sets and Systems*, (2):267-278, 1979.

5. Wood, K.L., and Antonsson, E.K., Engineering design-computational tools in the SYNTHESIS domain, The Study of the Design Process: An NSF Workshop, Coordinated by M.B. Waldron, the Ohio State University, 1987, pp. 75-95.

6. Clancey, William J., Heuristic Classification, in *Knowledge Based Problem Solving*, Janusz S. Kowalik, Editor, Prentice Hall, 1986.

7. Hayes-Roth, R., Waterman, D.A., and Lenat, D.B., *Building Expert Systems*, Addison-Wesley, 1983.

8. Chang, Sheldon S.L., Design Automation for the Component Parts Industry, 25th ACM/IEEE Design Automation Conference, June 12-16, 1988, Anaheim, California.

9. Zadeh, L.A., A Computational Approach to Fuzzy Quantifiers in Natural Languages, *Computer and Mathematics*, Vol. 9, pp. 149-184, 1983.

Form - Function Characteristics of Electro-Mechanical Designs

James R. Rinderle
Associate Professor

Eric R. Colburn
Stephen P. Hoover
Juan Pedro Paz-Soldan
John D. Watton
Graduate Research Assistants

Department of Mechanical Engineering
Carnegie Mellon University
Pittsburgh, Pennsylvania 15213

The Nature of the Mechanical Design Problem

Mechanical designs are solutions to multi-faceted problems. In this regard, mechanical design is no different from most other engineering design disciplines. However, in many design disciplines such as software design and circuit design, designs can be characterized as collections of weakly-interacting functional modules, each of which implements one of the functional requirements. In these domains, good designs can often be accomplished by successively decomposing requirements until the lowest level requirements match the behavior of some preexisting design component. Direct transformation and recombination can then be used to complete the design. On the other hand, good mechanical designs are often highly-integrated, tightly-coupled collections of interacting components because the cost, size, and weight of mechanical components makes a direct application of the decompose and transform strategy impractical. In well designed mechanical devices a simple correspondence between specific functional requirements of the product and individual components in the design does not usually exist. The converse is also true, i.e., a specific component does not contribute to a single function of a product.

We might summarize by saying:

- The form-function relations in mechanical design are complex.
- Function is not isomorphic to form.
- Design strategies based on functional decomposition and direct mapping into physical components are not generally applicable in mechanical design.

The absence of a close correspondence between functional and physical descriptions in mechanical design has important ramifications on the design process and on the efficacy of designs themselves. It is the designer's task to specify a *form* to satisfy constraints on *function* and *fabrication*. In light of this task, and the complexity of the form-function structure, we might ask what makes a good designer? One attribute, among many perhaps, is that good designers understand and take advantage of the subtle relationships which exist among function, form and fabrication for a mechanical device. If we are to reduce our dependency on the intuition of experienced designers we must also extract and utilize the relationships between form and function in designing a product.

Toward that end we assert that form-function characteristics of components and devices exist and can be identified from physical principles. Furthermore, we believe that design decisions are dominated by considerations having to do with relationships among form and function.

In our program of design research we have embarked on four areas addressing these basic questions. Firstly, we are seeking to identify the form-function characteristics of components and to abstract high level design relationships. Secondly, we are aggregating component level behavior to device level in a way that will consider the context and resolution of the design stage. Thirdly, we are seeking to identify the dominant design relationships which arise as a result of this aggregation and fourthly, we propose a synthesis strategy based on the opportunistic utilization of component form-function relations. Each of these four topics is discussed briefly in the sections which follow.

Identifying Form-Function Relations for Physical Systems

Much of the design process consists of selecting, sizing and configuring standard components into systems. During preliminary design, a designer reasons abstractly about the components, concentrating on high-level characteristics in order to evaluate alternative configurations. Consider, for example, the design of a print head positioning mechanism used in a dot-matrix printer as shown in Figure 1. To reason about the feasibility of the proposed configuration the designer does not refer immediately to a catalogue to select a particular motor, but instead relies on his experience to estimate the size, weight, torque, and power of the motor, ignoring other more detailed geometric and behavioral characteristics.

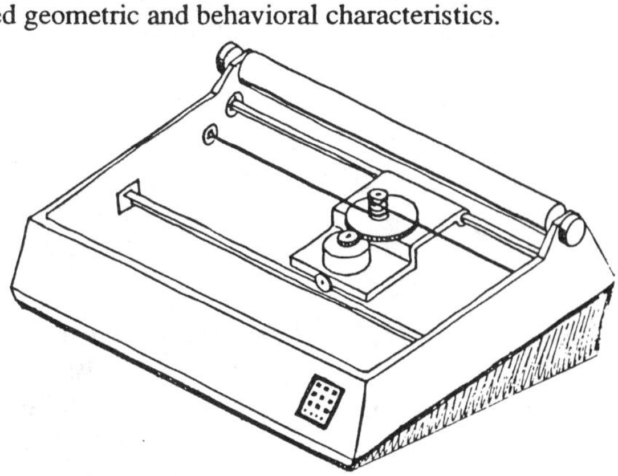

Figure 1: Print Head Positioning Mechanism

As the design is refined the designer will consider characteristics at many levels of abstraction. He reasons about high-level parameters such as weight and positioning speed because these are the terms which express the design specifications. He also employs intermediate-level parameters such as stress, aspect ratio and current density because these help him reason efficiently about the device, and he considers low-level parameters, for example, length, diameter and material since these are the design variables that he directly controls.

In order to reason about these various parameters designers use their experiential knowledge of the relationships between the form of a device and the behavior it exhibits. Knowledge of the relationship between weight and motor torque, for example, enables a designer to quickly evaluate the performance of the print head positioner. The configuring and evaluation of a complex design is aided by an understanding of these *form-function relations* which express inherent characteristics of the components that comprise the design [1, 2].

Although vast experience is beneficial it may not be sufficient to capture some relationships such as those between parameters of different levels of abstraction, for instance between the power consumption of an electric motor and the winding wire diameter. The two are related, but to a designer the form of the relation would not likely be obvious. Determining this sort of relation is useful to the designer because the design requirements are frequently given in terms of the high-level behavior desired, but the designer's task is to specify a detailed description of this component.

Form-function relations also exist between parameters that are at or near the same level of abstraction but that are normally associated with physically separate parts of the design, such as the rotor and stator of a motor, or parameters that exist in different energy domains, such as electrical and mechanical. Understanding these relationships aids the system designer by explicitly showing how high-level characteristics of the components are related to one another, allowing him to reason about tradeoffs between competing requirements or objectives, for instance the tradeoff between power and size of an electric motor. It is the consideration of these relations that dominate preliminary design decisions, for they are used to evaluate tentative configurations.

High-level form-function characteristics are a reflection of the underlying physics of the device and are abstractions of the complex interactions of low-level parameters. A change in the torque requirement for the motor may be met by adjusting the wire size, the number of poles, or a host of other parameters. However, other characteristics such as weight, will also change. Determining the resulting relationship between torque and weight is difficult to do directly. Instead designers usually gain this sort of knowledge empirically over years of design experience, acquiring a "feel" for the characteristic relations of a device. Empirical methods however, are subject to error due to hasty generalizations and narrowness of experience. In addition, these relations are dependent on the state-of-the-art, therefore major changes in technology can render the knowledge of particular form-function relations obsolete.

Motivated by the usefulness of form-function relations and the difficulty in obtaining them, we have developed a representation of electromechanical devices and a method for automatically identifying relevant form-function relations from it. The details of the representation and the method used are discussed in other papers [3, 2].

Device Representation

Our approach is based on the declarative representation of a device as a set of parameters related by constraints. The constraints arise from physical laws, spatial relationships and material limitations. Collectively these constraints define the space of acceptable designs[1].

In this model each parameter describes some characteristic of the form (such as a physical dimension or material density) or behavior (such as velocity, stress, or torque). The constraints relate the parameters typically through equalities or inequalities[2]. Equality constraints are relationships between parameters that always hold and may be the result of physical law (e.g., $f = ma$), may be imposed as a requirement of the design (e.g., $voltage = 12\,volts$) or may define a geometric relationship (e.g., $A = \pi D^2/4$). Inequalities are often used to express physical limitations (e.g., $temperature < melting\,temperature$), imposed requirements (e.g., $torque > 2.4 ft\text{–}lbs$) or spatial relations (e.g., $OD > ID$). The compositional nature of the constraints allows the model to be easily expanded to an arbitrary level of description by the addition of parameters and constraints.

A collection of these constraints forms a network [4] or a *bipartite graph* [5] with each node representing either a constraint or a parameter. In this graphical representation of the constraint model, each parameter node is linked to all of the constraints that it participates in, and each constraint is linked to all of its participating parameters. A constraint network for a brushless, unhoused d.c. electric motor under stall conditions is shown in Figure 2. Note that most of the inequality constraints and numerical limitations on the parameters are not shown in the figure for the sake of readability.

The constraint network itself represents a prototypical device or class of devices and *satisfying* the network of constraints by assigning a value to each parameter such that none of the constraints are violated, results in an instance of the class[3]. Thus we can view the satisfaction of a constraint network as analogous to parametric design. But our goal in developing this representation is not to support parametric design, it is rather to aid the designer by providing a way to automatically identify relevant form-function relations.

Design Context

A network of constraints represents the nature of a class of devices, but not a particular design problem. The specific requirements and objectives of a problem constitute a *design context*, which can be cast over the network to represent the given task. The context further constrains the space of the solutions to those that are acceptable for the particular situation and specifies the criteria necessary to identify the "best" design among these. Thus both a design and its corresponding set of form-function relations are directly related to the context under which it was created. For example one designer may be designing a motor for a given torque while trying to minimize weight, while another may be trying to maximize the efficiency while maintaining a fixed diameter. While both designers may be designing motors of the same class,

---

[1]Although both the parameterization and the choice of relevant constraints both influence the design space.

[2]We are currently using only these but our approach does not preclude other constraints such as boolean, discrete, differential equations, etc.

[3]Satisfaction of a constraint network is in general very difficult. As stated by Gosling in [6] "...boolean satisfiability can be cast in this framework and is NP-complete, so general satisfaction is at least as hard." But there have been many techniques developed and applied to this problem including constraint propagation, monotonicity analysis, relaxation and optimization.

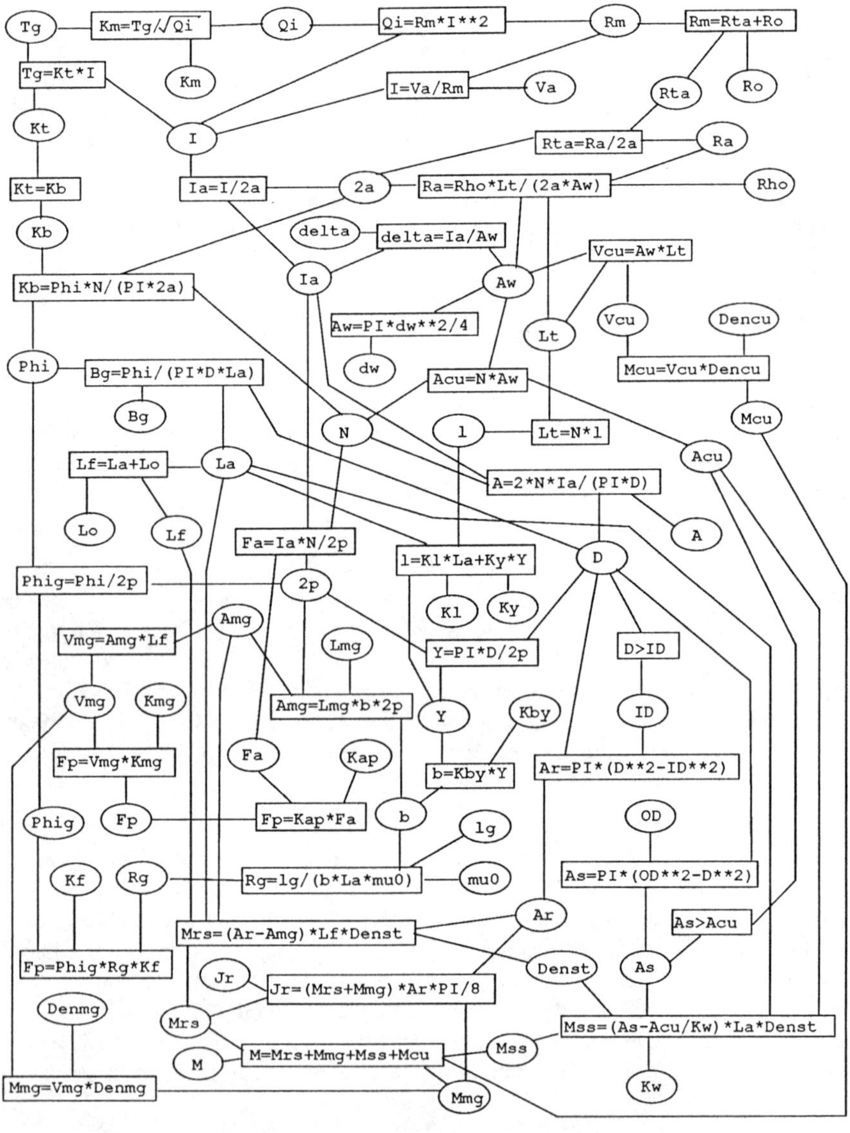

Figure 2: Example Constraint Network: D.C. Electric Motor

the first will produce a different design than the second, and the relations between characteristics such as torque and weight or efficiency and diameter will be different for the two contexts.

Approach

To aid the designer, we determine form-function relations from a constraint-based model of the underlying physics of a device. Our approach is based on solving a sequence of optimization problems, corresponding to a continuum of design contexts. This technique results in a series of optimas which can be plotted to show the relationships between various parameters[4]. The following example illustrates this approach.

Example: Brushless D.C. Motor

Consider the use of a frameless brushless d.c. motor in a printer as seen in Figure 1. The designer wishes to reduce the length of a standard motor to fit it in a particularly tight space. It is necessary to reduce the length as much as possible while maintaining the same torque and inside diameter and minimizing weight. The designer may be interested in increasing the outer diameter to effect this change, however, he will want to know how changing the diameter may affect some of the other aspects of the design.

Figure 3 shows the form-function relations obtained for this context from the constraint network shown in Figure 2. Each of the points marked on the curves represents a computer generated optimum of a prototype motor. Each curve then relates diameter of the motor to some other parameter. The relations are shown as relative changes in the outer diameter (measured along the abscissa) versus relative changes in the other parameters (measured along the ordinate). Thus the origin represents the standard motor (optimized for minimum mass), and the curves represent relative changes from that point.

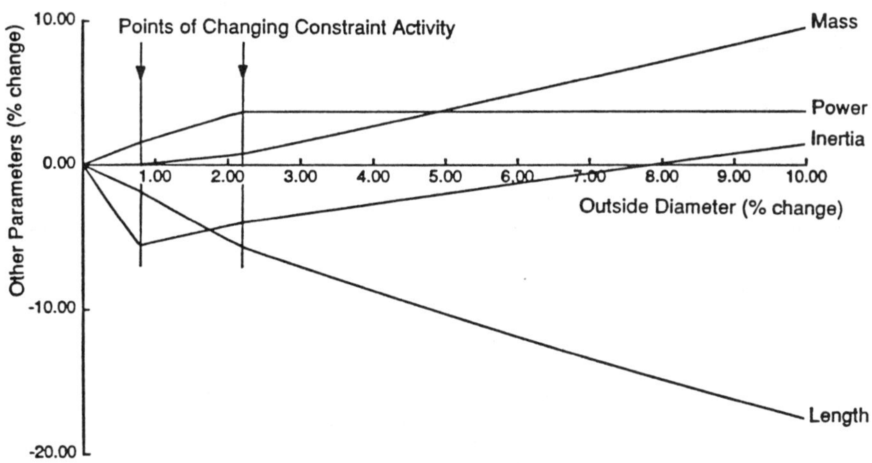

Figure 3: Form-Function Relations for a D.C. Motor

[4]This method is similar to what are called *interaction curves* in the optimization literature [7].

Constraint Activity

The relations shown are the result of the interaction of the defining constraints. In particular, the activity of the inequality constraints[5] shape the relations. We note that relationships among design parameters change dramatically with changes in the activity of the constraints resulting in the various cusps in the plots. Of particular interest in this example is the relationship between outer diameter and the motor inertia. Initially increasing the diameter actually decreases the inertia, but the rotor will reach a limit on strength and further increases in the diameter will call for a more robust and therefore higher inertia rotor. Understanding the nature of a design in this way can be very insightful. Thus determining the active constraints that reflect what is limiting the design can aid in reasoning about the design.

In general predicting the constraint activity is very difficult due to the complex, non-linear nature of the underlying equations and inequalities. The constraint activity depends on both the constraint network and the specific context of the design. For this reason it is not possible to determine *a priori* the active constraints, instead they must be determined for each context. Because they are difficult to predict and because they influence the form-function relations so strongly, the knowledge of active constraints is valuable to the designer.

An Environment for the Conceptual Design of Mechanical Systems

Component form-function relations can help the designer identify tradeoffs for classes of components such as motors, pulleys, masses and so forth, but ultimately the behavior of interest is that of the overall configuration, which is made up from an aggregation of components. Moreover, for mechanical devices the particular geometric interactions between components have to be taken into account to determine the overall system behavior. Reasoning about these behavioral and geometric relations can become a difficult task for the designer as he considers more numerous aggregations of components and as he refines the selected configuration. In this section we discuss the representation requirements for conceptual design environments that can support an interactive aggregation of standard components and the subsequent refinement of a selected configuration.

Representation of Component Aggregations

To allow the aggregation of components into a system, the design environment has to allow the *modular* aggregation of *behavioral* component models. Since mechanical devices include complex geometric relations among components, the internal component representation should include *geometric* component models as well.

Depending on the particular characteristics of the configuration, the designer may want to focus on specific aspects of component behavior. Consider the printer head drive configuration shown in Figure 1. For this configuration, the designer may want to investigate the effect of the linear inertia of the motor on the aggregate dynamic behavior of the drive system since the motor rides on the platform. To model adequately the behavior of the same component connected in different ways, the environment representation has to support several *functional perspectives* for a given component.

If the designer finds that the motor's rotational inertia has a much greater relative effect on the system dynamics than the its translational inertia, he may want to modify the resolution of the motor's behavioral model to investigate its electrical system dynamics. Alternatively, he

---

[5]An active inequality is one that is binding as a strict equality.

may want to assume an extremely simple motor model (e.g. an ideal source of torque) to establish a limit on maximum expected performance. To provide the designer the flexibility of varying the modeling detail of a component, the internal component model representation has to allow *variable resolution* of component models. The need for variable resolution also applies to the geometric model of components to support the gradual refinement of the geometric detail of the design.

Approach

The representation requirements of modular aggregation, functional perspectives and variable resolution have been achieved in an experimental implementation of an environment for conceptual electro-mechanical design. MEDA[6] is an interactive graphical environment in which component models can be aggregated modularly into a preliminary system description. Mechanical components can be connected by specifying kinematic relations between them. The designer can group connected components into subassemblies to consider the behavior of a meaningful part of the overall device. Once a subassembly is identified, the internal modular models of component behavior are automatically collected into a system description, with consideration of the functional perspective implied by the components' connectivity. Through the use of appropriate behavioral primitives, component models with variable resolution can be defined and stored in a component library.

The implementation is targeted to the design of electro-mechanical devices which may require complex dynamic analysis and therefore MEDA automates the process of dynamic model development. For this domain, the component connections are kinematic relations. The kinematic relations specified by the designer introduce constraints between behavioral component models when a subassembly system model is collected. After the behavioral component models are collected into a system model, the resulting system representation can be converted into a set of differential equations for subsequent numerical or symbolic processing.

MEDA is implemented in C as an extension to a commercial modifiable CAD platform [8] in a Sun™ 3/160 workstation. The details of the implementation and a more elaborate discussion of the requirements for conceptual design representations can be found in [9] and [10] respectively.

Automatic Identification of Critical Design Relationships

Once a design configuration has been determined it can be parameterized so as to obtain a set of algebraic constraint equations. These equations describe the functional relationship between the important behaviors of the design and the parameters describing the configuration. The ease with which quantitative and qualitative evaluations can be made depends, in large part, on the complexity of the design equations and constraints involved. Even very simplified design equations may be puzzling because changing the value of one of the design variables may influence many of the functional requirements. As a result, detailed analytical methods are often applied to any competing design configuration alternatives. The results of the analysis are used to judge the merits of the design configurations. Experienced designers, on the other hand, often shortcut the detailed analytical work by recognizing important relationships which govern the performance of the design configuration. This is accomplished by identifying important relationships among requirements and design variables, such as a critical ratio, a nondimensional

---

[6]MEDA stands for Mechanical Engineering Design Assistant.

parameter, or a simple difference; e.g. the column aspect ratio in structures, the Reynold's number in fluid mechanics, or the velocity difference across a fluid coupling. This achieves convenience and expediency in quantitative evaluations and enhances the qualitative physical reasoning associated with the design activity to better enable the designer to focus his creativity on the essential aspects of the proposed configuration.

The discovery of such critical relationships among parameters has been made on an ad hoc basis by experienced designers and engineers. Although certain nondimensional parameters are well known and methods exist for identifying such parameters, there are not, in general, strategies which assist the designer in identifying physically significant relationships which dominate the behavior of a particular design configuration. We seek to do this by establishing methods to identify physically significant new variables and to use them in performing a *transformation* of variable on the constraining design equations and inequalities. The terms *alternative formulation* or *reformulation* refer to a description of the same design configuration but with alternative design variables.

Two Bar Truss - Case Study

As an example of a variable transformation consider the design of a two bar truss [11]. A truss of the type shown in Figure 4 can be used to support both a vertical and horizontal load. The truss deflection must be limited and the truss must withstand the loads without yielding or buckling. The parameters describing the truss are height H, halfwidth B, tube diameter d, thickness t, and the modulus of elasticity, E. The relationships among actual stress, critical buckling stress, deflection and the design parameters are shown in Figure 5. Also shown in the figure is a transformation of variable applied to the truss equations. Changing design parameters does not change the truss design itself, only the form of the problem.

The Nature of Design Complexity and Coupling

Note, that each of the truss design relations depends on all four of the geometric design parameters making it difficult to reason about the design. When this occurs, we say that the desired functions are *coupled* or that the design is coupled. Simon [12], Preiss [13], Rinderle and Suh [14] are among those who commented on the nature of coupling in designs. The complexity due to coupling depends, at least in part, on the designer's selection of the product functional requirements and the design parameters. One structure of particular interest, which we refer to as serially decomposable, results from transformations that cause the design equations to become solvable without iteration. The transformation performed in Figure 5 makes the truss equations serially decomposed.

Generating and Choosing Useful Transformations

It is known that almost all of the nondimensional variables commonly used are one of four types: ratios of lengths, forces, energies, or properties [15]. This follows from the fact that all of the fundamental equations of continuum analysis can be nondimensionalized with these four types of nondimensional variables. In many cases the forces and energies are constructed from more primitive variables. For instance, Reynold's number is the ratio of fluid inertial force to fluid viscous force and in turn these forces are constructed from variables for fluid velocity, viscosity, length, and mass density. In fact, all the common nondimensional variables in fluid mechanics can be constructed from fluid force ratios. Rules for the construction of forces can be compiled for many mechanical engineering disciplines. In order to construct and use such rules, information other than fundamental units is required for each component variable.

The identification of dimensional variables is also enhanced using physical meaning rules. In the field of geometry, for instance, areas can be constructed from a radius squared but not

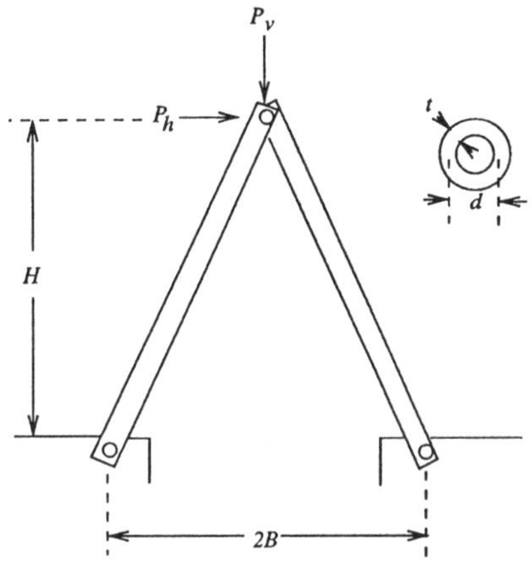

Figure 4: Two Bar Truss

The original design requirement equations are:

$$\sigma_{compression} = \frac{\sqrt{B^2+H^2}}{2\pi td}\left(\frac{P_v}{H}+\frac{P_h}{B}\right) \qquad = f_1(B,H,d,t)$$

$$\Delta\sigma_{buckling} = \frac{\pi^2 E(d^2+t^2)}{8(B^2+H^2)} - \frac{\sqrt{B^2+H^2}}{2\pi td}\left(\frac{P_v}{H}+\frac{P_h}{B}\right) \qquad = f_2(B,H,d,t)$$

$$\delta = \frac{P_v(B^2+H^2)^{1.5}}{2\pi tdH^2E} \qquad = f_3(B,H,d,t)$$

The following transformation:

$$d = d,\ H = H,\ R = B/H,\ A = \pi dt$$

produces an alternative form of the equations:

$$\sigma_{compression} = \frac{\sqrt{R^2+1}}{2A}\left(P_v+\frac{P_h}{R}\right) \qquad = g_1(R,A)$$

$$\delta = \frac{P_vH(R^2+1)^{1.5}}{2AE} \qquad = g_2(R,A,H)$$

$$\Delta\sigma_{buckling} = \frac{E(\pi^2d^4+A^2)}{8H^2d^2(R^2+1)} - \frac{\sqrt{R^2+1}}{2A}\left(P_v+\frac{P_h}{R}\right) \qquad = g_3(R,A,H,d)$$

where for both sets of equations the requirement constraints are:

$$\sigma_{compression} \le \sigma_{yield},\ \Delta\sigma_{buckling} \ge 0,\ \delta \le \delta_{max}$$

Figure 5: Two Bar Truss Equations

from a thickness squared. In the domain of rigid body dynamics forces or torques can be indicated as inertial or viscous and can be hierarchical in nature. For example, an inertial force may consist of the product of a mass and an acceleration, but acceleration in turn can be constructed according to other rules. In this way, for example, it is possible to generate both torque and energy variables (both of which may have units of foot-pounds) but to maintain the distinction between them and to use them appropriately. Construction techniques of this sort are chosen for general design domains and applied recursively. Spatial proximity methods are used to reduce the list of candidates prior to more complete evaluation based on design equation structure. In the truss example the transformation is to an area, A, and a ratio of lengths, R, while two parameters, d and H, remain the same. While the utility of this transformation is high, as will be seen in the next few subsections, the transformation itself is quite simple. Finding this transformation required a generate and test technique combined with goal oriented methods employing a measure of coupling. Details are described elsewhere [16].

Summary of Reformulation Advantages

There are several advantages to using less coupled, especially serially decomposed, design equations. The following ones will be discussed for the truss problem:

1. Enhanced ability to find numerical solutions.
    - Ease in determining satisfactory solutions.
    - Ease in making design changes to accommodate requirement changes with minimal iteration.
2. Identification of active constraints.
3. Symbolic computational benefits such as identifying form-function relationships.
4. Cognitive Benefits.

Numerical Solutions

Alternative variables with clear physical meaning allow, and perhaps even enhance, numerical value estimates. This is seen in the truss equations where a numerical solution of the original formulation could become an extended trial and error session of computations. This is because each expression relies on all of the design parameters making the problem one of three simultaneous nonlinear expressions in four unknowns (with additional inequality constraints). The reformulation on the other hand has been arranged in a serially decomposed form and a numerical solution can be easily found in a single iteration. This is facilitated by the fact that as one goes through the list of relations there is one additional parameter for each additional expression. The physical significance of each variable facilitates good estimates, for example the truss aspect ratio $R = 1$ is entirely reasonable. If for some reason in the future the requirements on any one functional requirement or design parameter change so that the design is no longer satisfactory then the serially decomposed reformulation has definite advantages. The designer will only have to loop back to the last equation which contains the design parameter or functional requirement which makes the design unsatisfactory. For example, if after a satisfactory solution is found it becomes necessary to change pipe diameter then only the equation for $\Delta\sigma_{buckling}$ need be consulted in the reformulation when making that change in pipe diameter.

Identifying Active Constraints

For a particular design configuration defined by parametric constraints there may not be a solution space for the given specifications. This would eliminate that configuration from consideration as a viable alternative, or promote a respecification, or modification of the configuration [17]. At this stage of preliminary design it may be difficult to determine if there is a simple design limitation and what constraint or constraints constitute it. Alternative reformulations facilitate the identification of limiting constraints considerably, especially if the problem is made nearly monotonic. Furthermore, reformulations often make it possible to employ more powerful symbolic techniques, such as those discussed by Agogino [18] and Choy [19] which are most useful when the design equations are monotonic and serially decomposed.

Considering again the truss problem, now with the objective of minimizing weight. Weight is a function of the original and new parameters respectively and is proportional to volume, $V = 2\pi dt\sqrt{B^2 + H^2}$ or $V = 2AH\sqrt{R^2 + 1}$. The reformulation makes it clear that the value of d plays no part in minimizing volume. It influences the satisfaction of the buckling constraint only. The reformulation also makes it clear that the buckling constraint can be satisfied after all the others and that there is an active constraint on compressive stress ($\sigma_{compression} = \sigma_{yield}$) or an active constraint on maximum deflection ($\delta = \delta_{max}$) depending on the numerical values involved.

Symbolic Computational Benefits - Form-Function Relationships

Very often it is desirable to know how some aspect of a design's form, other than the design parameters themselves are influenced by changes in the functional specifications of a design. The problem with developing such relationships analytically is that most design problems are underconstrained, but in some cases a degeneracy exits that will overcome this problem. Identifying such degeneracies so as to develop analytical form-function relationships can be greatly enhanced by a transformation of variable. Consider the truss problem where the expression for volume in terms of new design parameters does not depend on d. This eliminates the expression for $\Delta\sigma_{buckling}$ from consideration but still leaves an underconstrained problem. If we set P_h to zero then the two remaining expressions are simplified. From these simplified expressions it is possible to determine that volume is proportional to δ if the active constraint is on deflection and inversely proportional to $\sigma_{compression}$ squared if the active constraint is on compressive stress.

Cognitive Benefits

Cognitive benefits are those which help the designer reason about the design problem. His previous experience, current frame of mind, degree of concentration, etc., will interact with the problem representation and formulation to influence his understanding. Any insights he might have will depend on that understanding. With respect to the truss problem, the ability to easily find numerical solutions enhances the designer's thinking by supplying quick design parameter value estimates. In addition for the truss, insights on active constraints, and relationships between form and function were facilitated by the reformulation. By the time the designer has gone through the exercises above he will look at the truss problem from a different perspective[7].

---

[7]In the words of Simon [12] "All mathematical derivation can be viewed simply as change in representation, making evident what was previously true but obscure. This view can be extended to all of problem solving - solving a problem simply means representing it so as to make the solution transparent".

Results

A testbed computer program, called EUDOXUS has been implemented to test and demonstrate the techniques discussed for identifying critical design relationships. The program has been used in the design of a Geneva mechanism, a simple suspension system, an impact energy absorber, a helical coil spring, a heat exchanger, and the simple two bar truss.

A Structured Synthesis Strategy for Mechanical Devices

During the design process the designer transforms an abstract functional description of a device into a physical description which satisfies the requirements. In this sense, design is a transformation [20] from the functional domain to the physical domain.[8] However, the basis for selecting favorable transformations and methods for accomplishing transformations are not well understood. Our work is an effort to discover the desirable characteristics of these transformations and develop a structured approach for transforming a device specification into a physical description, thereby creating design alternatives. We contend that form-function relations provide a strong basis for selecting favorable design transformations and that by combining these with the simple guideline of integrating functionality, a robust and useful structured approach to design can be defined. This approach can be used to design a device configured from classes of known components.

A Strategy for Mechanical Design

The direct functional decomposition of a set of device specifications and subsequent one-to-one matching of individual functional requirements to physical components results in weak designs for two reasons. The first of these is that by matching individual functions to some collection of components we forsake the opportunity to integrate functions into more compact or economical collections of components. Secondly, components provide not only the desired function but also many additional, unintended behaviors. The following simple examples illustrate the problems with direct decomposition and demonstrate one method that helps resolve them.

Consider the design of a speed reducing device. Specifications for the device are given as:

- The ratio of input to output speed must be 8:1.
- The input and output shafts must lie at right angles to each other.

This set of specifications is already functionally decomposed into two independent functional requirements, the 8:1 reduction and the right angle requirement. The most direct solution is to match each of the individual requirements with a separate piece of physical hardware. A spur gear set may be selected because its behavior matches the functional requirement of an 8:1 reduction. Then a bevel gear set may be selected because its behavior matches the functional requirement of a right angle between the input and output shafts. The resulting physical description then consists of a spur gear pair connected to a bevel gear pair. So, by matching between the decomposed functional description and the behavioral descriptions of known components a physical configuration can be generated. However, this simple case demonstrates that one-to-one matching usually results in a poor design because the resulting device is more complex than is necessary. If instead the matching is done so that functional integration is emphasized, then more compact and economical designs result. Here, by using the form-function relations for bevel gears, it can be determined that a bevel gear provides both of the functions required, a reduction *and* a right angle between the input and output shafts. So, matching *both* functions to a bevel gear results in higher degree of function integration and a more economic execution of the design.

---

[8] It is important to note that the word transformation is used here in a different sense than that expressed in the previous section on Automatic Identification of Critical Design Relationships. Here, the transformation is not applied to a set of equations which describe an already configured design, but instead to a set of specifications which are transformed until they correspond to a complete physical description of a device.

This example illustrates that combining functions in a single device will often result in a superior design. The difficulty lies in properly matching the functions in the specification to actual components. Form-function relations of the components provide a means to this end because they express the relation between behaviors and physical form. However, matching the functional specifications to the physical form of the components is not always as straightforward as in the previous example. The following example demonstrates some of these complexities and some methods for accomplishing an economic execution of the design.

Consider the design of another speed reducing device. Specifications for the device are given as:

- The ratio of input to output speed must be ***40:1***.
- The input and output shafts must be at right angles.

If the catalogue of available items consists only of straight bevel and spur gear sets, we find that there is no single component which is capable of providing the 40:1 speed reduction. So an immediate matching between these specifications and a physical description is not possible. Instead the specifications must be *transformed* into a form that allows for matching between the specifications and device behavioral characteristics. In this case we apply a transformation which decomposes the specified functionality into a collection of equivalent functions; for example, a 5:1 speed reduction followed by an 8:1 speed reduction and a right angle between the input and output shafts. Note that this transformation preserves the overall functionality expressed in the original specifications. This is an important characteristic of all transformations; that they be function preserving, otherwise the completed device will not meet the specifications. The decision to decompose the specifications in this manner is guided by the form-function relations for the components. They represent the fact that no single spur or bevel gear pair can have a reduction ratio greater than 8:1. Therefore, this particular form-function relation is used to guide the decomposition of the 40:1 reduction ratio into two elements neither of which exceeds an 8:1 reduction ratio. After this decomposition, it is possible to match each of the individual functions, the 8:1 reduction, 5:1 reduction and right angle, into a physical component; however, direct matching would result in a design which is large, costly and complex as was shown before. Alternatively, we can seek *groups* of functional elements which closely correspond to available physical components. In this case, by considering the form-function relations for the known components, bevel and spur gears, we find that we can group or associate the 5:1 speed reduction with the right angle function. So we can view the specifications as consisting of two groups; a 5:1 reduction coupled with a right angle, and an 8:1 reduction. We now find that we can match each of these *groups* with a physical component; specifically, a bevel gear set can satisfy both the right angle requirement and one of the reductions, and a spur gear set can satisfy the second reduction requirement.[9] In this case we have not only achieved multiple functions with a single component but we have identified a particularly favorable selection of components by grouping individual functions into collections that closely corresponded to real physical devices.

This example illustrates the integration of multiple functions into single components and the value of function preserving transformations intelligently applied to the device specifications. It also demonstrates the role of form-function relations in guiding these processes. However, the approach used above will not always be clear or unambiguous. The matching between specified functionality and component behavioral characteristics is not always as perfect. Physical components have many behaviors and an exact match between specifications and component behavior will not always be possible. If this is the case, unintended behaviors will be introduced into the device. Consider if the lexicon of known components used in the example above included worm gears. A worm gear introduces an offset and a right angle between the input and output shafts. It also can provide a 40:1 reduction ratio. Therefore, a worm gear could be used to integrate all of the desired functionality, 40:1 reduction and right angle, into one component resulting in a lighter, simpler device. But, it also introduces a behavior not required in the specifications, an offset between the input and output shafts. This

[9]The relative configuration of the two components can be selected to maintain coplanarity of the shafts or to achieve an offset. The way in which components are configured relative to each other is critical to both function and geometry.

configuration is desirable only if this behavior is acceptable. So, the specifications must be transformed and matched to the components incrementally and the the desired functionality of the device must be maintained after each transformation, grouping or matching of the specifications.

Results

We are developing a design methodology that will both enable the integration of functions into single components and will take advantage of, or compensate for, unintended functionality of components by appropriately utilizing the form-function relations of the components used in the design process. The methodology is suitable for designing mechanical devices composed of standard components. In this case, by design we mean generating the configuration, i.e. the type and arrangement of standard components utilized in the device. The methodology requires a representation of the form-function relations of the standard components, and a representation of the specifications for the device. Also required is the ability to transform the representation of the specifications in a function-preserving manner. Using these representations of the design requirements, the available components, their form-function relations and the function preserving transforms, the most desirable transformations can be identified and applied until a complete physical specification of the device is determined. This methodology has been developed to be applicable to a broad range of design domains. A trial study of its effectiveness has been completed for the example domain of single speed, single input-output geared transmissions. A discussion of its application to this domain and a description of a computer program which utilizes the methodology can be found in [21].

Summary

Form-function characteristics of mechanical components and devices exist and can be identified from physical principles. Furthermore, design decisions are dominated by considerations having to do with form and function. The four areas in our program of research demonstrate that progress is being made to address the absence of a close correspondence between functional and physical descriptions in mechanical design. Tools and methods were described that will reduce our dependency on the intuition experienced designers use to extract and utilize relationships between form and function.

References

1. Rinderle, J.R., ''Function and Form Relationships: A Basis for Preliminary Design'', *Proceedings from the NSF Workshop on the Design Process*, Waldron, M.B. ,ed., Oakland, CA, February 1987, pp. 295-312.

2. Colburn, E.R. and Rinderle, J.R., ''Identifying Design Relations'', Internal Lab Report CMU-MEDL-88-25, Department of Mechanical Engineering, Carnegie-Mellon University, 1988.

3. Colburn, E.R. and Rinderle, J.R., ''Supporting Design Reasoning with a Constraint-Based Representation'', Internal Lab Report CMU-MEDL-88-22, Department of Mechanical Engineering, Carnegie-Mellon University, 1988.

4. Gross, M.D., *Design as Exploring Constraints,* PhD dissertation, Department of Architecture, Massachusetts Institute of Technology, 1986.

5. Serrano, D., *Constraint Management in Conceptual Design,* PhD dissertation, Department of Mechanical Engineering , Massachusetts Institute of Technology, 1987.

6. Gosling, J., *Algebraic Constraints,* PhD dissertation, Department of Computer Science, Carnegie Mellon University, 1983.

7. Siddall, J.N., *Optimal Engineering Design,* Marcel Dekker, New York, NY, 1982.

8. *M/P/E: Modifiable/Portable/Extendable, Reference Manual,* Unicad, Inc., Boush & Plume Streets, P.O. Box 1839, Norfolk, Virginia 23501, 1987.

9. Paz-Soldan, J.P., "Conceptual Design Environments for Opportunistic Refinement", Internal Lab Report CMU-MEDL-88-24, Department of Mechanical Engineering, Carnegie-Mellon University, 1988.

10. Paz-Soldan, J.P., "Abstraction and Structure in Conceptual Design Environments", Internal Lab Report CMU-MEDL-88-21, Department of Mechanical Engineering, Carnegie-Mellon University, 1988.

11. Parkinson, A.R., Balling R.J., and Free, J.C., "Exploring Design Space", *Computers in Mechanical Engineering*, March 1985, pp. 28-36.

12. Simon, H.A., *The Sciences of the Artificial,* second edition, M.I.T. Press, Cambridge, MA, 1981.

13. Preiss, K., "Data Frame Model for the Engineering Design Process", *Design Studies*, Vol. 1, April 1980, pp. 231-243.

14. Rinderle, J.R. and Suh, N.P., "Measures of Functional Coupling in Design", *Journal of Engineering for Industry*, Vol. 104, November 1982, pp. 383-388.

15. Kline, S.J., *Similitude and Approximation Theory,* Springer-Verlag New York Inc., 1986.

16. Rinderle, J.R., and Watton J.D., "Automatic Identification of Critical Design Relationships", *Proceedings International Conference on Engineering Design*, Boston, MA, August 1987, pp. 512-525.

17. Dixon, J.R., "Iterative Redesign and Respecification: Research on Computational Models of Design Processes", *Design Theory '88: The 1988 NSF Grantee Workshop on Design Theory and Methodology*, June 1988.

18. Agogino, A.M., "AI in Computer-Aided Design: Qualitative Reasoning and Symbolic Computation", *Proceedings from the NSF Workshop on the Design Process*, Waldron, M.B. ed., Oakland, CA, February 1987.

19. Choy, J.K. and Agogino, A.M., "SYMON: Automated Symbolic Monotinicity Analysis For Qualitative Design Optimization", *Proceedings ASME Computers in Engineering Conference*, Chicago, July 1986.

20. Mostow, J., "Towards Better Models of the Design Process", *The AI Magazine*, Spring 1985.

21. Hoover, S. P. and Rinderle, J. R., "A Structured Synthesis Strategy for Mechanical Devices", Internal Lab Report CMU-MEDL-88-23, Department of Mechanical Engineering, Carnegie-Mellon University, 1988.

STRATEGIC CONCEPTUAL DESIGN IN MECHANISM SYNTHESIS

Kevin Otto
Donald R. Riley
Arthur G. Erdman
Department of Mechanical Engineering

Albert Esterline
Department of Computer Science

Productivity Center
University of Minnesota

ABSTRACT

Research into the conceptual stage of design was motivated by the difficulties users experience with theoretically based CAD tools. Based on studies of protocols of mechanical design experts, we have developed a control strategy to aid problem formulation. This representation can accommodate the methods and operators needed for original mechanism design. We are now formulating these methods and operators and are developing techniques to implement them.

1. Introduction: Conceptualization Problem with Advanced Kinematic Design Methods

Research into the conceptual phase of mechanical design was motivated by problems we have encountered with CAD tools and knowledge-based systems. We are investigating what is commonly referred to in the literature as *Original Design*, and, in particular, original design in the field of mechanisms. The interest is not in searching for some component and varying it to fit a new application. Rather, the interest is directed toward specifying a task and an environment, and then designing a new solution to this.

There exists a semantic gap between most currently available CAD synthesis tools and their intended users. The concern is when CAD systems require a set of inputs which are often unintelligible or unnatural to potential users and thus not used. Unfortunately, this is usually the case with CAD tools intended for industrial use, be it actual software, or just the design algorithms, such as optimization techniques. Convinced that powerful design tools must include facilities to aid the user in formulating problems, we undertook a study of the problem formulation methods of industrial kinematicians and design engineers.

2. Formulating Mechanical Conceptual Design Expertise

How does one generate software which can assist in problem formulation? The direct answer is to observe a human expert who is capable of formulating problems, and then try to mimic that performance in software. This task was undertaken using a well established, proven knowledge acquisition methodology [1] which used protocol analysis and was developed at the University of Minnesota. This methodology, however, was aimed at traditional AI heuristic classification problems. The method was modified substantially to suit mechanical engineering design [2], by developing a top level control structure consisting of three *spaces*, operating in parallel: a *task decomposition, operating environment,* and *problem formulation* space. A space represents a coherent level of design knowledge. This completely circumscribes the area of design performance we have observed and wish to implement to some level of detail.

The task decomposition space covers the generation of sub-problems, and the dependence of the solution to some sub-problems on solutions to others. The operating environment space elaborates the working environment of the mechanism and thereby imposes constraints. The problem formulation space covers the development of a specification of the problem into terms which can be solved by mechanism design modules.

Each space consists of *states* representing the aspects of the design task relevant to the definition of the space. From the protocol analysis it was found that when particular design strategies are employed, only certain *transitions* from one set of states to another occur. In the resulting model, only certain states are *activated*, or capable of doing computation. Only the states with possible transitions from active states can become active, and only if certain conditions exist do these transitions occur. Thus, this provides a guidance structure for conceptualizing and designing a mechanical assembly.

The spaces operate simultaneously on a global data structure which is based on an entity-relationship model and which we call the *cue network*, as *cues* found in the network are abstracted into the conditions that sanction transitions within the spaces. The states which become active by virtue of these transitions in turn place more cues on the network. Cues consist of *tasks,* as well as *relations* among and *properties* of *entities* for a specific problem being worked on. For a complete description, the reader is referred to [2], but an illustrative example is presented.

```
1.   I would say that if I lifted up the applicator 4", I would have no problem
     clearing the application area.
2.   So my Y movement, let's assume that's 4" then, and no X movement, 0".
3.   I'm inclined to just mount the applicator on a slide rail, so I can move
     straight up and down.
```

Figure 1: Transcript of a Protocol Fragment

Figure 1 is a fragment from a transcript of a protocol. This is well into the protocol, and the expert is about to present a solution to an assembly line problem of gluing boxes. To conduct the analysis, the entire protocol is divided into minimal coherent units, or *episodes.* A transition occurs during an episode. For each episode, the data that cued the episode is recorded. What was determined in the episode is also recorded. And finally, the point of the episode, or the *goal,* is also recorded. In Figure 1, lines 1 and 2, which records the determination of how to move the glue applicator, constitute an episode. Line 3 starts a new coherent unit, hence it is a new episode. This information is shown in Figure 2 for the problem formulation space. Similar information is recorded for

REF	EPISODE	GOAL	DATA CUES	GENERATED CUES
1,2	Propose	Propose Feature (Output Motion)	Applicator out of Appl. Area Gluing region extends 4".	Applicator motion 4" in Y direction.
3	Propose	Propose Feature (Position)	Relative locations of cartons, gluing task	Appl. position above center of box

Figure Two: Problem Formulation Encoding of the Protocol Fragment

competence in the task decomposition and operating environment spaces. Also, the design methods used are recorded into a separate space, and are incorporated as design modules.

This information is then further compiled into the state-spaces previously described. The states are determined through abstractions of the goals and cues. The transitions between states occur as the procession of intervals in the protocol. A fragment of the problem formulation space is shown in Figure 3. The states are represented as ovals, and transitions as the interconnecting lines. For each interval, the goals are represented as "clouds" and the conditions as boxes. The conditions are abstractions of the data cues. The protocol fragment in Figures 1 and 2 demonstrates the transition between the states Output Motion and Position in Figure 3.

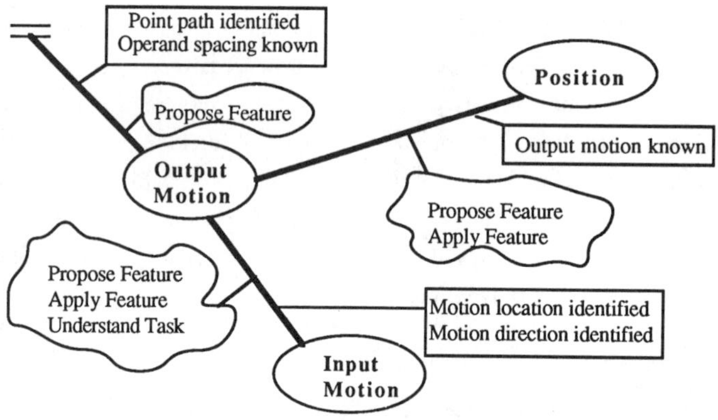

Figure 3: Problem Formulation Space Fragment

3. Salient Features of The Representation Derived from the Expert Protocols

We shall first discuss what is required of an intelligent CAD tool and then discuss salient features of our representation of mechanism design expertise. We have determined that an intelligent CAD tool must include methods:

1) to identify what entities or components of entities must be designed, that is, to identify design tasks.
2) to design these actual entities, that is, to fulfill the task requirements.
3) to determine which of the available design methods from 2) should be used.
4) to select among the results from the design methods from 2), and to determine how completely these designs solve the task presented; and
5) to combine designs.

Most of the current CAD tools include only methods falling in either the second or fourth category. Our three spaces address all five sets of methods and global control over them. The spaces also directly reflect the design expert's control over these methods[2].

This representation provides a control framework which can accommodate all of the above methods. For each transition enabled by active states in a space, the cue network is examined for aspects of the design being worked on, and these are matched to the set of conditions regulating the transition. If the conditions hold, the transition may be fired. Notice that this provides tightly focused computation, as opposed to production rule systems, which fire across the entire system. Also notice that this representation is modular: each state is really an object with its own methods for updating the design and the design requirements (via the cue network) as it sees fit.

State-space control, however, would still be non-deterministic. What will occur when several transitions could fire? To accommodate this, scripts[11], or stereotypical patterns of action, are being formulated. Such a script corresponds to an abstraction of the problem as initially presented. Thus, different design approaches, or strategies, are accommodated. If a designer chooses to work at a local level, the script selected will direct transitions to states of a local character. If a different designer chooses to work at a more global level, a different script, directing transitions to states with more global focus, is selected. For example, if a problem is amenable to designing part by part, the script would direct transitions to states of local character (Position and Output Motion, for example). If, on the other hand, the problem is amenable to specifying the number of links, the script would direct transitions to states of more global character. A range of approaches was observed in our protocol analysis. It is difficult using protocol analysis to achieve a representative number of strategies, as they are evoked by the problem formulated by the expert, and they tend to be idiosyncratic to the expert. To overcome these difficulties, techniques similar to those discussed below are being used.

We thus have a representation of problem formulation as we observed it done by our experts. This can then be used to formulate problems for advanced kinematic design techniques. Burmester precision point theory[4], cognate methods[5], catalog interpreters[6], and optimization modules[7] can all be accommodated. We are also developing design databases[8] of non-analytically defined components as used by some of the observed experts. For example, no analytic techniques exist for components such as turn stops (small levers which hold gears in place as a safety device). Yet these components can readily be modeled using qualitative simulation. It was concluded that qualitative simulation techniques[9] and common sense reasoning[10] provide a means to capture the functionality and qualitative, geometric descriptions which characterize methods for designing such components.

4. Development of Design Nets for Our Representation

A structure has been developed for guiding the design process and conceptualizing the problem into applicable design methods as the problem formulation and solution proceeds, much as we saw our experts doing. Current work is now concentrating on various categories of methods required. Heuristics for selecting among these are also being formulated.

A related area is redesign[12], and what Ullman calls patching[13], that is, combining designs at the same level of abstraction. What sort of representation allows this to be implemented? How can a system determine when and with what methods a patch or redesign is necessary in a new problem?

Protocol analysis alone will not answer this question. Protocols deal with specific models in the expert's mind, and elicit only the specific traversals instantiated by the problem presented by the knowledge engineer. For example, why would an expert discuss knowledge of cams if the problem presented does not elicit cams? Since the expert will not, to assume protocol analysis will provide methods to make generative patches is to assume the knowledge engineer can determine problems a priori which completely cover the set of methods for all required patches.

To provide these models and methods, a variation on the critical incident technique[14] is used to analyze hundreds of mechanism designs, and methods for arriving at them. This is done by means of a structured interview, asking questions on the relative merits of particular mechanism topologies, methods of synthesis, and similar considerations. This knowledge is then formed into nets which represent the critical dimensions found. This does NOT mean an exhaustive catalog of solution instances is sought; rather, dimensions are formulated such as design methods, kinematic types, and structural motion characteristics. Nets of elements for each dimension are found, as well as the compatibilities and conditions among them. These are arranged for selection and pruning in a network fashion. This work is still in progress. Notice, however, that this information is easily verifiable scientifically. One need only present the experts with the preconditions of a hypothesis, and see whether they indeed follow the hypothesis. For example, if it is hypothesized that a Watt I six bar mechanism topology is the best to pursue for a particular set of conditions, these conditions can be presented to the experts to see whether they indeed pursue a Watt I.

5. Conclusion

The goal of the research is the development of an intelligent assistant to the mechanical design process. A framework has been developed for controlling the design process which can accommodate different design strategies and many design methods. Compiled mechanical design knowledge is being developed in the form of nets. Also, scripts of stereotypical behavior are being developed for initial problem definitions, which will provide guidelines for proceeding with a design. When the nets and scripts are formulated, a truly generative design assistant, capable of operating in novel situations, will be possible.

Bibliography

1. Johnson P., Zualkernan I., and Garber S., "Specification of Expertise", *International Journal of Man-Machine Studies 26*, 1987, pp. 161-181.

2. Esterline A., Rosen D., Otto K., Nelson L., Hessburg T., Riley D., and Erdman A., "A Methodology for Capturing Mechanical Design Expertise", *Proceedings of the 1988 ASME Computers in Engineering Conference*, Vol. 1, San Francisco, 1988, pp. 47-56.

3. Chase T., *Burmester Theory for Four Precision Positions: An Extended Discourse with Application to the Dimensional Synthesis of Arbitrary Planar Linkages*, PhD Thesis, 1984.

4. Freudenstein F., and Maki E., "The Creation of Mechanisms According to Kinematic Structure and Function", General Motors Research Publications, GMR-3073, September 1979.

5. Cayley, A. "On Three Bar Motion", *Proceedings of the London Mathematical Society*, Vol. 7, 1876, pp. 136-167.

6. Kota S., Erdman A., Riley D., "MINN-Dwell - Computer Aided Design and Analysis of Linkage - Type Dwell Mechanisms", *Proceedings of the 1987 ASME Computers in Engineering Conference*, Vol. 2, New York, 1987, pp. 351-358.

7. Paradis M., Willmert K., "Optimal Mechanism Design Using the Gauss Constrained Method", *Journal of Mechanisms, Transmissions, and Automation in Design*, Vol 105, June 1983, pp 187-196.

8. Kannapan S., and Marshek K., "Design Synthetic Reasoning: A Program for Research", *Mechanical Systems and Design Technical Report 202*, 1987.

9. de Kleer J., and Brown J., "A Qualitative Physics based on Confluences", *Qualitative Reasoning about Physical Systems*, D. Bobrow ed., MIT Press, 1985.

10. Hayes P., "The Second Naive Physics Manifesto", *Readings in Knowledge Representation*, Brachman R., and Levesque H., ed., Morgan Kaufmann, 1985.

11. Schank R., and Abelson R., *Scripts, Plans, Goals, and Understanding*, Lawrence Erlbaum Publ., Hillsdale NJ, 1977.

12. Kannapan S., and Marshek K., "Design Methodologies: A New Perspective on Approaches and Tasks", *Mechanical Design Technical Report 201*, 1987.

13. Ullman D., and Dietterich T., "Mechanical Design Methodology: Implications on Future Developments of Computer Aided Design and Knowledge-Based Systems", *ASME CIE, 1986*, pp. 173-180.

14. Flanagan J., "The Critical Incident Technique", *The Psychological Bulletin*, July 1954, pp. 327-359.

15. Ericsson K., and Simon H., *Protocol Analysis: Verbal Reports as Data*, The MIT Press, Cambridge Mass., 1984.

OPTIMALLY-DIRECTED INNOVATIVE DESIGN

Alice M. Agogino, P.E., Principal Investigator, Associate Professor
Jonathan Cagan, Graduate Research Assistant
Department of Mechanical Engineering
University of California, Berkeley, CA 94720

Sponsored by NSF PYI grant # DMC-8451622

Abstract

Innovative designs, having some relation to previous design prototypes, are derived by manipulating the design space to generate new design features or prototypes. Our research effort [1,2] is aimed at developing a computer-based innovative design methodology known as 1stPRINCE (FIRST PRINciple Computational Evaluator) which incorporates the qualitative, functional, and numerical levels of reasoning outlined by Agogino [3]. These methods utilize monotonicity analysis, which observes the sign of the algebraic first derivative of design variables, to make qualitative decisions that direct globally optimal behavior and detect flaws in the problem formalization. Mathematical functional information is used by computer algebra routines as deemed important from the monotonicity analysis. Monotonicity analysis and functional backsubstitution are implemented in the SYMON-SYMFUNE programs [4,5]. If constraints cannot be met or a designer wishes to investigate alternative designs, 1stPRINCE goes one step further in modifying the design space by breaking integrals into smaller ranges to create novel design prototypes. Algebraic solutions to the design problem are reduced by computer algebra routines which can be solved numerically to obtain specific solutions and quantitative comparisons between solutions. 1stPRINCE has been applied to a beam under torsion load and a beam under flexure load. In each case the system was told to minimize the weight of the beam subject to various physical constraints (maximum stress, deflection, weight, etc.) By starting with a solid round beam as the only structural prototype for the torsion load, 1stPRINCE derives a hollow and composite rod. For the flexural beam, a series of stepped beams are created, approaching a tapered beam. Some designs proposed by 1stPRINCE may be unusual, but all geometric modifications that lead to new design prototypes satisfy all assumed constraints and are *optimally-directed* within those constraints. Because 1stPRINCE treats material properties as design variables, combinations of various materials lead to optimally-directed composite beam configurations.

Publications

[1] Cagan, J. and A.M Agogino (1988a), "Innovative Design of Mechanical Structures from First Principles," *in press: AI EDAM: AI in Engineering, Design, Analysis and Manufacturing*, V. 1 (3).

[2] Cagan, J. and A.M. Agogino (1988b), "Reasoning about Mechanical Structures from First Principles, *Proc. of the 12th IMACS World Congress on Scientific Computation*, V. 4, pp. 278-282.

[3] Agogino, A.M., "AI in Computer-Aided Design: Qualitative Reasoning and Symbolic Computation," *The Study of the Design Process*, (ed., M. Waldron: Ohio State University), pp. 263-294.

[4] Choy, J.K. and A.M. Agogino (1986), "SYMON: Automated SYMbolic MONotonicity Analysis System for Qualitative Design Optimization", *Proc., ASME 1986 International Computers in Engineering Conference*, ASME, V. 2, pp. 305-310.

[5] Agogino, A.M., Almgren, A.S. (1987), "Techniques for Integrating Qualitative Reasoning and Symbolic Computation in Engineering Optimization", *Engineering Optimization* V. 12 (2), Sept./Oct. 1987, pp. 117-135.

Chunking Structural Design Knowledge as Prototypes

by

John Gero
Architectural Computing Unit,
Department of Architectural Science,
University of Sydney 2006 Australia

Mary Lou Maher
Weiguang Zhang
Department of Civil Engineering,
Carnegie Mellon University,
Pittsburgh, PA 15213 USA

A prototype is a generalization of groupings of elements in a design domain which provides the basis for the commencement of a design. A prototype carries with it the required interpretations or requirements, a vocabulary set of elements, knowledge about that vocabulary and its relation to the requirements and some parameterized design descriptions. The statement "design a rigid frame" brings forth the prototype associated with the label "rigid frame". This prototype includes the requirements of lateral load resistance, open bays, rigid connections, and the design parameters of bay width and height, material and size of beams and columns, and so on, even though none of these was mentioned explicitly in the prototype's label. A prototype based structural design knowledge base has been implemented. The implementation supports the design process through the application of these prototypes to specific design problems. Examples of prototypes and their application are described.

Engineering Design Synthesis: A Domain Independent Representation

by

Mary Lou Maher
Department of Civil Engineering
Carnegie Mellon University
Pittsburgh, PA

The knowledge used in the synthesis of engineering systems includes understanding systems and their components and the implications of design decisions on other decisions on further problem decomposition. In using a knowledge based approach to synthesis, design knowledge can be characterized as planning knowledge, including design goals and planning or ordering of goals, and design knowledge, including alternative solutions for each goal and constraints on the selection of a solution for a given goal. This approach is implemented in a program called EDESYN (Engineering DEsign SYNthesis), developed using a frame-based reasoning tool. EDESYN provides a domain independent enviornment for building design expert systems.

PART 4

ENVIRONMENTS, LANGUAGES, REPRESENTATIONS, AND DATA BASES FOR DESIGN

Automatic Composition in Design

by Charles M. Eastman
Graduate School of Architecture and Urban Planning
University of California, Los Angeles, CA. 90024

1. Introduction

About forty years ago, an important revolution took place in engineering design and education. Until that time, engineering practice relied on empirical approximations and rules of thumb for determining design performance. This revolution linked the performance of materials to physical science principles and led to the development of engineering science.

There is a need for a second revolution. This time the focus must shift from analysis to synthesis, to methods for composing a design. Methods for composing a design are at the same general state that analysis of a design was 40 years ago. The goal is to move design synthesis from a set of practices derived from experience and art and to that of a science based on mathematical and logical principles.

This paper approaches design problems at an abstract level, attempting to identify common ground among different domain dependent issues. The perspective taken summarizes an emerging paradigm, contributed to by many people and reinforced by many of the papers in this volume. It outlines answers to a set of questions about the nature of design[+] as an intellectual activity: How may we characterize the space of design alternatives? Given a design space, how can it be systematically searched or enumerated? How can the alternatives that are generated

---

[+]Throughout this paper, I will use the following working definition of design:

> "design is the specification of an artifact (or change in an artifact), in sufficient detail to both construct it and to determine its performances in all dimensions of interest".

be restricted to realizable and meaningful ones? In answering these questions, my intention is to relate current work with previous efforts and to sketch out a conceptual framework that others can extend.

2. The Structure of Design Problems

Some approaches to design attempt to characterize the solution space explicitly, as a constrained set of points within n-dimensional space. In design, however, the dimensionality of the space is not fixed at the beginning; dimensions are defined as decisions are made. A decision-theoretic approach avoids this issue.

Any engineering, architectural or other designed product involves millions of variables. Some are <u>continuous</u>, others <u>discrete</u>. Discrete variables include a wide range of structures: networks, technologies, shapes, mechanisms etc. Continuous variables involve sizes, flow rates and other properties of the structures. Prescriptive design typically involves a mixture of the two; it involves an outside iteration process that selects candidate discrete structures, then search or optimization over the continuous variables defined by those structures. A solution is the best combination of structure and continuous variables.

This is called a <u>single level design problem</u>. See Figure One. There are many examples of single level design involving mixed discrete-continuous search or optimization [Kirsh,81; Gero,85]:

- the selection for a process in a chemical plant versus the sizing of the components and flows within the process [Kocis and Grossman,88];
- the topological structure of rooms in a building versus the dimensioning of those rooms [Mitchell et al,76];
- the network definition of a structural system versus the sizing of the members [Sheu and Schmidt,72].

In each case, a discrete set of structures defines alternative parametric designs. The result is a discrete OR-tree, each branch with a continuous space below it.

Even small single level design problems are immense. A one level composition problem to arrange a set of ten elements in a space, such as a furniture layout problem, is estimated as having a tree with 10^{13} leaves[++]. The point is that in design, exhaustive enumeration is not possible; the one best solution is seldom identifiable. Design is a hard problem.

Real design problems combine single level OR-trees with other branch structures that define systems and subsystems that make up the total artifact. See Figure Two. These AND-trees characterize a technology or system in terms of its components. Each of the components are resolved with its own OR-tree. Design, then, is characterized as a mixture of AND- and OR-trees. AND-trees are sometimes characterized as "frames" within the AI paradigm [Winograd,75] and by others as "prototypes"[Gero, Maher and Zhao,88].

The resulting structure is not neatly hierarchical. Leaves may be attached to multiple branches. A pump, for example, may be part of the hydraulic AND the electrical distribution structures. It is often not possible to order the hierarchy so that all dependencies apply downward.

Such semi-hierarchical structures have been proposed in a wide area of design studies, including the structure of human design process [Waldron et al,88], the structure for CAD databases [Eastman,78] and as the structure for expert systems in design [Mahar,88]. Many of the papers in this volume assume or develop such a structure.

A design space, then, consists of multiple OR-trees combined with alternative AND-trees. Branch structures (both AND- and OR-trees) are added during decisionmaking. At no time is the total tree structure apparent because there is never the time or resources to generate it. The general formulation of the composition problem is the dynamic definition and search of such a tree.

[++]Each object has locations defined by other objects (next to them, etc.). Thus for 10 objects, there will be about twenty locations for each, or 20^{10} or 10^{13} total.

DISCRETE-CONTINUOUS STRUCTURES IN DESIGN

OR-Tree: EXAMPLES:

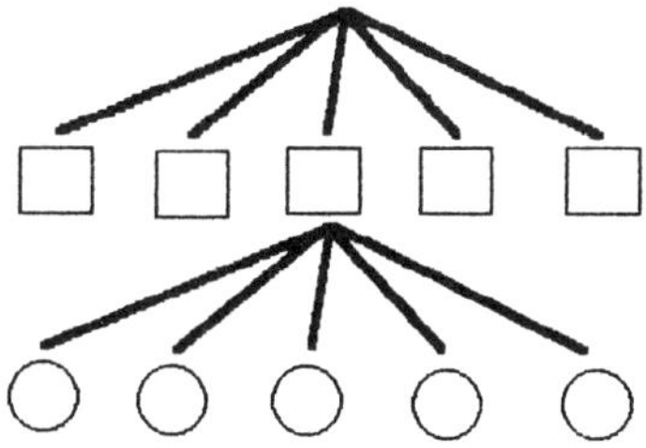

Structural Frame — Chemical Process

Member Sizes — Sizing of Material Flows

Figure One

AND-Tree:

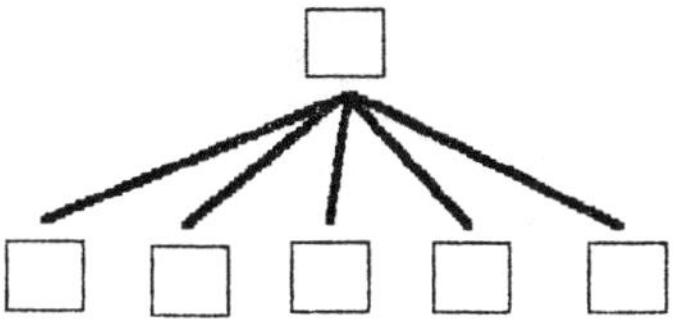

Building:

Site, Structure, HVAC, Exterior Cladding, Assigned space, Public space, Vertical Circulation etc.

Figure Two

3. Generating Designs

One set of issues involve enumerating the range of scalars in the continuous domain. Linear [Dantzig,65] and geometric [Duffin et al,67] programming are sometimes appropriate, and if not, hill-climbing [Wilde,64] is possible if the goals of the model can be defined as a differentiable function.

The more challenging area is in the discrete domain. Within a type of structure, distribution system, or technology, there is a wide range of configurations possible. Needed are systematic methods to enumerate these discrete structures.

Methods to enumerate discrete structures include:

1. enumerate the alternatives manually; this is called <u>morphological analysis</u> [Zwicky,69; see also Arciszewski, 88]. Morphological analysis is usually done in an ad hoc manner and there is the question whether the enumerated set is complete. Only very small spaces may be enumerated. Taken alone, morphology seems a poor method of enumeration. However, morphology is the primary method for identifying the alternative base technologies (the AND-trees) used in design. Once invented, a candidate type of structure or technology is recalled by name, which allows it be defined morphologically[+++].

2. other design problems can be characterized as the <u>selection of members of a finite set</u>. Set enumeration relies on a fixed superset whose members are themselves members of sets, or

[+++]It is always possible to design recursively, defining a new problem whose solution is the desired technology. The recursive problem's solution may be the invention of a new technology, with a new AND-tree.

$$S = \{s_1, s_2, .. s_i, ... s_n\}$$

where $s_1 = \{m_{1,1}, m_{1,2}, ... m_{1,p1}\}$
where $s_2 = \{m_{2,1}, m_{2,2}, ... m_{2,p2}\}$
. . .
. . .

These sets may be any finite set, including scalars. The matrix representation of a graph with a finite number of nodes is an example. Location and routing problems are often amenable to this formulation [Ford and Fulkerson,63]. Because there are finite choices for each member of the set, methods to enumerate alternatives are easily defined. Integer programming, using branch-and-bound techniques [Lawler,66], or applying integer constraints to the convex hull in linear programming [Balinski and Speilberg,69] may be applicable to this method.

3. in most other cases, alternatives can be developed by <u>a set of operators</u> that transform an initial design state into alternative states. The range of states possible with the operator set is the range of designs. This method can theoretically enumerate very large design spaces. Systematic enumeration without repetitions is difficult, however, and sometimes not possible. Various forms of heuristic search are examples of this method [Newell,69; Eastman,73] as well as formal grammars [Stiny,80]. Design is usually associated with problems where some effort is required to generate candidate points within the solution space. It is this last method that seems to best characterize this condition. It also has similarities with how humans design [Eastman,69; Akin,86; Ullman and Dietterich,88].

Different structures or technologies (OR-trees) justify having their own enumeration methods, based on their unique organization and constraints.

Operator sets for generating design alternatives themselves involve the design of a representation of the space of solutions for a single level design problem. The range of operator-based approaches includes various automata [Hopcroft and Ullman,79] and graph and shape grammars [Stiny,80]. Theoretically, these representations too can be enumerated and evaluated against the constraints of the problem domain. This is easier said than done, of course; defining such representations and operators is an important area of design research.

4. The Structure of Physical Objects

Given a means for generating candidate compositions, many may be meaningless, or if not meaningless, violate necessary properties of a solution. Each of the means to enumerate alternative designs outlined above rely on different mechanisms for dealing with the semantics of the design, eg. those that are meaningful versus meaningless.

Certain compositions are meaningless because physical reality does not allow them to exist. The physical world imposes limitations on the combinations of values allowed in a composition. Some combination of floor layouts and total areas in a building are impossible, by summation and equivalence. Physical definitions impose equivalencies across sets of variables, such as mass-density-volume or volt-amp-watt. The definitional well-formedness of certain classes of descriptions, such as shape geometry [Requicha,80] or a planar graph (Kirchoff's Laws) impose other restrictions. Many combinations of properties do not co-exist in known materials. These definitions and relations impose restrictions on the combinations of values allowed in the definition of a single design object. I call these <u>realizability constraints</u>++++.

Another set of constraints deal with the relations between pairs of entities. While certain entities themselves may be meaningful, their combinations are not; attached surfaces must meet, objects must be supported against gravity, flow properties between adjacent pairs of objects must be consistent. Another class of unrealizable conditions exist if pairs of solids occupy the same location in space at the same time, eg. collisions. I call constraints between pairs of objects <u>compositional constraints</u>.

++++Constraints are considered here in the standard artificial intelligence sense: as relations over a set of terms, without the fixed and unfixed terms being distinguished. Constraints may be solved for different terms, depending on the sequence of decisions [Sussman and Steele,78; Sutherland,63; Gross et al,88].

At another level, the variables may describe a realizable composition, but the performance may be unacceptable. At the simplest level, it may not support its own weight or survive natural forces. At another it may not maintain necessary operational conditions within the range of environmental conditions expected. These are called performance constraints. Tests to accurately determine performance is traditionally called analysis and is the best developed area of design. An expanded set of analysis criteria are needed, however, for maintenance, assembly, testing and other conditions that must be supported during the artifact's lifetime.

I have grouped constraints according to the localization of variables to which they refer. This classification of constraints is motivated by the desire to associate constraints with objects using object-oriented programming. Configurational and performance constraints are managed by objects that are aggregates of others. Other organizations may be more appropriate, depending upon the constraint management scheme envisioned.

Sometimes, it is desired to anticipate constraint behavior between high level entities, before the low level technologies have been generated. See for example Arciszewski [1988]. This is a form of look-ahead based on experiential probabilities of conflict. Such an approach to constraint definition and resolution usually masks a decision across multiple technologies or structures, each with different discrete variables. Thus specific constraints cannot be identified. The probabilities are based on previously used methods of enumeration and evaluation. If these methods are the same as those used in the current problem, then the probabilities may have validity. Thus explicit control of the design method during the collection of probabilities is necessary for them to have more than crude heuristic value.

Constraints are a general characterization of design knowledge. In most design domains, formal definition of the applicable constraints has not been attempted. Such efforts are necessary, however, if we are to develop new representations of design problems. Constraints, if they have been specified at all, are defined in a representation-dependent manner. That is, only those constraints that are not embedded in the representation are considered. The approach taken here attempts to avoid the distinction between constraints encoded in the representation and those that must be treated explicitly. Selection of a representation then can be approached formally.

5. Representation of Design Problems

Given a problem defined as a set of possible objects to be composed and a set of constraints across object variables, the search for a combination of variables that satisfy realizability, compositional and performance constraints can be a formidable task. The power of design tools in this task is largely determined by the representation used. The relations between operators and constraints within a representation largely determine the <u>power</u> of the representation.

I am aware of three means of relating constraint information to search operations:

First, the representation and operators may have rules built into them (explicitly or implicitly) that only allow realizable or satisfactorily performing compositions. In such representations, all expressible compositions are automatically feasible (in some respects). Examples are constructive solid modeling [Requicha,80; Eastman and Preiss,84] and operators to generate planer floorplans [Flemming,86; Galle,86]. The author and a group of students developed a simple piping design system with operators that automatically maintained expressed flow conditions (by keeping pipe diameters variables and re-computing them as the layout was generated) [Eastman,81]. Thus performance constraints can be dealt with in operators as well as realizability constraints. Borning's Thinglab has the goal of developing computer environments that simulate a range of realizability constraints [Borning,78]. Several graphically oriented constraint satisfaction schemes have been developed for embedding into a representation. The earliest was by Sutherland [1963]. See also Gosling [80]. I call constraints embedded in the representation <u>universal</u> constraint satisfaction.

It may be thought that a goal of computer modeling of physical systems is that all constraints should be handled automatically by the representation. In practice, such a representation would be terribly limited. I can conceive of a representation that would always maintain structural integrity. In such a representation, one would have to prop up the structure in order to move a column or wall. One would have to go through all the steps of structural support that would have to be done on the real object. Such a representation would be very odious and inefficient to use.

Rather, I wish to violate constraints for periods of time, then apply them to candidate designs incrementally, as decisions are made. This applies for example to shape definition; non-manifold geometries are needed during design [Weiler,85]. In practice, only the simplest constraints should be embedded in the representation to be always satisfied. Equivalency and consistency across data values are strong candidates for universal constraint management.

Second: in other cases, the representation does not guarantee the satisfaction of certain constraints, but there is a match between operators and constraints. Application of the operator satisfies the constraint. That is, there is a mapping from evaluation to resolution, which in heuristic search methods is defined as "means-ends-analysis" [Newell,69]. High level examples in design include: provision of doors to provide access between rooms, the addition of a valve to provide control of flow in piping, provision of a drain for fluid drainage, and so forth. Low level examples include constraints over continuous variables and solving for a specified variable. I call these <u>associated</u> constraints.

Third: for many types of constraints, no operator mapping can be found that is meaningful. The most common cause is that the constraint is not local to a fixed set of objects, but rather global to the configuration. Cost is a prime example. In other cases, the constraint is a function over several objects and there is a wide combination of actions over the set to change the constraint test. In such cases, the information available to guide a search process toward an acceptable solution is very limited and trial and error methods predominate. I call this constraint condition an <u>unassociated</u> constraint.

The power of a representation in composition is a function of the relations that can be utilized between problem knowledge and actions that can be taken. These three means order how constraints and design actions may be associated.

6. Discussion

A very important research undertaking is to analyze current design domains, define their constraints and develop representations and operator sets that can enumerate the domains. These formulations encode the logical structure of conventional design issues and potentially such issues as fabrication, testing, maintenance and reliability. They provide a formal model of a design domain with much of the semantic knowledge embedded within the model. Of course, the constraint set is always open-ended; some of importance in a particular problem will be exogenous to the system. Associated with such single level solution schemes is the task of searching the domain to solve particular problems and placing value on the alternatives generated. These should remain exogenous to the structuring of the design domain.

Various disciplines are slowly making progress on the formulation and structuring of single-level design problems. Such design domain formulations have been developed or are evolving for: pumps, fluid valves, electric motors, different types of piping systems, standard structural systems for buildings, heating systems for buildings, and many other domains. Eventually, single level design systems will encode expertise that exceeds all but the most expert human designers in the domain. Incremental improvement of these design systems will allow them to be perfected, in the same manner in the same manner that allows refinement of scientific theories. As this work proceeds, it will be important to define standard interfaces for the domains, allowing the procedures to be embedded into larger, multi-level systems.

I expect to see multi-level system development proceeding in parallel with the single level systems, focussing on such issues as integration, constraint propagation across domains and between levels, search strategies, and concept-to-schematic-to-detail design development.

These structures allow or will allow well understood domains of design knowledge to be used widely by an organization or society. It provides a means to organize and better utilize expert design knowledge. By routinizing such knowledge, human design skill will be freed to focus on the never-ending issues of complexity, invention of new technologies or structures, improved performance and integration. It is interesting that today, the knowledge and structures of similar design problems are re-invented over and over throughout the world.

7. Summary

In this review, I have attempted to articulate a paradigm that is quickly evolving in many different design domains. I have suggested how such a paradigm contributes to the practical solution of design problems and its logic. I have focussed on single level design problems, not because of their inherent importance, but because they are the best developed. Multi-level designs and the integration of different single-level design systems are a significant challenge.

I have avoided discussion of search strategies and whether these should be manually controlled or automated. This has been done on purpose, because the paradigm applies regardless and mixed manual and automated multi-level systems are likely to become the eventual standard. I have also not meant to imply any order of processing. Design may or may not proceed in a top-down manner; that is, low-level structures may be specified before the high-level structures they are part of. Design development strategies are another discussion.

Today, design experience is often more important than schooling. Design is currently an art; designers re-construct from mostly personal experience the structures and technologies, constraints and enumeration methods used in design. This says a lot about the level of formal knowledge that exists today regarding design.

References

Akin, O. The Psychology of Architectural Design, Pion Press, London, (1986).

Arciszewski, T. "Stochastic form optimization", J. Engr. Optim., 13:1 (Jan. 1988).

Balinski, M.L. and K. Spielberg, "Methods of integer programming: Algebraic,combinatorial and enumerative", Progress in Operations research Vol.III, J.Aronofsky (ed.) Prentice-Hall,(1969).

Borning, Alan, "Thinglab: A Constraint Oriented Simulation Laboratory", Computer Science Research report No. STAN-CS-79-746, Stanford University (1979).

Dantzig, G. B. Linear Programming and Extensions, Princeton University Press, Princeton, N.J. (1965).

Duffin, R., E. Peterson, C. Zener, Geometric Programming: Theory and Applications. Wiley, N.Y. (1967).

Eastman, C. "On the analysis of intuitive design processes", in Emerging Methods of Environmental Design and Planning, G. Moore, (ed.), M.I.T. Press, (1969).

Eastman, C. "Automated Space Planning" J. of Artificial Intelligence, 4(1973), pp. 41-64.

Eastman, C. "Representation of design problems and maintenance of their structure", in Artificial Intelligence and Pattern Recognition in Computer Aided Design, J.C. Latombe (ed.), North-Holland Press, N.Y. (1978).

Eastman. C. (1981) "Recent developments in representation in the science of design", Proc. 18th Des. Auto. Conf. ACM, IEEE, (June, 1981).

Eastman, C. and G. Lafue, "Semantic integrity transactions in design databases", in File Structures and Databases for CAD, J. Encarnacao and F.L. Krause (eds.), North-Holland Press, N.Y. (1982).

Eastman, C. and K. Preiss, "A review of solid shape modeling based on integrity verification", Computer Aided Design, 16:2 (March, 1984), pp.66-80.

Flemming, U. "On the representation and generation of loosely packed arrangements of rectangles", Environ. & Design, 13(1986), pp. 189-205.

Ford, L.R. and D.R. Fulkerson, Flows in Networks, Princeton University Press, Princeton, N.J. (1963).

Freeman, P. and A. Newell, "A model of functional reasoning in design", Proc. Second Intern. Conf. A.I., British Computer Society, London, (1971).

Galle, P. "An algorithm for exhaustive generation of building floor plans" Commun. ACM, 24:12, (1981), pp. 813-826.

Gero, J., M.L. Maher and F. Zhao, "Chunking structural design knowledge as prototypes", this volume, (1988).

Gero, J. (ed.)Design Optimization, Academic Press, Orlando, FL. (1985).

Gosling, James, "Algebraic Constraints", Computer Sscience Department Report No. CMU-CS-83-132, Carnegie-Mellon University, (May, 1983).

Gross, M., S. Erwin, j. Anderson and A. Fleisher, in Computability of Design, Y. Kalay (ed.), Wiley, (1988).

Kirsh, U. Optimum Structural Design - Concepts, Methods and Applications, McGraw-Hill, N.Y. (1981).

Kocis, G. and I. Grossman, "Solution of mixed-integer nonlinear programming problems in engineering design with DICOPT", this volume (1988).

Kutay, A. and C. Eastman, "Transaction management in engineering databases", SIGMOD Conf. on Engr. Design, (1983), pp. 73-80.

Lawler,,E.L. "Branch-and-bound Methods: a survey", Opns. Res. 14(1966), pp. 699-714.

Maher, M.L. "Engineering design synthesis: a domain independent representation" this volume, (1988).

Mitchell, W. P. Steadman, R. Liggett, "Synthesis and optimization of small rectangular floorplans", Eviron. & Plnng. B, 3(1976), pp. 37-70.

Newell, A. "Heuristic programming: Ill-structured problems", Progress in Operations research Vol.III, J.Aronofsky (ed.) Prentice-Hall, 1969.

Requicha, A.A.G., "Representations for Rigid Solids: theory, methods and systems", ACM Comput. Surv. 12:4, (Dec.,1980), pp. 437-464.

Sheu, C. and L. Schmidt, "Minimum weight design of elestic redundant trusses under multiple static loading conditions", AIAA J. 10, (February, 1972), pp.155-172.

Steadman, M. Architectural Morphology, Pion Press, London, (1984).

Stiny, G. and J. Gips, Algorithmic Aesthetics: Computer Models for Criticism and Design in the Arts, University of California Press, Berkeley, CA. (1978).

Stiny, G. "Introduction to shape and shape grammars", Environ. and Plnng B, 7(1980), pp. 343-351.

Sussman, G. and G. Steele, "Forward reasoning and dependency-directed backtracking in a system for computer-aided circuit analysis", Artif. Intell. 9(1977), pp. 135-196.

Sutherland, SKETCHPAD: A man-machine graphical communication system", Ph.D. dissertation, MIT, (January, 1963).

Ullman, D.G. and T. Diettrich, "Progress in understanding the process of mechanical design", this volume, (1988).

Wheeling, R.F. "Heuristic search: structured problems", in Progress in Operations Research Vol.III, J.Aronofsky (ed.) Prentice-Hall, (1969).

Waldron, M. K. Waldron and D. Owen, "Use of systematic theory to represent the conceptual mechanical design process" this volume, (1988).

Weiler, Kevin, "Edge based data structures for solid modelling in curved surface environments", IEEE Comp. Graphics and Applic., 5(1985), pp. 21-40.

Wilde, D.J. Optimum Seeking Methods, Prentice-Hall, Englewood Cliffs, N.J. (1964).

Winograd, T. "Frame representations and the declarative/procedural controversy" in Representation and Understanding: Studies in Cognitive Science, D. Bobrow and Collins (eds.) Academic Press, N.Y. (1975).

Zwicky, F. Discovery, Invention, Research Through the Morphological Analysis, Macmillan, N.Y. (1969).

Formal Devices for Design

by

George Stiny
Graduate School of Architecture and Urban Planning
University of California
Los Angeles, California 90024-1467

1 Introduction

Many of the formal devices used in mechanical design are also used in architecture. This is probably no accident. These subjects -- mechanical design and architecture -- have parallel histories that can be traced at least to Vitruvius. Vitruvius counts mechanical design as one of the three departments of architecture, and devotes an entire book to the design and construction of machinery. The importance of this exposition is confirmed by Ernst Mach, who turns admiringly to Vitruvius in *The Science of Mechanics* for an account of ancient practice. History, however, may be gratuitous. Substantive links between mechanical design and architecture are more compelling, and are not hard to find. Both subjects are concerned with making things that are described according to form and function.

Practice today in mechanical design and in architecture suggests that describing form and function requires schemata, features, hierarchies, form-function relations, calculi, and grammars. Such formal devices are the subject of this essay; they are examined separately to establish their special characteristics -- some good and some bad -- and together to show how they combine and interact in some important ways.

Architecture is the source of examples, not of buildings or of building plans but of labeled shapes. These arrangements of lines and symbols, their algebras, and shape grammars defined in these algebras to carry out computations have been used for more than fifteen years now in an ongoing effort to formalize architecture and related areas of design in the fine and applied arts. This effort to formalize is not without difficulties. The way these have been resolved provides insights that are perhaps of application in mechanical design.

2 Schemata

What, then, is a schema? A schema defines representations of entities of some kind. It is comprised of variables in sets, or in combinations of sets. Values are assigned to these variables by functions to define representations. The values assigned by a function may be required to satisfy certain explicitly stated conditions and constraints, thereby restricting representations in some desired way. The entities corresponding to these representations are determined according to an appropriate semantics.

Representations of labeled shapes are defined in a simple kind of schema given by a pair $<A, P>$. The components of the pair are sets. The set A may contain sets of the form $\{x, y\}$, where the variables x and y take distinct points as values. And the set P may contain pairs of the form $<x, y>$, where the variable x takes points as values, and the variable y takes symbols from a specified vocabulary.

The semantics of representations defined in a schema $<A, P>$ is also straightforward. If a function g assigns values to variables, then the elements in the set $g(A)$ determine straight line segments of nonzero, finite length. Taken together, these lines form a shape. In a like fashion, elements in the set $g(P)$ determine labeled points that are arranged spatially. Lines obtained from $g(A)$ and labeled points obtained from $g(P)$ determine a labeled shape. Sometimes, the set A or the set P is empty. In these cases, a shape, an arrangement of labeled points, or even a blank space is determined, according to which set contains elements.

Take, for example, the schema

$$<\{\{x_1, y_1\}, \{x_2, y_2\}, \{x_3, y_3\}\}, \{<x_1, y_4>\}>$$

If values are assigned to variables according to the function given in this table

Variable	Value
x_1	(0,0)
y_1	(4,0)
x_2	(4,0)
y_2	(0,3)
x_3	(0,3)
y_3	(0,0)
y_4	A

then a representation of a three-four-five right triangle with A at its right angle vertex is defined

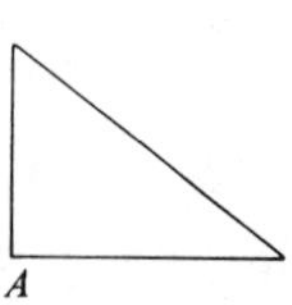

Of course, different functions define representations of different labeled shapes

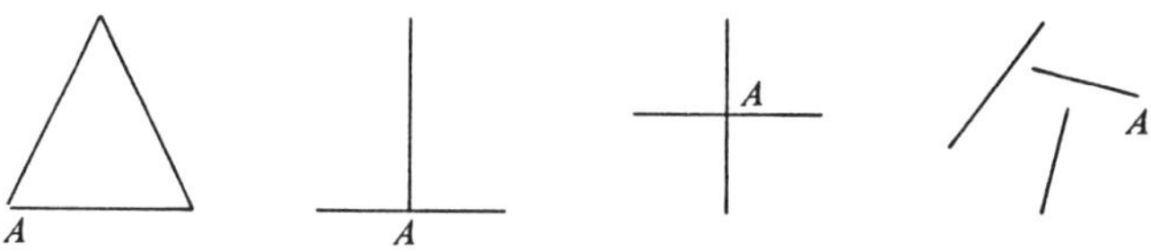

If conditions are imposed, so that functions assign colinear points to variables, then representations are defined for labeled shapes of a more specialized kind

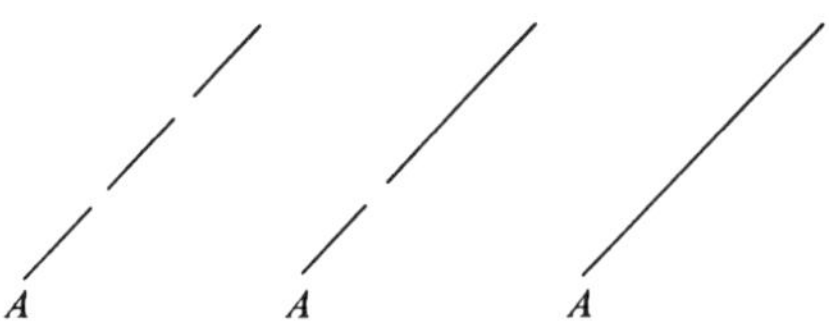

And conditions may even require that constant values be assigned to variables to define a single representation for a single labeled shape.

The symbols in a labeled shape may be just that, in which case, they identify points that are of use spatially, say, to modify the symmetry of a shape, or logically, say, to classify lines in it. But the symbols in a labeled shape may also have their own semantics allowing for symbols to carry information about function, material, and so on.

Schemata with conditions and constraints are meant for optimization. When is one entity determined by a representation defined in a schema better than another one? And what relationships among variables account for this difference? Answers to such questions are important in design. But reciprocal questions about identity are of prior logical concern. When are entities determined by representations defined in schemata the same? And what criteria can be used to establish this relation?

Answers to these questions depend on how entities are represented. Certainly, two entities are the same if they have the same representations, even if these are defined in different schemata or in the same schema by different functions. But this approach is sometimes too stringent. For example, different functions apply to the schema

$$\langle\{\{x_1, y_1\}, \{x_2, y_2\}, \{x_3, y_3\}\}, \{\langle x_1, y_4\rangle\}\rangle$$

to define different representations of a line with A at an endpoint

as these diagrams illustrate

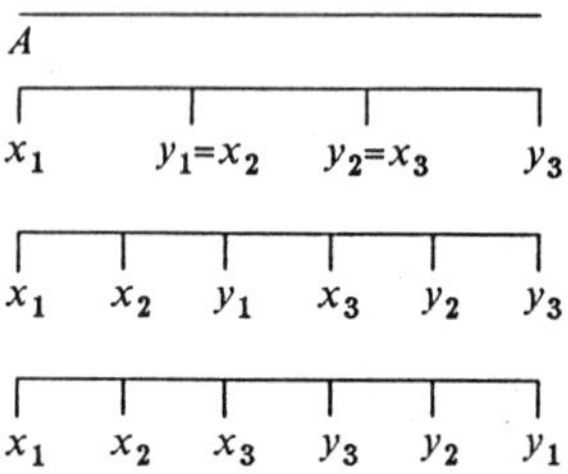

And schemata in which arbitrarily many lines are determined also define representations of this labeled shape, as different combinations of lines obtained in this way show

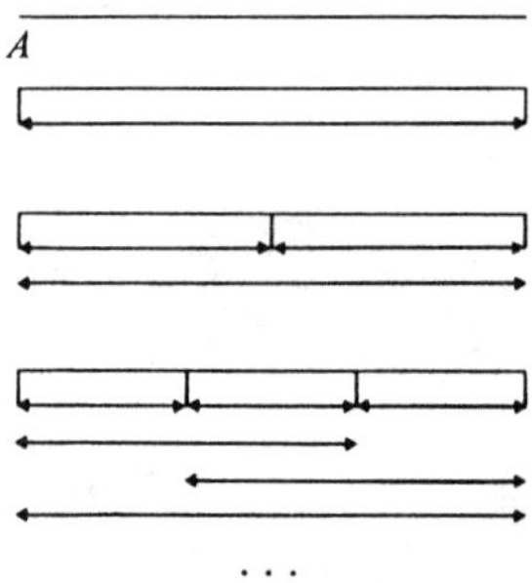

Why does one labeled shape have so many different representations, and how are all of these representations to be handled?

Labeled shapes have multiple representations because lines can be subdivided into segments in any way whatsoever. The segments of a subdivision are recorded separately in a set, but they fuse seamlessly when they are combined to determine a line. Reduction rules are required to decide whether different subdivisions determine the same line, and so to decide whether different representations determine the same labeled shape. In this way, an equivalence relation is defined on representations of labeled shapes.

Reduction rules apply to a representation of a labeled shape to change the lines determined in it into maximal lines. Every labeled shape corresponds to a unique set of maximal lines. These are the longest lines that combine to form a shape, and no smaller collection of lines does so. For example, these shapes are formed by one, two, three, and six maximal lines apiece

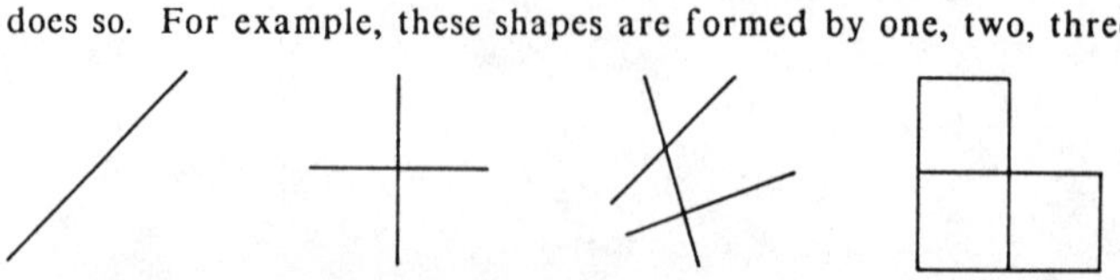

Two representations determine the same labeled shape whenever the reduction rules apply to both with identical results, and both contain the same labeled points.

Reduction rules for identity do more than define an equivalence relation on representations of labeled shapes; they also determine a sum operation that makes any two labeled shapes one. And additional reduction rules can be given to determine a difference operation. Together, these operations lead directly to an algebra for labeled shapes, or, to use the old-fashioned term, calculus. Reduction rules are used to carry out computations in this algebra and to ensure that computations are effective.

In the algebra for labeled shapes, the operations of sum and difference combine labeled shapes, and labeled shapes are related according to whether one is a subshape of another. For example, two squares have the sum

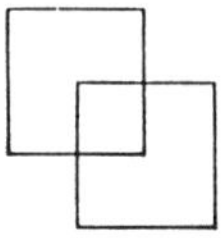

and are subshapes of it. But notice that this sum has infinitely many other subshapes, including the emergent, central square and less obvious subshapes, too

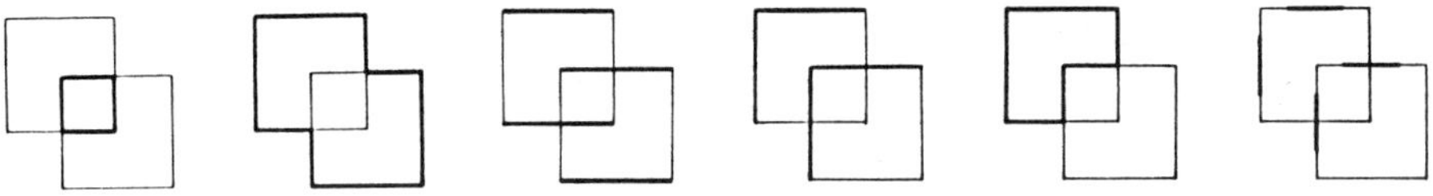

Technically, labeled shapes and the operations of sum and difference form a Boolean ring. Or in an equivalent formulation, the subshape relation is a partial order on labeled shapes that forms a relatively complemented, distributive lattice with a zero and no unit.

3 Features

Entities obtained from representations defined in schemata have features. Features are perhaps mythical; they defy definition by example and by description. Sometimes, though, features are thought to be named entities with attributes of both form and function.

Straightforwardly then, labeled shapes are features of labeled shapes. This may seem tautologous but is not, reflecting instead the closure of the algebra for labeled shapes. If several labeled shapes are summed to make a labeled shape, then they are features of it, along with other subshapes of the labeled shape. But genuine features are named. Does a name for a labeled shape make a difference?

Names may merely designate. For this purpose, names need not be familiar, and can even be obtained according to formula, perhaps from this series

feature1, feature2, feature3, . . .

These names provide access to features any time any one determines labeled shapes that answer to it. Sometimes, pointing is used to achieve this end: name a labeled shape and tell how to transform it into other labeled shapes that have the same name. Transformations may require identity, congruence, geometrical similarity, origin in a common schema, or connection in some other way, say, membership in a language (set) defined in a shape grammar.

Once a name is bestowed to acknowledge a feature, the feature may be extracted or used to design. Feature extraction proceeds when a labeled shape or a transformation of it is determined to be a subshape of a given labeled shape. And designing with features proceeds analogously when transformations of labeled shapes are combined in sums and differences.

It is sometimes claimed that extracting features is more difficult than designing with them. This claim is almost certainly idle. Extracting features involves recognition, whereby selected entities are identified as parts of a given entity, and designing with features involves operations -- appropriate versions of generic ones like add, modify, and delete -- that combine selected entities to make them parts of a new entity. But if the part relation and operations are related in a Boolean fashion, then the methods used to design with features can be used to extract them. And, in fact, this is the case for labeled shapes. Either the reduction rules for sum or the reduction rules for difference can be used to check the subshape relation.

Of course, the transformations applied to entities may complicate feature extraction in a variety of ways. For example, an indefinite number of different transformations may change one entity into parts of another one, either because there are infinitely many such transformations or because there is no way of deciding whether there are more transformations than the ones already found for this purpose. The first problem arises for similarity transformations and labeled shapes, but is tractable. Cases where infinitely many transformations may change one labeled shape into subshapes of another labeled shape can be distinguished from cases where this is not so. And in these finite cases, transformations can be enumerated exhaustively and without repetition.

But don't forget names: they frequently do more than designate. Names are meant to describe and to classify, and to provide heuristics for design. This is a seductive intrusion. Names convey information that may be useful in extracting features or in designing with them; at least, they give the impression of knowledge and control. Still, names may limit imagination; they suggest that a single description will do even when others are available, and they exclude original gizmos from design precisely because they cannot be named in an established taxonomy.

Entities given by name are apt to blunt imagination in a more serious way, too: there is an irresistible temptation to combine them by name and not by entity. This leads ineluctably to a tinker-toy kind of design. Where names are disjunctive in combination, announcing insistently that they designate separate things, entities may be conjunctive in combination, fusing to create new things other than the entities given by name -- and groupings of them -- in disjunction.

Thus, for example, this *square* and that *square* in combination are just that, two *squares* without any possibility of being anything else. This outcome is predictable and boring because of it. However, translations of this square

in this sum

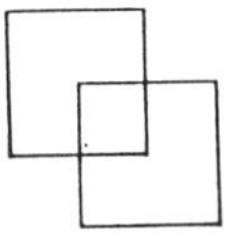

are much more than names (descriptions) suggest

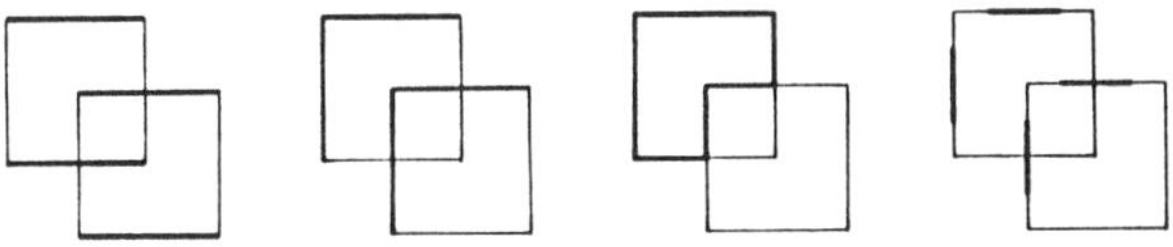

They form horizontals and verticals, Italian quotation marks pushed together to interfere, and L-shaped regions that enclose a new square; and they even have subshapes without familiar names. But surely, to handle these possibilities requires that they be anticipated. From where are such predictions to come?

Oracles are out of the question, and are not necessary anyway. Predicting and recognizing are independent goals. Knowing what features someone will use to describe a labeled shape requires that the decision be made in advance, or special insight that relies heavily on psychology or clairvoyance, and little on design. But neither of these approaches is likely to sanction arbitrary features. The preferred approach that makes this possible is already at hand; it requires pointing, transformations, and the subshape relation.

4 Hierarchies

People who like to name entities like hierarchies. Both names and hierarchies work in comparable ways; they stamp entities with preferred structures, thereby excluding other possibilities. Even so, hierarchies are worth considering in their own right; they are invoked widely and used with approval in many approaches to design.

A hierarchy is a description; it decomposes an entity into others that are parts of it -- larger composites into smaller components -- to form a tree. This allows for entities to be broken down independently, each with its own parts that are not shared with others and that interact with them, if at all, only in actual combination.

Hierarchies may be defined directly for entities, as is this one

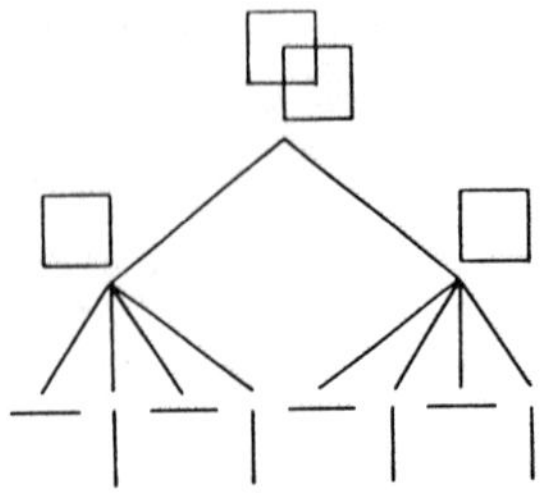

in which the shape

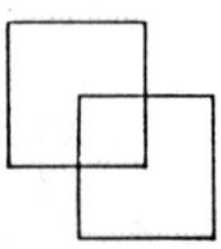

is divided into squares and sides of squares. Such hierarchies are embedded in the lattice defined by the subshape relation on labeled shapes, but they barely begin to suggest the possibilities for description to be found in it. Alternatively, hierarchies may classify entities by name

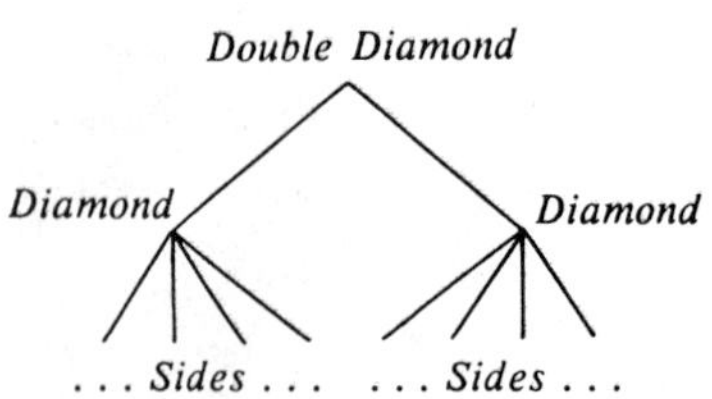

A hierarchy is rarely unique. Every labeled shape with at least one line as a subshape is described in infinitely many hierarchies. For example, these hierarchies

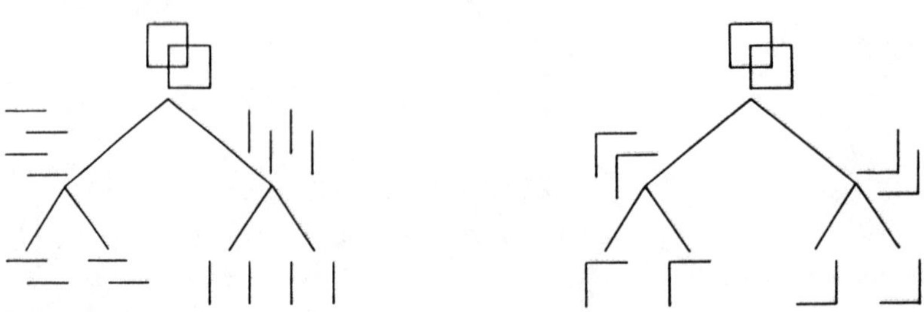

also describe the shape

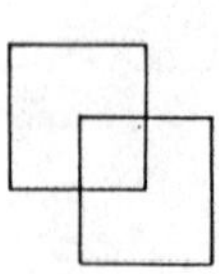

And if the shapes in these hierarchies are classified by name, then other hierarchies are also forthcoming

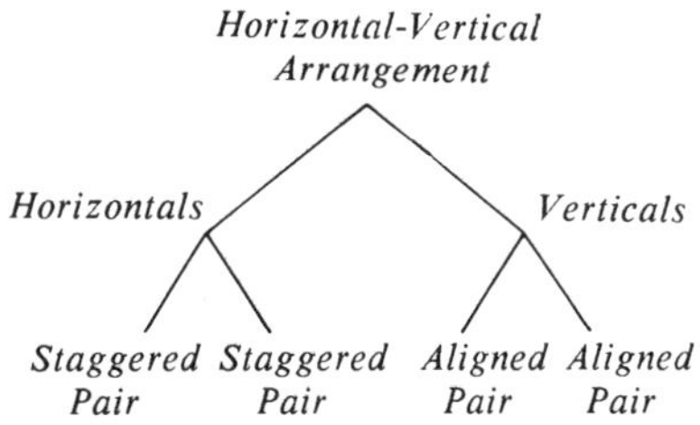

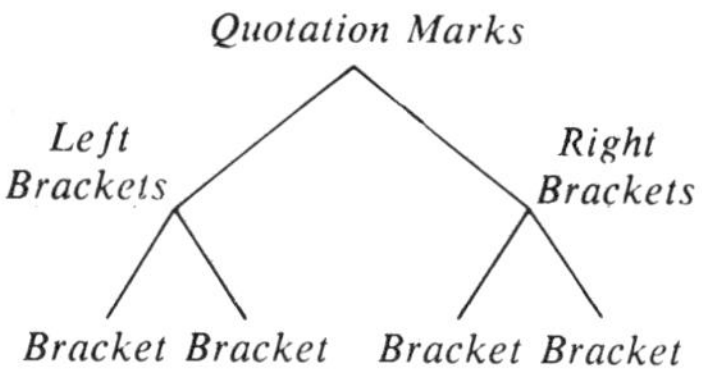

In most cases, two entities corresponding to different nodes in a hierarchy are discrete, so long as there is no descending path from one node to the other. But this rule can be broken for overlapping entities if their common parts are viewed as distinct copies that do not interfere with one another. Thus, for example, in the hierarchy

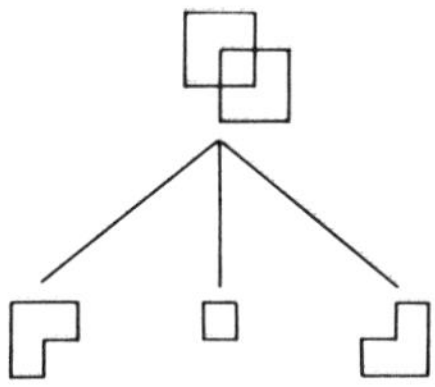

the sides of the square and corresponding sides of the L's are all separate entities.

Hierarchies are a convenient way of distinguishing the parts of an entity, and showing relationships among them. Still, without constant vigilance, one is apt to forget that hierarchies are selective, describing entities from a single point of view among a host of alternatives. Like names, hierarchies are useful in design, but they rarely show everything and are sometimes misleading. Like names in combination, hierarchies show how entities combine disjunctively without the slightest hint of conjunctive possibility.

5 Form-Function Relations

A form-function relation is a set of entities with attributes of both form and function, say, a set of labeled shapes. The definition of such relations is at the root of design. Form-function relations contain designs: their elements describe objects for making, and tell how they work.

In mechanical design, it is sometimes claimed that universal, physical laws are enough to determine form-function relations, but experience in architecture suggests otherwise. Style intervenes to provide the impetus for practice. But this is an old and involved argument that need not be rehearsed here. For now, it is enough to consider how the formal devices at hand can be used to define form-function relations.

One approach uses hierarchies or even more complicated graphical devices such as finite lattices and networks to describe entities in terms of their structures. An entity belongs to a form-function relation if it has a structure of a certain, specified kind. A structure may show how entities are broken down into others, or classify entities in this decomposition by name. Either way, the definition of preferred structures may involve grammars, at least something like context free grammars to obtain hierarchies, or stronger computational devices to obtain lattices and networks.

An ambitious but sometimes vague application of this approach is developed by Alexander [2] with devices like those used by Fu [4] to define context free pattern grammars. In Alexander's pattern language, patterns of space determine the shape of buildings and towns. Patterns of space are relationships of others; they determine rules $X \rightarrow r(X_1, \ldots, X_n)$ allowing for composite patterns X to be broken down into independent components $X_1, \ldots, X_n$ in an approved arrangement r. As components may also be composite patterns, a hierarchy is defined

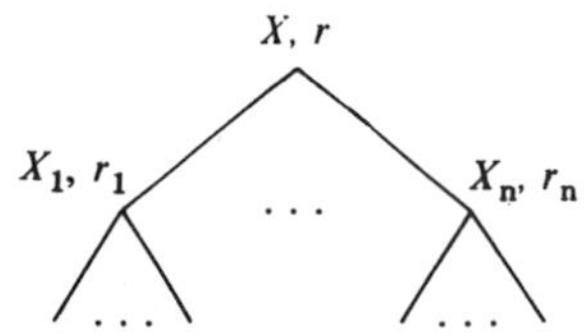

or, by fiat, a network

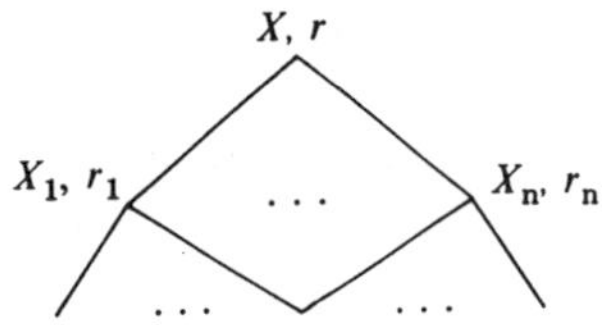

in those cases where some instance of a pattern corresponds to several nodes in the hierarchy. In each such structure, patterns of space are classified by name, and associated with patterns of events to provide the link between form and function.

Alexander's work has been influential, perhaps because he is sympathetic with the view that universal laws control form-function relations, but more likely because he relies on hierarchies and names to describe patterns of space for buildings and towns. Alexander's pattern language may be useful; still, like other form-function relations defined in the same manner, it never gets beyond the limitations of disjunctive combination. For this purpose, alternative devices are required.

6 Algebras and Grammars

The algebra for labeled shapes encourages conjunctive combination. Shape grammars define computations in this algebra to determine languages of labeled shapes. These languages are form-function relations, and may be used to obtain designs, for example, for buildings of different types and styles.

In a shape grammar, computations are carried out by following rules. A rule is comprised of schemata that are used with functions and similarity transformations to determine pairs $a \rightarrow b$ of labeled shapes a and b. If any pair is such that a is a subshape of a labeled shape c, then the rule applies to c to replace a with b: another labeled shape is produced in the sum of b and the difference of c and a. Rules are used recursively in this way to obtain the labeled shapes in the language defined in the grammar.

The character of computations defined in shape grammars is clear in a very simple example. Suppose a rule determines the pair of shapes

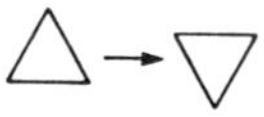

and other pairs formed in similarity transformations of this one. The rule replaces any equilateral triangle with its 60° rotation about its center, so that this point is always fixed. The rule can be applied to the shape

and then to shapes produced in previous applications of the rule to obtain the shape

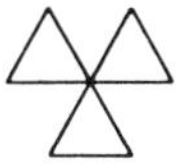

At first sight, this computation may seem elusive. The second shape is a 60° rotation of the first shape about its center. This point is fixed, but the centers of the equilateral triangles in the first shape change their positions as the shape is rotated. How can a rule that leaves the centers of equilateral triangles alone make these changes?

The computation is easy to see. The rule can be applied to the shape

to replace each triangle separately, thereby providing the new shapes in this series

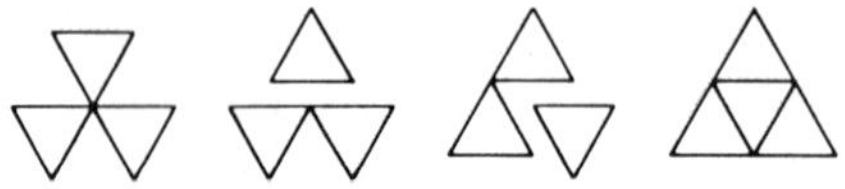

The second shape in the series is obtained from the first shape by replacing the uppermost triangle with its 60° rotation. This process is repeated for the leftmost triangle in the second shape to obtain the third shape, and for the rightmost triangle in the third shape to obtain the fourth shape.

Two equilateral triangles -- the large, outside one and the small, inside one -- in the shape

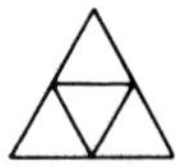

have sides formed from sides of the three equilateral triangles replacing those in the shape

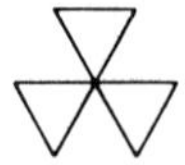

The rule can be applied to replace each of these two emergent triangles in turn in the same way as before to produce the shapes in the series

The second shape in the series is obtained from the first shape when the large, outside triangle is replaced with its 60° rotation, and the third shape is obtained from the second shape when the small, inside triangle is replaced with its 60° rotation.

Now, three equilateral triangles -- one at each corner -- in the shape

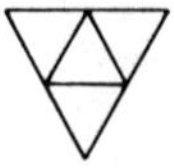

have sides formed from segments of sides of the two equilateral triangles replacing those in the shape

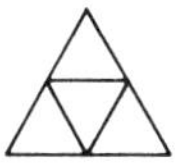

The rule replaces each of these three emergent triangles in the prescribed way to produce the shapes in the series

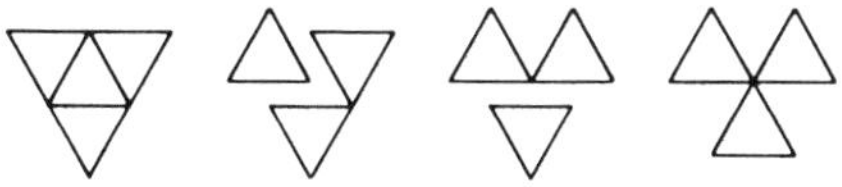

The shape

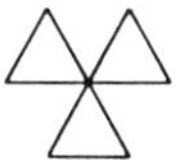

is so obtained from the shape

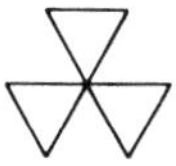

In computations like this one, rules exploit the basic property of labeled shapes: their subshapes may be distinguished at will, and repeatedly so. Subshapes are not fixed -- as in a hierarchy -- for the labeled shape at the beginning of a computation or for any other labeled shape produced during the computation; they are free to vary for each of these labeled shapes, being determined by a rule and where it applies.

The rule used above replaces each of the three equilateral triangles in the shape

with its 60° rotation to obtain two equilateral triangles -- the large, outside one and the small, inside one -- in the shape

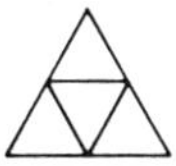

And each of the triangles in this pair is replaced in the same way by the rule to obtain the three equilateral triangles at the corners of the shape

Only when equilateral triangles are distinguished freely in applications of the rule can the shape

be obtained from the shape

in a computation.

Of course, designs are likely to be more than arrangements of triangles. And, in fact, shape grammars have been used to define a host of languages in architecture. There are shape grammars for Palladio, Wren, Wright, and Terragni, and for vernacular styles from different periods and places. Shape grammars provide designs for churches, villas, houses, multi-storied structures, and buildings of other types, for ornament to decorate them, for furniture to fill them, and for gardens in which to put them. New shape grammars are always available; they have been defined from scratch to explore fresh ideas, and by combining or transforming shape grammars already in use to profit from past experience with known types and styles. Shape grammars furnish more opportunities for design in another way, too.

On close examination, it is easy to see that shape grammars perform parallel computations in two different algebras -- one for shapes and one for arrangements of labeled points -- to obtain designs. The algebra for shapes U^* provides for descriptions of form, and the algebra for arrangements of labeled points V^* for descriptions of function. These algebras are combined in a Cartesian product U^*V^* to form the algebra for labeled shapes. And grammars used to carry out computations either in U^* or in V^* can be combined in products in an analogous fashion to form shape grammars to carry out computations in U^*V^*.

In a shape grammar, computations with descriptions of form and computations with descriptions of function influence one another according to how rules in a grammar in U^* comprised of schemata for shapes and rules in a grammar in V^* comprised of schemata for arrangements of labeled points are paired and variables are shared. As a result, dependencies between descriptions in designs are established and vary as computations proceed. And conditions and constraints placed on one description can be propagated to the other as these conditions and constraints are met. Designs may so be viewed to follow from considerations of form, of function, or of both, depending on how the rules in the shape grammar are defined, and on what conditions and constraints are given and how they are used.

This idea of computing in the Cartesian product of two algebras can be extended to Cartesian products of multiple algebras. In this way, n-ary relations containing designs comprised of descriptions in appropriate domains of interest can be defined, either in shape grammars when the algebra for labeled shapes is used repeatedly to form Cartesian products $U^{*} \ldots U^{*}V^{*}$, or in description schemes when other algebras are used as well.

A Cartesian product $U^{*} \ldots U^{*}V^{*}$ provides for n-ary relations among labeled shapes to be defined. Relations of this kind are used throughout design; for example, they may hold for plans, sections, and elevations of single components, or for multiple components in assemblies. Other algebras are used in Cartesian products to provide descriptions of additional kinds. Labeled shapes themselves may be described in multiple decompositions. These may be hierarchies or arithmetic-like expressions in which labeled shapes are combined in sums and differences, and are classified by name. It is so possible to provide different semantics for labeled shapes, and to reason about them in different contexts. Or labeled shapes may be components in larger designs in which they provide descriptions that are connected with others to establish relationships among form, function, manufacture, use, and so on.

In *The Kinematics of Machinery*, Franz Reuleaux argues that machines have been treated scientifically to the extent that mathematics has been applied to their analysis, but that every question about machines is not thereby answered.

> "The mathematical investigations referred to bring the whole apparatus of a great science to the examination of the properties of a given mechanism, and have accumulated in this direction rich material, of enduring and increasing value. What is left unexamined is however the other, immensely deeper part of the problem, the question: How did the mechanism, or the elements of which it is composed, originate? What laws govern its building up? Is it indeed formed according to any laws whatever? Or have we simply to accept as data what invention gives us, the analysis of what is thus obtained being the only scientific problem left -- as in the case of natural history?"

Reuleaux believed that mechanical design could be scientific and tried to make it so. This belief more generally -- that design can be formalized -- motivates work of the kind described in this essay. As more and more formal devices are suggested, tried, and proven in design, this will surely show that Reuleaux was right.

7 Background

The types of formal devices examined in this essay were culled, at least by name, from a number of recent sources describing research supported by the Design Theory and Methodology program in the Division of Design, Manufacturing, and Computer Engineering in the Engineering

Directorate of the National Science Foundation. Schemata are used in Gero, Maher, and Zhao [5] to characterize prototypes. Features are the province of Dixon [3], where shape grammars are also mentioned. Hierarchies are used to very good advantage in Spillers and Newsome [8], and are invoked conspicuously in Agogino and Guha [1] and in Dixon [3]. Form-function relations correspond closely to features and are treated in Rinderle, et al [7]. And a calculus for design is proposed in Agogino and Guha [1].

The formal devices used as examples in this essay come from work in shape grammars. An informal survey of the subject is given in Stiny [11]. Important technical references include Stiny [9, 10, 12] and Knight [6].

References

[1] A. M. Agogino and R. Guha, "Object-Oriented Data Structures for Reasoning about Functionality, Manufacturability, and Diagnosability of Mechanical Systems", *Design Theory '88: The 1988 NSF Grantee Workshop on Design Theory and Methodology*, RPI, Troy, New York, June 3-5, 1988.

[2] C. Alexander, *The Timeless Way of Building*, Oxford University Press, 1979.

[3] J. R. Dixon, "Feature-Based Design: Research on Languages and Representations for Components and Assemblies", *Design Theory '88: The 1988 NSF Grantee Workshop on Design Theory and Methodology*, RPI, Troy, New York, June 3-5, 1988.

[4] K. S. Fu, "Syntactic (Linguistic) Pattern Recognition", *Digital Pattern Recognition*, edited by K. S. Fu, Springer-Verlag, 1976.

[5] J. Gero, M. L. Maher, and F. Zhao, "Chunking Structural Design Knowledge as Prototypes", *Design Theory '88: The 1988 NSF Grantee Workshop on Design Theory and Methodology*, RPI, Troy, New York, June 3-5, 1988.

[6] T. W. Knight, *Transformations of Languages of Designs*, Ph.D. Dissertation, UCLA, Los Angeles, California, 1986.

[7] J. R. Rinderle, et al, "Form-Function Characteristics of Mechanical Designs -- Research in Progress", *Design Theory '88: The 1988 NSF Grantee Workshop on Design Theory and Methodology*, RPI, Troy, New York, June 3-5, 1988.

[8] W. R. Spillers and S. Newsome, "Some Attributes of Design Theory", *Design Theory '88: The 1988 NSF Grantee Workshop on Design Theory and Methodology*, RPI, Troy, New York, June 3-5, 1988.

[9] G. Stiny, *Pictorial and Formal Aspects of Shape and Shape Grammars*, Birkhauser, 1975.

[10] G. Stiny, "A Note on the Description of Designs", *Environment and Planning B*, Vol. 8, pp. 257-267, 1981.

[11] G. Stiny, "Computing with Form and Meaning in Architecture", *Journal of Architectural Education*, vol. 39, no. 1, pp. 7-19, 1985.

[12] G. Stiny, *Shape: Algebra, Grammar, and Description*, manuscript, 1988.

FORM + FUNCTION + ALGEBRA = FEATURE GRAMMARS

Steven N. Longenecker and Patrick A. Fitzhorn

Department of Mechanical Engineering
Colorado State University
Ft. Collins, CO 80523

Abstract

Much of the current research and development in feature-based design of discrete objects is both pragmatic in nature and opportunistic in approach. We develop a formalism for features that is consistent in both enumeration and recognition as well as extensible enough to handle the complex relationships between function and form. Using the ***descriptive, parametric shape grammars***, this paradigm is easily expanded to exploit feature set descriptions and manufacturing process planning in a unified and formal framework.

1 Introduction

"Autant de langues que l'homme sçait parler, autant de fois est il homme."

Charles V (1337-1380)

Languages, either formal or natural, provide an excellent basis for the study of descriptive paradigms for object representation. The beauty and elegance of these systems is evident in the succinct, consistent way they can both *enumerate* an infinite set of sentences in some language of interest, as well as recognize or *parse* each sentence back into its component elements. Of course, the real beauty of formal language theory rests in the infinite descriptive power of a finitely describable grammar. Many different methodologies exist to represent and classify languages and their generating grammars. The more prevalent grammar forms are nicely summarized in [7] including the one of particular importance here; the *descriptive, parametric shape grammars* of Stiny [18,20]. The name given to this linguistic paradigm is quite graphic in fact. We use this grammar here for the following reasons:

- *descriptive*, to provide a placeholder for functional descriptions along with shape
- *parametric*, to yield geometric equivalence classes for non-nominal geometry
- *shape grammars*, to capture the formalism and algebra of shape as a major component of feature-based design and manufacturing

The disassociation of design and manufacturing is one of the current, major problems in CAE. While this problem has plagued the entire manufacturing process for dozens (if not hundreds) of years, it is no doubt based on the division of labor and knowledge manifested in tasks associated with each side of the process. Currently, designers create an idea and develop it to completion in function and detailed form. Typically, it is only then that the design is "tossed over the wall" to the manufacturing side of the house. At this point, the manufacturing personnel must determine not only the designer's original intent, but also the manufacturability of the final design. This problem is exacerbated by many object representations and exchange standards, which carry product form (typically in terms of design geometry and possibly topology) without intent, function or manufacturing/assembly information. In fact, a consistent methodology for the union of both form and function is difficult to capture because of the diverse requirements of design and manufacturing. Ideally, a design by constraint approach would give the designer flexibility in function and form and would require relevant specifications and constraints about the design. Such a system might use variational geometry [1,15] and a high-level object representation while alleviating messy details such as specific dimensions required during the design phase and an irrelevant and weak set of modeling primitives like edges or holes.

2 Feature definitions and use

Since typical, evaluated object representations consist of points, lines, surfaces and their connectivity, it is believed that more abstract, functional representations can be used to depict various manufacturable objects. Much of the current work centers around the concept of feature-based design representations. One description of features is [5]:

> "...a vocabulary of familiar geometric features. This vocabulary of features is similar to the terms with which one designer would describe a design to another (designer). A feature is a geometric form or entity:
>
> - whose presense or dimensions are relevant to one or more CIM function(s) (graphics, analysis, process planning, manufacturability evaluations, etc), or
> - whose availability to a designer as a primitive facilitates the design process."

In this particular description of features, Dixon explains that problems in casting are closely related to certain geometric features of design, such as slabs, bosses and holes and thus knowledge of these features and associated problems is helpful when designing a casting. He states that "it is important that the features logically represent the domain for the designer" (and of course the manufacturer as well).

The paper continues with various descriptions of ***primitive features*** [22], ***macro-features*** [17], ***micro-features*** [13] and ***co-features*** [5] all developed for different purposes and describing different aspects of manufacturable objects. For example, ***macro-features*** include boxes, L-brackets and U-channels, while ***micro-features*** are walls, intersections and fillets. This brings up interesting questions concerning feature definition and standard feature set construction. Note that each of these "features" evokes a typical shape, probably parameterized in some geometric class, that carries descriptive (non-geometric and non-topologic) information. While most current research in feature-based representation is both pragmatic and thought-provoking, it seems somewhat incomplete. It provides new methods in shape representation, topology and functional structure but typically fails to adequately bridge the design/manufacturing gap with a consistent formalism.

One of the first interpretations of features was introduced by Henderson [10]. This work presented feature extraction for representing, recognizing and extracting volumes from a solid geometric model in a form suitable for input to a process planning system. In this context, features were generic volumes which defined cavities described by facts and rules in a logic programming language. Feature extraction was the process of instantiating these appropriate higher-level data representation (Hendersons's *feature tree*). Others [2,3] have also showed how such features could be used in automatic process planning. Currently, it seems as though the major drawback to these previous ideas lies in the definition of the feature itself. Each of these systems is centered around a specific set of features and various associations to describe the overall function/form of the final product. For example, the basic features might consist of such things as through-slots, holes, countersunk holes and planar faces, while spatial relationships might be perpendicular, parallel, sharing, on, etc. Since most object representation forms use a variety of primitives like circles, lines, points, B-spines, relationships such as intersection, tangency, and parallelism and the set-theoretic operations of union, difference and intersection, clearly there are some major discrepancies between the two forms. While some objects may already be represented similarly in the design and manufacturing realms, there is clearly not a common medium in which both arenas can be described. What is needed is a specific *language* of object description for both function and form which can be understood by both the manufacturers and designers alike. It should be complete for all designs, unambiguous, reliable and accurate.

3 Production systems and shape grammars

The notion of a production system was first introduced by post and later demonstrated in the work of Chomsky's weak generative capacity grammars [4]. Chomsky developed grammars that generate sequences or 'strings' of symbols. Other types of production systems generate arrays, graphs, trees and shapes. These grammars have been applied in a number of areas including linguistics, pattern recognition, cognitive psychology, architecture and spatial design. A production system in general consists of symbols of a specified vocabulary. The production rules are then used to change the current string into a new one. The generation of a string using a Post production system begins with an initial string and proceeds by repeatably applying productions. The shape grammars differ from other production systems in that they replace sub-shapes with other shapes as described below. For the introduction to shape grammars and their applications, see [18].

3.1 Structure and form of shape grammars

Any design can be understood in a syntactic or compositional manner as a collection of shapes and relationships between these shapes. In a more formal sense, this understanding corresponds to viewing a design in terms of a *vocabulary* of shapes in the artifact, along with the *spatial relations* between them. Here, a shape might be defined as a finite collection of lines, a vocabulary as a set of unique shapes, and a spatial relation (or element of a *shape grammar*) as the arrangement of such shapes in a particular design. Then given some shape grammer G, the use of *features* as description mechanisms can be precisely characterized as being:

1. *recognizable* in sentential forms of sentences in G, and

2. *generable* by sentence enumeration from the grammer.

From this, it is possible to generate a concise, encompassing definition of a feature as a defining or shape entity in a language of objects as follows. A shape grammer is defined as a four-tuple $G = (N, \Sigma, R, I)$ for which

N is a finite set of non-terminal shape elements (also called markers),

Σ is a finite set of pattern primatives or symbols called terminal shape elements,

R is a finite set of replacement shape rules given as a binary relation with elements (Θ, Ψ), and

I is the initial shape containing element Θ such that (Θ, Ψ) is in R.

A shape grammer is similar to a string grammar in that it is required that N and Σ be disjoint. A rule (Θ, Ψ) in R is written $(\Theta \Rightarrow \Psi)$ to indicate that the replacement of the left-hand shape Θ by right-hand Ψ is allowed. To generate shapes then, one begins with I and recursively applies rules in R according to the following specifications:

1. Locate a part of the current shape that has a one-to-one correspondence with Θ in a rule $(\Theta \Rightarrow \Psi)$. There must be an exact correspondence between the terminals and markers in Θ and the terminals and markers in the part of the current shape to be replaced.

2. Find the geometric transformations necessary to make Θ identical to this part of the current shape. These transformations may be adjustments of scales, rotations, translations or formation of mirror images.

3. Apply the same set of transformations to shape element Ψ in rule $(\Theta \Rightarrow \Psi)$.

4. Replace the part of the shape by the transformed version of Ψ.

The *shape language* generated by G is the set of all shape sentences generated by G. A *sentential form* of G can be defined inductively as follows; (*i*) the start symbol I is a sentential form, and (*ii*) if xyz is a sentential form and $(y \Rightarrow w)$ is a production in R, then xwz is also a sentential form of G. Discussion of shape generation, with several excellent examples, can be found in [16]. A discussion of shape parsing in syntactic image processing is detailed in [8].

3.1.1 The parametric shape grammars

Parametric shape grammars are an extension of shape grammars in which shape rules are defined by filling in the open terms (right hand side) with a general schema. Thus, shape descriptions become Euclidean points in space and a set of governing algebraic equations describing the relationships between these points. A square, for example, could be represented in a non-parametric shape grammar as a set of four predetermined points connected by four lines of equal and predetermined length whose adjoining angles are all 90°. In a parametric representation, the same square could be represented as the four perimeter points $\{(x1, y1)(x2, y2)(x3, y3)(x4, y4)\}$ and several spatial relationships that determine its exact description. Some of these might be that $(x1, y1)$ defines the square's origin, and $x1 = x2$ and $y2 - y1$ define the length of all sides. Then the definition of the square becomes non-monotonic in a parametric shape grammar and requires various specifics (i.e. length of sides, origin) at the point in which a corresponding replacement rule will apply.

Thus, parametric shape grammars can be used to define languages of shape with proportional relationships determined in any given manner. One such manner might be the use of variational geometry using explicit dimensions as symbolic variables. These variables would then be assigned values as the last step in defining a particular geometry for the shape model. For a general overview of variational geometry, see [14].

3.1.2 The descriptive shape grammars

Given a language of designs defined by a parametric shape grammar, the intended functions or descriptions of these designs can be explicitly constructed using a similar recursive schema based on the grammar. Even though designs are shape given by a set of lines, it is possible to describe them in functional terms not directly related to their component spatial elements. Design descriptions are defined in shape grammars as follows [20]:

> "The way a language of designs is understood depends on two main factors:
>
> 1. the rules given to compose the spatial elements in these designs, and
> 2. the rules given to describe these designs in other terms pertaining to purpose, function, and use, meaning, type, and form.
>
> Typically, rules of the first kind are specified by a shape grammar G. Rules of the second kind can be specified by a *description function* $h : L_G \Rightarrow D$ ..."

The description function h maps shapes or designs in the language generated by the shape grammar G into descriptions in the description set D. This function provides a formal placeholder to investigate and merge shape form and function in either design or manufacturing.

Thus the formal concept of a manufacturing or design feature within the parametric shape grammar can be elegantly exploited. Just as the shape grammar is applied to the spatial representation of the design, these description functions apply changes to the corresponding functional components in the description. Depending on the design domain, these functions relate the spatial representation of the artifact with various recognizable features such as doors, windows, and walls in an architectural environment or gears, shafts, and motors in a mechanical one. Further, the description functions would likely be applied from the result of direct parsing of the design's spatial representation at any given time during its development. This block-by-block description of the artifact's history provides the basis for determining the properties of the particular design.

> *"This simple idea of recording the history of a design as it is generated can be invoked for any shape grammar. Given the generative history of a design, a line-by-line description of it can always be derived, and spatial elements in it can be distinguished in terms of the sequence of shape rules used to compose them. Thus, if descriptions of designs in languages*

defined by shape grammars depend on their constituent spatial elements and the way these are combined, then description functions of the kind envisaged here can be defined" [20, page 265]

3.2 Shape grammars in design

From this, one can then generate a concise, encompassing definition of the concept of a *feature* as a defining or shape entity in a language of shape elements as follows:

Definition 1 (Feature set) *For the descriptive, parametric grammar G with alphabet Σ, α is a simple feature of G iff $\alpha \in \Sigma$. The feature set of G is the set of all sentential forms of G.*

That is, all members of the defining alphabet of some domain are simple features, and the feature set of the domain is the set of all right-hand sides of the defining productions. Note that this may contain "meta-alphabet" descriptors, or non-terminals. Thus in a language characterized by straight lines, one can generate a circle feature by parametric, iterative applications of sines and cosines. In this trivial example, the circle is a member of the feature set, and the lines would be simple features. Feature function is then carried by the descriptor functions.

This definition is quite powerful in several different aspects. Discussion of features has centered, as of late, on the development and use of different covering sets of features that are capable of describing the same set of objects. As a direct result of the definition here, evaluation of this problem becomes straightforward. First, it is well known that one can generate sets grammars whose languages are equivalent. It is difficult at best however, to actually *prove* language equivalence for any but the simplest grammars. Explicitly, only the question of equivalence for the regular languages in the Chomsky hierarchy (the grammars with the least descriptive power) is known to be decidable [11]. The general solution for the remainder of the grammars is either unknown (in the case of deterministic context free languages) or undecidable. Thus, the *general* answer to feature set equivalence will, no doubt, also be undecidable. Another question with some interesting positive results is whether or not a word is an element of some language. In the domain of feature descriptions, this question could be reworded as follows; is a particular feature an element of a shape language? For all but the recursively enumerable grammars (the language in the Chomsky hierarchy with the most descriptive power), the question is decidable.

As should be expected, these grammars provide formal placeholders and terminology for Dixon's definition of features (discussed in the previous section). They capture both implicit shape and explicit function as necessary.

3.3 Design formalisms

The use of formal rules in the design process has several distinct advantages:

1. Rules are usually much less complicated than the designs they produce. They can be expressed in relatively simple terms and then easily enumerated many times by a computer. Thus, design can be reduced to an algorithm solved by a *divide and conquer* paradigm.

2. Rules increase the designer's power of observation, open up new avenues of design direction and formalize thoughts and initial design requirements. This does require that these constraints be

represented in an explicit and detailed form. In this way, the designer acquires command of a language of designs which is accessible via these defining rules.

3. The emphasis in design is shifted away from individual designs already in existence to the general language itself. A designer can concentrate on the constructive basis of the design and an explicit awareness of its properties and structure by expressing them in specific rules. The understanding of a particular design is complete since it is defined solely on the rules and not by chance or human intuition. This opportunity has not yet been possible when individual designs are modified in an ad hoc manner to obtain variations. Without such rules, the designer often settles for initial ideas or early variations, even when better designs are within grasp.

4. Rules can be modified systematically to define new languages of design to reflect changing circumstances or to incorporate new ideas. The designer is able to define new languages of design based on construction rules and not on any individual members.

Reduction of design to an algorithm is certainly not a new idea. The *design machines* of Stiny and March [21] are an interesting formalism of this concept. Design as an abstract computational model is suggested by Fitzhorn [6] which suggests that design has an underlying algorithmic formalism that can be studied and quantified as necessary. Concerning the concept of a design machine or automaton,

> *"The common structure of all design machines should provide the basis for a future science of design."* [19, page 461]

4 Concluding remarks

This approach to the definition of features and their constituents is quite powerful and alleviates many of the problems of previous research, two of which are

1. informal, conflicting terminology and definitions, and

2. lack of an algorithmic approach to tightly couple both design and manufacturing process planning

Several other researchers have noted the connection in general design (see [9,12]) or enumeration. Discussion of features has centered on the development and use of different covering sets that are capable of describing the same set of objects. As a direct result of the use of a shape grammer, evaluation of this problem becomes straightforward.

Another question with some positive results is whether an object is an element of some language. In the domain of feature descriptions, is a particular feature an element of a shape language? For all but the recursively enumerable grammers (the language in the Chomsky hierarchy with the most description power), the question is decidable.

No doubt the most interesting fact is that grammars are useful for both sentence *enumeration* as well as *parsing*. Thus, given two grammars whose languages can be conjectured to be equivalent, one could be used to generate sentences (as in the design process), while the other could be used to parse the result of the first, giving us a manufacturing process description of its contents. This paradigm is obviously easily expandable to exploit feature set descriptions and recognition for design and manufacturing process planning in a unified and formal framework. The use of a unified shape feature methodology (based around parametric shape grammars) would allow multiple sets of features to be used in both the design and manufacturing of discrete products. This work proposes two sets of grammars, the first (*i*) for design to recursively enumerate sentences (objects) in its language. These grammars could be optimized for a particular domain or class of objects in practice. Given the shape productions and operators, this paradigm should prove to be quite powerful. Finally, once the object is designed (*ii*) a second grammar of manufacturing features would then be used to parse the current design into its component manufacturing features.

5 Current work

Research in feature-based design in the Design Automation Laboratory at Colorado State University is currently exploiting the connection between features and shape grammars. We are concerned with three general areas in this work:

1. creation of a LISP-based descriptive, parametric shape grammar enumerator
2. definition of feature-based, domain dependent object representation schemes exploiting both form and function
3. developing parametric, shape-based meta-productions to exploit high-level feature paradigms

Our goal is to represent discrete mechanical objects in such a manner that could be both designed and recognized easily, concisely and unambiguously through the use of a formal, parametric shape grammar. This underlying formalism is necessary to a science of design since it allows one to state formal questions, and receive provable answers.

Acknowledgements

This material is based upon work supported by the National Science Foundation under Grants No. DMC-8611220 and DMC-8714173. P.A. Fitzhorn gratefully acknowledges National Science Foundation support as a Presidential Young Investigator. Any opinions, findings, and conclusions or recommendations expressed in this publication are those of the authors and do not necessarily reflect the views of NASA or the NSF.

References

[1] P. Beaty and P. Fitzhorn. Driving surface shape definition using variational geometry. *IEEE Computer Graphics and Applications*, 1988. (in review).

[2] K. Chang and A. Bond. Feature-based process planning for machined parts. In *ASME Computers in Engineering*, Los Angeles, CA, November 1987.

[3] B. Choi, M. Barash, and D. Anderson. Computer recognition and extraction of form features. *Computers in Industry*, 5(4):329–339, 1985.

[4] N. Chomsky. *Syntactic Structures.* Humanities Press, Atlantic Highlands, NJ, 1957.

[5] J. Dixon and M. Simmons. Expert systems in mechanical design. In *Proceedings of the 1985 ASME Design Automation Conference*, Cincinnatti, OH, 1985.

[6] P. Fitzhorn. A computational theory of design. *Design Computing*, 3(1), 1988.

[7] J. Gips and G. Stiny. Production systems and grammars: a uniform characterization. *Environment and Planning B*, 7:399–408, 1980.

[8] R. Gonzalez and M. Thomason. *Syntactic Pattern Recognition*. Addison-Wesley, Reading, PA, 1978.

[9] I. Ham and S. Joshi. Features in process planning. February 1988. Abstract presented to the 1988 NSF Workshop on Features in Design and Manufacturing, UCLA.

[10] M. Henderson. *Extraction of feature information from three dimensional CAD data*. PhD thesis, Purdue University, May 1984.

[11] J. Hopcroft and J. Ullman. *Introduction ot Automata Theory, Languages, and Computation*. Addison-Wesley, 1979.

[12] G. Jared. Shape features in geometric modeling. In *Solid Modeling by Computers: From Theory to Applications*, Plenum Press, New York, 1984.

[13] E. Libardi, J. Dixon, and M. Simmons. Designing with features: design and analysis of extrusions as an example. In *Proceedings*, Mechanical Design Conference, Chicago, Ill, March 1986.

[14] R. Light. *Symbolic Dimensioning in Computer-aided design*. Master's thesis, MIT, Cambridge, MA, 1980.

[15] V. Lin, D. Gossard, and R. Light. Variational geometry in computer-aided design. *Computer Graphics*, 15(3):171–177, 1981.

[16] E. Lord and C. Wilson. *The Mathematical Description of Shape and Form*. Ellis Horwood Limited, Chichester, England, 1986.

[17] S. Luby, J. Dixon, and M. Simmons. Creating and using a features data base. *Computers in Mechanical Engineering*, 5(3):25–33, 1986.

[18] G. Stiny. An introduction to shape and shape grammars. *Environment and Planning, B*, 7:343–351, 1980.

[19] G. Stiny. Kindergarten grammars: designing with Froebel's building gifts. *Environment and Planning, B*, 7:409–462, 1980.

[20] G. Stiny. A note on the description of designs. *Environment and Planning, B*, 8:257–267, 1981.

[21] G. Stiny and L. March. Design machines. *Environment and Planning, B*, 8:245–255, 1981.

[22] M. Vaghul, J. Dixon, G. Zinsmeister, and M. Simmons. Expert systems in a CAD environment: injection modeling part design. In *Proceedings ASME Computers in Engineering Conference*, Boston, August 1985.

"Design Theory: A Model for Conceptual Design"

W.R. Spillers
Professor of Civil Engineering

Sandra Newsome
Assistant Professor of Psychology

Rensselaer Polytechnic Institute, Troy, NY 12180

Abstract

This paper addresses the problem of "conceptual design" in which the designer begins with some high level of ambiguity in terms of both goals and the means to achieve them. A "dictionary" approach is used in which an object is described by listing its properties, which are in turn described by listing their properties, ... This leads to a hierarchical structure of modules which can be described using the methods of partially ordered sets (combinatorial mathematics). The relationship of this approach to knowledge representation is discussed. Examples are included.

Introduction

As the papers in this volume attest, there are many opinions concerning what a "design theory" is to be about. There are equally many histories of how we got to where we are.

Our particular history goes back to 1974 when the National Science Foundation supported a symposium on Basic Questions of Design Theory at Columbia University (1). Roughly speaking, that symposium brought together engineers, architects, urban planners, mathematicians, and behavioral scientists in an attempt to explore the common ground which occurs in engineering and architectural design when these activities are approached on an abstract level. That common ground turned out to involve issues of discrete models, heuristics, "wicked" problems (an invention of architects to describe difficult, open-ended, design situations), linguistic methods, behavior (particularly as it relates to housing layout), formal methods of composition, and a remarkable discussion by Stiny and Gips (2) of of questions of aesthetics. Having identified these common interests within the design community, there was no attempt made at that time to push through anything that might be called a "theory of design".

The recent work in design theory is said to have originated in about 1984 with discussions within the National Science Foundation concerning the lack of "order" in the design process: Why is design not done in a more scientific manner? Indeed, what is the science or theory of design?

It should be stated at the outset that, at this moment, thee is no theory of design. But there are many discussions of what might comprise such a theory. In one such discussion (3) we have attempted to review the literature of systems theory (if that is to be the designation given to basic studies concerned with formal systems, computer organization, knowledge representation, automata, ...) as it might relate to design theory. Two simple points can be made about this literature. First of all it is robust. Second, it can be argued that rather than being the "odd fellow", design theory research is quite compatible with many areas of contemporary research within mathematics and computer sciences:

- Algebraic Methods. There is a strong segment within the computer science literature concerned with the use of formal algebraic methods to describe both computer architecture and software semantics.
- Lattice Theory. Birkhoff's work in the 1940's has had considerable application within the computer sciences. Most recently it is being used to discuss questions like the "expressiveness" of a language, the properties of entities obtained by combining known entities, valid statements about entities, ...
- Inductive Inference. System identification is an area of active resarch these days. Workers in this area might deal with questions such as "Given computer output, what is the program?" (Why not then "Given some prose, what is the grammar?", "Given some designs, what is the design theory?", ...).
- Artificial Intelligence. Methods of expert systems are now actively being pursued in areas such as software development.

In historical terms, a major worker within this systems literature is the British mathematician R.H. Atkin (4-6). Atkin has spent a good deal of time working with models of sophisticated events which span most of the arts and sciences. In terms of the above classifications, Atkin's work is algebraic: He begins by constructing a hierarchical decomposition of an entity and then studies the topology induced by this decomposition.

Having argued that design theory is wide-ranging, the dilemma is then how to approach it or even which aspect of it to approach. In the material which follows we have chosen (arbitrarily) to discuss what it is commonly called "conceptual design" in which the design goals and the methods to achieve them both contain high levels of ambiguity. It is, for example, at the conceptual design level that the designer can be creative. We argue, in fact that design contains a heuristic* component. It is these types of arguments which result in a strong tie between design theory and the behavioral sciences.

In the work which follows we begin by listing some properties which help to define conceptual design, at least as it will be dealt with in this paper. That is followed by a section which applies some of the mathematics of partially ordered sets to design problems. (A chain decomposition, for example, will be used to provide semantics for discussing designs.) Finally, the relationship between design theory and knowledge representation will be considered.

In terms of approach, we have tried to go back to basics. There is, for example, nothing more basic than listing properties of an object in an attempt to describe (or design) it. The mathematics of partially ordered sets then provides some structure with which meaning can be assigned to design representations. Finally, we propose an open-ended approach in which additional structure can be added as appears appropriate to enrich the theory.

Design, Design Theory, and Conceptual Design

First of all, a philosophical point about design. We are now in a dynamic environment in which more and more of what we used to call design is being automated. However, one could ask whether an activity that involves no more than pushing a button for a set of blueprints (as would occur for the completely automated project) is really design. Here we assume that the answer is no! In the same vein, if at some point I know exactly how some design is going to turn out (perhaps the details exist only in my mind), beyond that point I cease to do design. In the old days we might have talked about routine design versus creative design. In any case this type of argument leads to some general attributes of design:

- Design must have a heuristic* component. If button pushing is not to be design, certainly one way to exclude it is to require some heuristics of design.
- Optimization is not design. As a corollary of the above, once a design has been reduced to parametric studies of a given model, by virtue of the push of a button, design activity has ceased.
- Enumeration is not design. We would argue less strongly that once the design

*In this paper the term heuristic implies the intentional use of a vast amount of background information and experience. The term cognitive heuristics might be substituted for it.

process has been reduced to simply examining all possibilities, design activity has ceased. This argues against the use of grammars of some kind to generate designs, at least to the extent that these grammars do not respond to some sort of demand. Put another way,

- Design implies a demand.
- Design theory is a metatheory. It is assumed arbitrarily that the existing theories of structures, machine design, aesthetics, ..., stand alone and that "design theory" is a theory about these theories or a metatheory.

Heuristics and/or creativity by definition are difficult to analyze formally. By default then, arguing in favor of heuristics in design argues in favor of the role of the behavioral sciences in design since they offer the only available systematic approach to human behavior.

The term "conceptual design" is used in this paper to describe the design process at its most abstract. For the structural engineer this is probably the time at which he sits down with a new client to discuss a structure that he has never built before; for the inventor, this is the point at which he first begins to get an idea for a new project. Design at this level is characterized by a high level of ambiguity. The designer, at the initiation of conceptual design, does not know exactly what he is going to do; at this stage in the designer/client relationship, not only do they both not know how the design is going to evolve, they do not understand each other completely and will have to go through a process of conflict resolution.

It is common (3) to model situations of abstract designing using a hierarchical tree structure in which properties (modules) are sequentially assigned to (other) modules. An example of this is indicated in Figure 1. This figure attempts to describe, simply, a rather remarkable event of the late 1960's. At that time the structural engineer David Geiger designed the first large pavilion to be covered by an air-supported, fabric covered, cable net. (Geiger went on to build some 12 major structures of this type in the following 10-15 years and is now regarded to be the father of this type of structure.) As an aside, he first designed a conventional structure for this pavilion which could not be made to come in under budget. Out of desperation he then produced a truly creative design.

The logic of Figure 1 goes something like this:

1. The federal government initiates a design competition for the US Pavilion at the 1970 Worlds Fair in Osaka.
2. It is stipulated that there are floor space requirements, a fixed and restrictive budget, and considerable interest in the "image" projected by the structure.

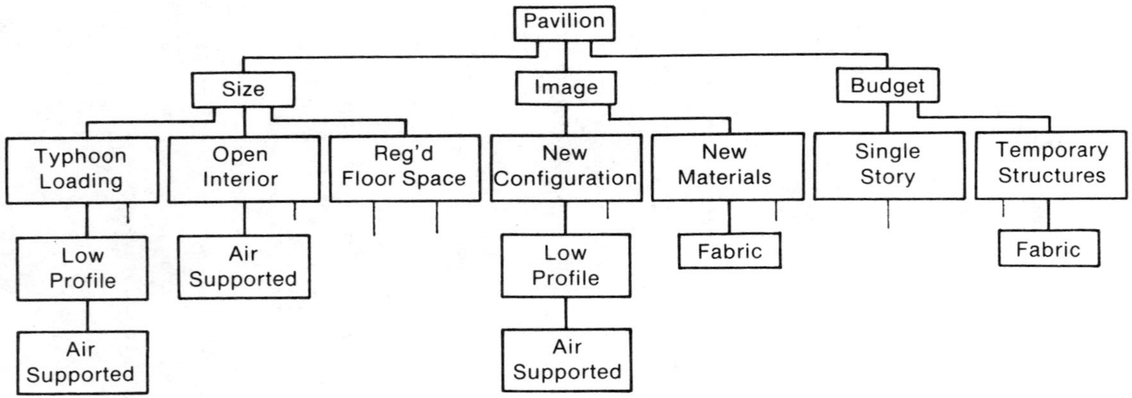

Figure 1
A Hierarchical Tree Structure

3. Size invokes considerations of environmental loading, floor space requirements, and the advantages of an unobstructed interior.
   Image invokes considerations of new materials and new configurations.
   Budget invokes considerations of the low costs of single story and temporary structures.
4. ...

Figure 1 introduces two ideas, the hierarchical tree structure and the "dicctionary" approach of describing an entity (module) by assigning properties (other modules) to it. At this basic level, the only mathematics involved is that of partially ordered sets. (The fact that A is a property of B is written as $A \prec B$.) The next section formalizes this type of representation of designs, adds some additional features, and describes how combinatorial mathematics can be used in the design process.

A Formal Model for Conceptual Design

The work of this section owes much to the remarkable book of Ford and Fulkerson (7), Flows in Networks. A more contemporary view of the combinatorial mathematics used here can be found in Fraisse's new book (8).

The basic element of the model proposed in this paper is the hierarchical tree structure which is used to define any concept or design. (See, for example, Figure 1.) Clearly, with this type of structure, as you move downward through the hierarchy the level of detail of the description increases.

This tree structure can be augmented in many ways. First of all it is natural to assign weights to each of the branches to indicate the importance of the element to follow in the description of the design. It is furthermore convenient to have the weights coming out of any component sum to one. Since there is a unique path from the top of the hierarchy (the root of the tree) to any component, the weight of any component within a design can be defined as the product of the weights associated with each branch in this path. It follows that if all component descriptions are continued (perhaps artificially) to all levels, the weights of all the components at any level must also sum to one. More generally, arbitrary annotations can be attached to nodes and branches of this tree. (This will be discussed in more detail later.)

Various operations over the hierarchical tree structure can be defined:

- Simple search procedures. It is a simple matter to calculate the "role" or "effect" of a specific component in a gien design by summing component weights which occur each time a component appears in the tree. Similarly, the "function" and "context" of a component are expressed by the branches leaving and entering it.

- Contracting the system. The role of a component within the system is obscured by its occurrence at more than one location in the tree. When these locations occur at the same level it is a simple matter to "contract" the system by removing the extra components and adding the appropriate branches to the system. (At this point the tree structure is lost.) When the same component occurs at different levels within the hierarchy, the structure of the system can be maintained by keeping the component at its lowest level and bringing the other components artificially to this level. In this manner, properties of elements always lie <u>below</u> them in the tree.
- Representations of the system. Partially ordered sets can be decomposed into "chains" (7). Let the components of the hierarchy be called C_1, C_2, ..., C_n and let $C_i \prec C_j$ describe the fact that C_i is a property or element of C_j. A chain is then a sequence of one or more elements

  $$C_{i_1} \succ C_{i_2} \succ C_{i_3} \cdots \succ C_{i_n}$$

  A decomposition of a design is a partition of it into independent chains (independent chains do not share any elements); a minimal decomposition has the smallest number of chains.

  Two elements C_i and C_j are said to be "unrelated" if neither $C_i \succ C_j$ nor $C_j \succ C_i$. Dilworth's theorem states that the number of chains in a minimal chain decomposition equals the maximum number of unrelated elements in the design.
- Induced structure. A more detailed analysis of the effect of a given component within a design can be achieved through studying structure which the component itself imposes upon the hierarchy. This can be done using the idea of a "serial decomposition" of a graph into subgraphs. In a serial decomposition of a graph, the first subgraph is arbitrary and can be taken to consist of all occurrences of the particular component which is under study; subsequent subgraphs are constructed by including all elements which are connected to the last subgraph constructed and which have not been included in any subgraph already constructed. (Note that it is probably not appropriate to carry this construction beyond the root of the tree.)

This is, perhaps, a good place to describe some applications of the analysis just presented. Figure 2 indicates a typical hierarchical description of a project. (This type of hierarchical description might be of use to an engineer or architect as he or she negotiates with a client.)

Figure 3 represents a different example that indicates how chain decompositions can be used. In this case an attempt is being made to discuss the human factors of

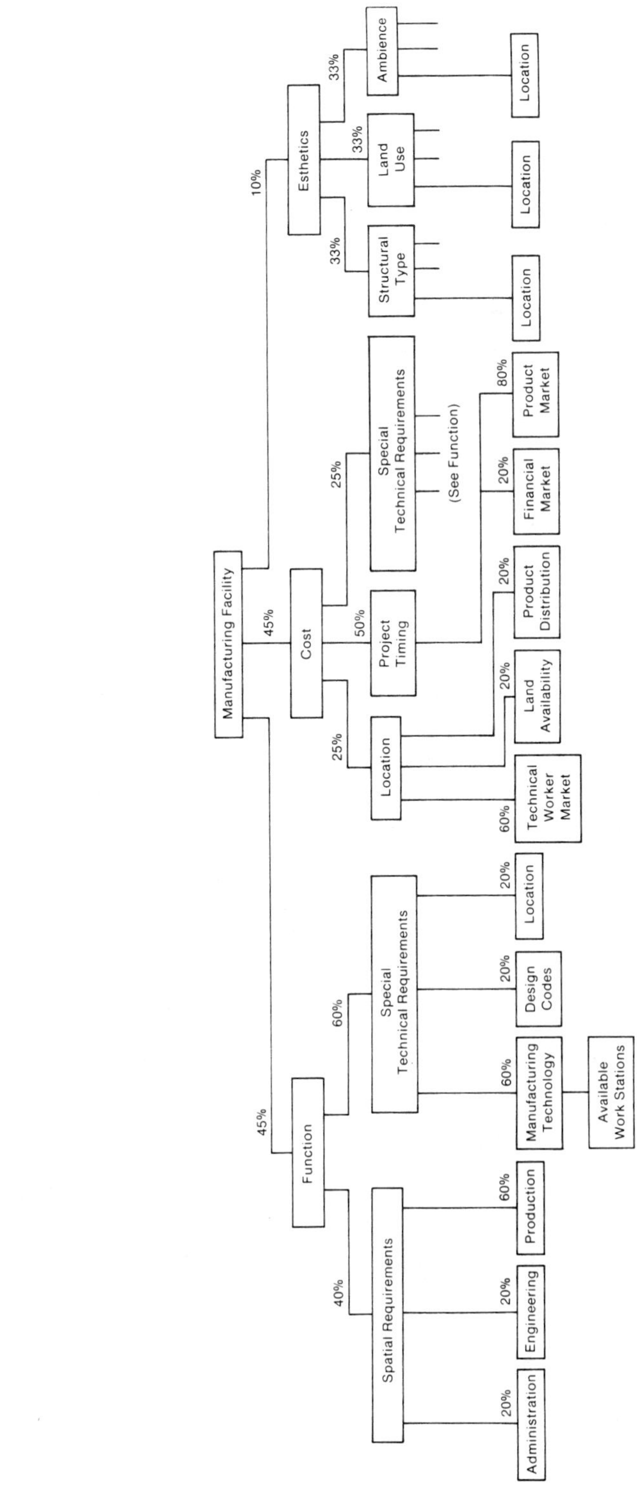

Figure 2
Hierarchical Description of a Typical Project

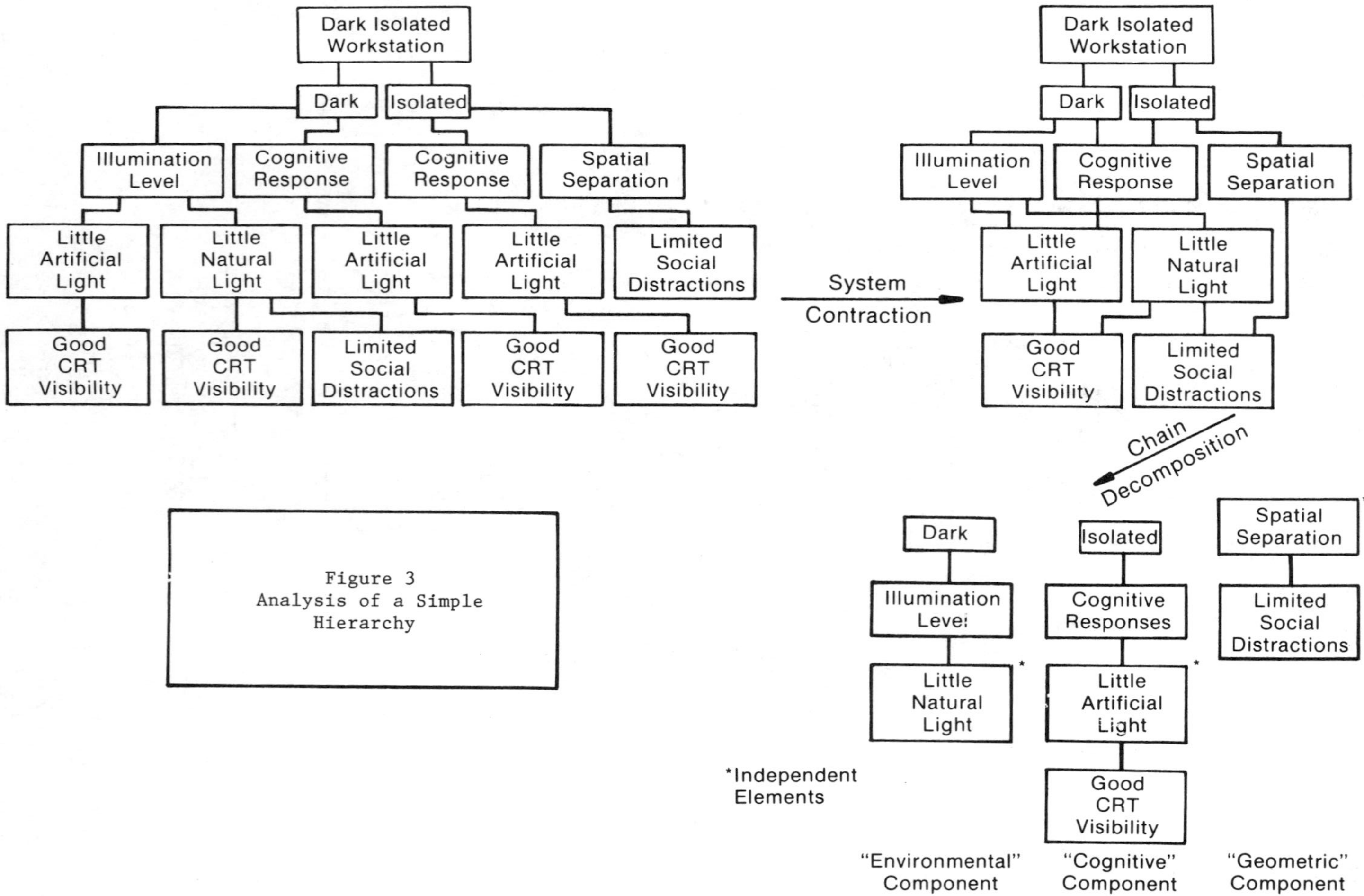

Figure 3
Analysis of a Simple Hierarchy

a dark, isolated workstation environment. (It might be supposed, for example, that while a dark, isolated workstation environment might lead to high productivity in the short run, it might also carry with it long term deficits such as increased sick leave and a high rate of employee turn-over). In any case, Figure 3 first of all describes a dark, isolated workstation:

1. Dark invokes concerns of both illumination and cognitive response. (It is most easy to read a CRT in a dark room.)
   Isolated again invokes concern over cognitive response and is of course concerned with spatial separation.
2. The illumination level has both natural and artificial components.
   Cognitive response is argued here to be primarily concerned with the level of light.
   Spatial separation in this case implies limited social distractions.
3. . . .

Figure 3 is next concerned with a system contraction in which repeated modules are combined. The figure finally indicates a minimal chain decomposition. In this case there are 3 chains in the decomposition and (via Dilworth's Theorem) 3 independent elements. The chains have been assigned names in this figure and provide a simple semantic device for discussing the original concept of a dark, isolated workstation. (To the extent that a chain decomposition is not obvious, the reader is referred to (7); for clarity, no weights have been associated with the branches of Figure 3.)

Figure 4 illustrates the idea of induced structure. In this case the dark, isolated workstation of Figure 3 is viewed from the point of view of cognitive response using the serial decomposition of a graph described above. In this case some of the ordering of the modules is lost but could be salvaged by assigning arrows to the branches.

Design and Knowledge Representation

It is a central theme of this paper that one of the issues designers share across disciplines is the manner in which ambiguity is dealt with. On the surface, a VLSI designer shares little in terms of technology with, for example, an architect. They are at the opposite ends of the scale in terms of size and even the common ground of spatial packing or network flow becomes strained in these cases. To what extent then is a theory of design to subsume both of these activities?

We have argued above that design, certainly conceptual design, should be defined as an activity which contains a high level of ambiguity. When that has been

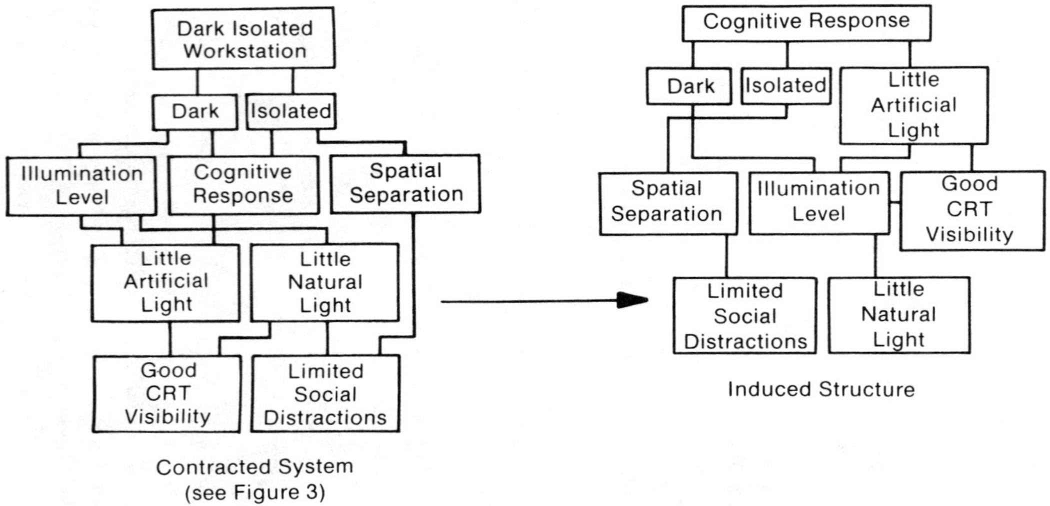

Figure 4
A Dark Isolated Workstation as Viewed from the Cognitive Response Point of View

done, design becomes an activity of dealing with ambiguity and the necessity of dealing with ambiguity is something all designers share. This argument takes design theory into the realm of knowledge representation. And the chain decomposition of partially ordered sets provide a "representation" or semantics for this discussion.

In this section some additional applications of this representation will be considered. They are intended to explore the use of chain decompositions in both the design process and other areas in which knowledge representation is a significant issue. It will be noted that a chain decomposition of a partially ordered set may not be a unique representation of it and may not even describe the partial order of the set. But in these cases the chain decompositions allow additional structure to be superimposed on the system under discussion.

First of all, consider another example of creative design by the structural engineer David Geiger. In a 1986 article in Engineering News Record (9) Geiger describes how he came up with a new type of structure for the forthcoming olympic games in Korea:

- Image. There were pressures from the project architect to develope "innovative concepts".
- Air supported, fabric covered cable nets. While air supported structures had served Geiger well, they had some nagging mechanical problems involving both maintaining internal pressure and melting snow. How then could the mechanical problems be eliminated while maintaining the other features of air supported, fabric-covered cable nets?
- Tensegrity. Geiger was familiar with the work of the inventor Buckminster Fuller.

Figure 5 attempts a hierarchical description of this discussion. (We have again omitted the weightings on the branches.) The advantages of the chain representation are clear here to the extent that the chains allows the original hierarchy to be discussed in terms of only 4 chains. These chains then play the role of subcomponents within the design and may be given names if that is of interest. It is equally clear that a) the chain decomposition is not typically unique and that b) it does not include all the information of the original partial order. (That is, it is not possible to go back to the original partially odered set given the decomposition.) These two properties actually provide additional degrees of freedom for the system which can be used to support additional structure just as redundancy in mathematical programming allows additional constraints. That will be the subject of future work.

Finally some comments on the relationship between knowledge representation and design theory. Conceptual design, as it is discussed in this paper, is a problem of

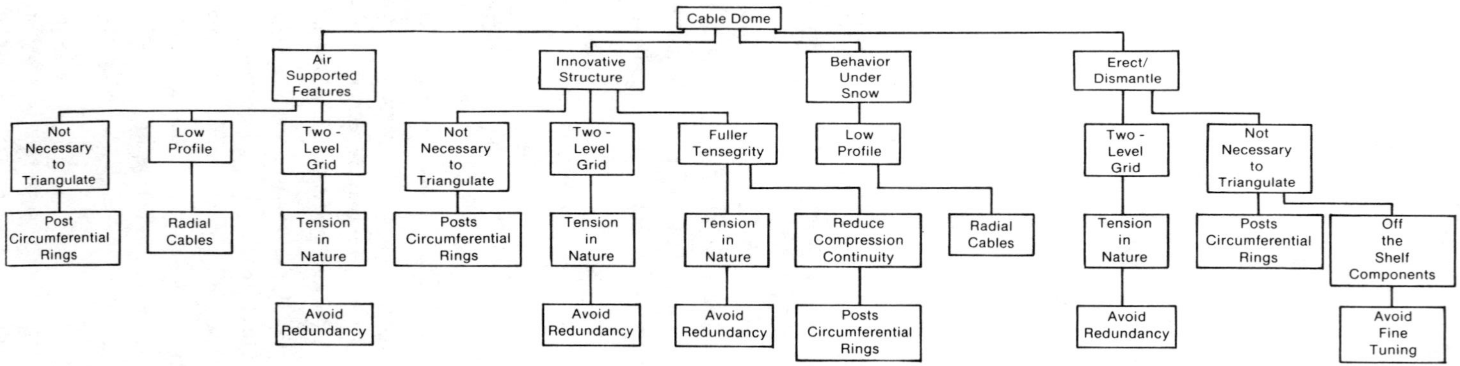

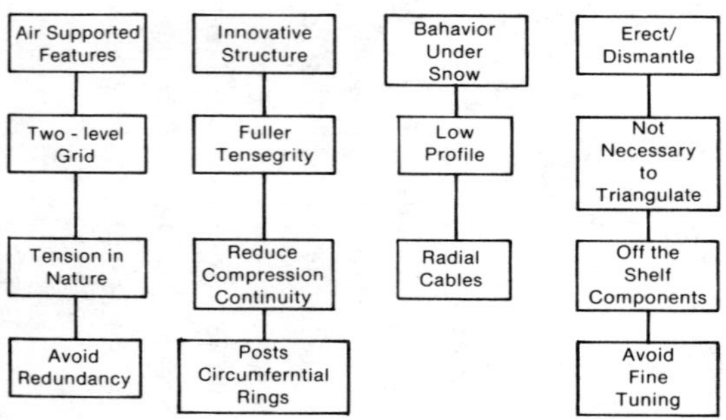

FIGURE 5
DESIGN OF CABLE DOME
Tree Hierarchy and Chain Decomposition

resolving ambiguity. That there exists an infrastructure which carries various types of technology to be used during this process of resolution only adds complexity but does not change the basic activity. Clearly talking about design is equivalent in these terms to talking about problem solving in any domain. This places design theory well within the more broad area of knowledge representation. Then techniques of design theory should have a more broad application to other areas of knowledge representation.

We end with an application which is not concerned with engineering design but with management. In Figure 6 which has been taken from the book of Margulies and Raia (10), the attempt is made to "diagnose" alienation in the workplace. What that means is that the authors are trying to describe (understand) the concept of worker alienation by listing components or factors in this very complex figure. Figure 7 shows this same concept described using the hierarchical approach of this paper. It certainly provides a simplified discussion of the concept of alienation (which may even be arged to be more meaningful). In fact the discussion of Figure 6, with its highly ambiguous tree roots and fruit and arrows, is reduced to four chains in this representation.

Concluding Remarks

We have explored the use of some ideas from partially ordered sets in problems of conceptual design while leaving many issues unsettled. We feel that this is appropriate given the fact that design theory is in its early stages of development.

In terms of our own research we would now like to see if an automated version of the work described here can actually help a designer with conceptual design and perhaps at the same time investigate the needs of creative designers. We would hope that the intersection of these activities turns out to be interesting.

Acknowledgement

This work has been supported by the National Science Foundation under its program in Design Theory and Methodology.

References

1. W.R. Spillers (ed), Basic Questions of Design Theory, N. Holland, Amsterdam, 1974. Proceedings of the 1974 NSF Symposium on Basic Questions of Design Theory. (These proceedings are summarized in W.R. Spillers, "Design Theory", IEEE Transactions on Systems, Man and Cybernetics, 7, 1977, pp. 201-204).

2. George Stiny and James Gips, Algorithmic Aesthetics, University of California Press, Berkeley, 1978.

3. W.R. Spillers and Sandra Newsome, "Some Attributes of Design Theory", submitted for publication.

4. R.H. Atkin, Mathematical Structure in Human Affairs, Crane, Russak and Co., NY, 1974.

5. R.H. Atkin, Combinatorial Connectivities in Social Systems, Birkhauser, Basel, 1977.

6. R.H. Atkin, Multidimensional Man, Penguin Books, Harmondworth, 1981.

7. L.R. Ford, Jr. and D.R. Fulkrson, Flows in Networks, RAND Corp. Report, 1962, Santa Monica.

8. R. Fraisse, Theory of Relations, Studies in Logic and the Foundations of Mathematics, Vol. 118, North-Holland, Amsterdam, 1986.

9. "Olympic Domes First of Their Kind," Engineering News Record cover story, March 6, 1986, pp. 24-27.

10. Newton Margulies and Anthony P. Raia, Conceptual Foundations of Organizational Development, McGraw-Hill Book Co., NY City, 1978.

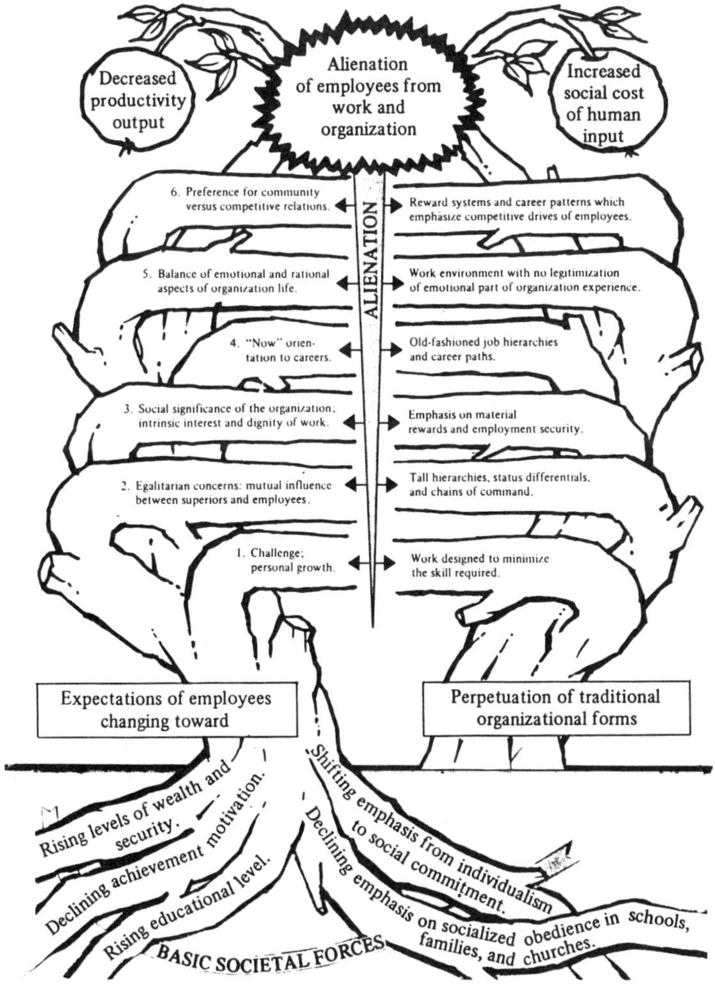

Figure 6
"Diagnosis of Alienation"
(Margulies and Raia (10))

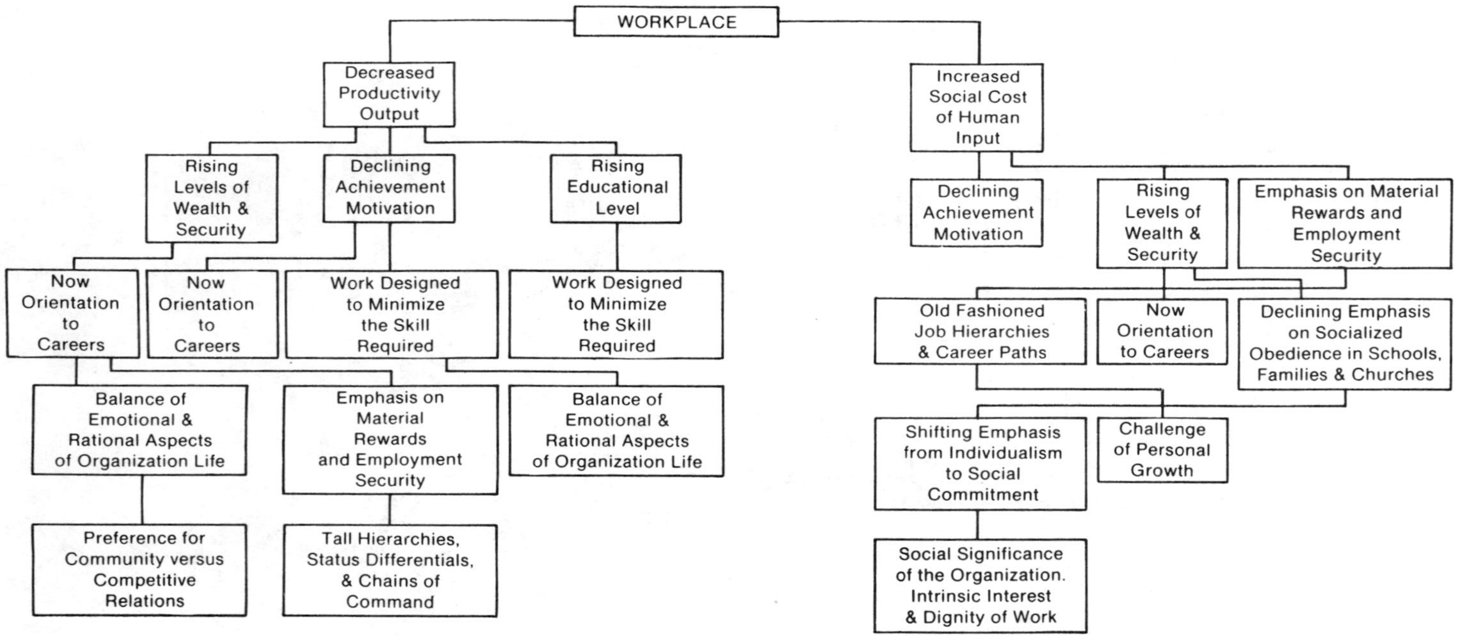

Figure 7
Hierarchical Representation of Figure 6

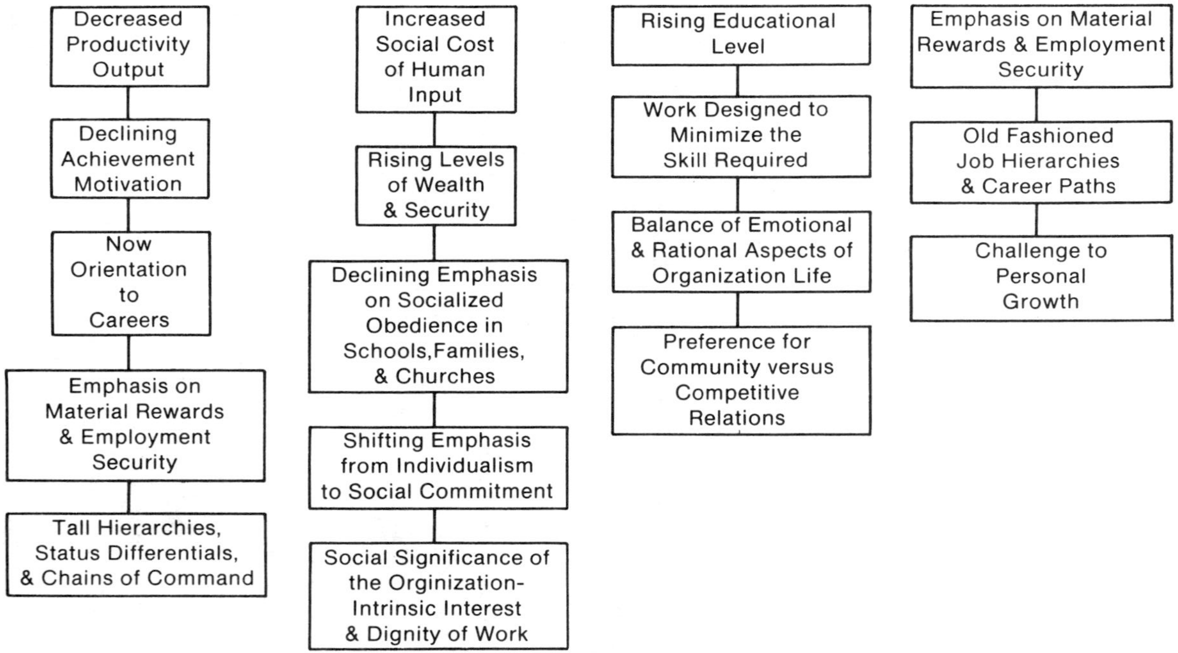

Figure 8
Chain Decomposition of Figure 7

Development of Design Methodologies

Jo Dale Carothers and Harvey G. Cragon

Dept. of Electrical and Computer Engineering
The University of Texas at Austin
Austin, Texas 78712

I. Introduction

In our study of the design process, the goal has been to develop methodologies that will form a foundation for computer design at the architectural and implementation levels. The architecture of a computer system is defined by Blaaw and Brooks to be "its functional appearance to its immediate user; that is, its conceptual structure and functional behavior as seen by one who programs in machine language." The implementation, on the other hand, is defined by them to be "the logical organization of its data flow and controls." [1]

It is known that computers can serve an important part in the design of new computers. Our aim has been to outline the role of computers and determine the characteristics that a high-level design system should possess. These studies have been conducted in our Symbolic Engineering Laboratory with the use of Texas Instruments Explorers. Our current findings verify earlier predictions [5] and further extend the work.

As stated in [5], it is seldom that a new concept is developed. In general, an innovative architecture results from the creative and skillful blending of known ideas and concepts. The present design approach often depends on the memory of a few individuals on the design team. Thus, the richness of the complete body of knowledge is usually not available as a new architecture is being designed. Our intent is to combine the full body of knowledge, in the form of computer design systems, with the experience of an expert designer(s) to produce more accurate, complete designs at an early stage in the design process.

II. Taxonomy and Pruning

We have found that one of the fundamental requirements in design is the ability to define the design space, including the minimum and maximum complexity boundaries. The development of taxonomies has proven to be an excellent method for defining this design space for a given problem. These taxonomies include:

1. Conventional Instruction Set Architectures
2. Object Oriented Instruction Set Architectures
3. Disk System Configurations
4. Data and Resource Tagging Methods
5. Branch Handling Strategies

After successfully defining the design space, it is necessary to explore that space by evaluating options, estimating performance, estimating cost, and finally producing a complete specification for the resulting design. This is a complex process that will benefit from the use of computer tools. The Artificial Intelligence community has provided us with tools such as LISP, PROLOG, and expert system development capabilities, to name a few.

Considering and evaluating the design options is a crucial part in this process. We are taking an approach called pruning. Pruning consists of presenting the known alternatives of a taxonomy to the designer, who prunes the alternatives down to an acceptable set. In many cases, presenting all alternatives will be impractical. In those cases a pre-pruning, or automatic pruning, mechanism is needed. This pre-pruning step would limit the number of options presented to the designer based upon user-defined design goals and requirements. These pre-pruning algorithms would automatically perform this pruning step during, or immediately after, the formulation of the design alternatives. The purpose of this pre-pruning is to ensure that the design options presented are reasonable given the context of the current design goals and that there are a reasonable number of options presented to allow the designer to give proper attention to each relevant option.

The taxonomies and menus are multi-level. That is, the top-level design decisions lead to lower level taxonomies. For example, if a top-level decision is to use a memory address, the lower level taxonomy will list the options of memory addressing such as direct, indirect, base plus displacement, etc.

The creator of the design system must determine when a leaf node has been reached in the taxonomy and provide, as will be discussed later, an analytic performance model, a cost model, and the documentation of the design choice.

It is necessary to find the proper balance between the system guiding the designer and the designer guiding the system. For example, if the system dictates the complete design sequence, flexibility is not available to allow for a wide range of design situations, and the expert designer's experience is neglected. On the other hand, if the designer "blindly" guides the system, then the potential of the design tool will not be fully utilized, and the goal of an accurate, complete design specification will probably not be met. The proper balance will be obtained by the system providing the design alternatives, performance and cost estimates for various alternatives, and suggested guidance on possible design paths to follow from that point. The designer will choose from the alternatives based upon his/her own knowledge and the statistics that are provided.

It is essential to provide feedback about the current state of the design to the expert throughout the process. We feel that this close cooperation between the designer and the design system will ultimately lead to more accurate and complete designs in a shorter time than is currently possible in a "first pass" design. Thus, cost and time are both reduced.

We have found that the determination of the design alternatives requires an expert system, an algorithmic approach, or a combination of both depending upon the goals. Expert systems have proven to be very useful when there is a fairly static taxonomy to describe the design space. However, in some cases, such as intraconnection minimization, an algorithmic approach is more appropriate since the taxonomy describing the design space is entirely dependent upon the given set of data. In other words, the type of reduction that can occur is not known until the data is provided. That is, the taxonomy is dynamically determined.

III. Performance Models

An essential element of a design system is the ability to provide a performance and cost analysis of a high-level design. This requires the development of cost and performance models. Determination of valid models is not a trivial task. In some cases, currently defined models can be used. However, in most cases, satisfactory models have not been developed. This is an area open for further research.

Evaluation of these models requires either analytical techniques or simulation. It is crucial to determine which of these methods are appropriate for a given design situation. At a high level there are several inherent problems with simulation. First, simulation is often a lengthy, time-consuming process in which it may be necessary to specify certain parameters. Determination of these parameters may not be an easy task. In addition, it is essential to be able to show that a simulation is valid. This is not always possible. One validation technique is to show that a simulation is valid for an existing computer by comparing it with real statistics which can be obtained. Then there would be reason to accept a simulation of a modified version of that same computer as valid. However, when designing new systems, similar computers do not always already exist. Therefore, there is nothing with which to compare, and validation is that much more difficult. In spite of these negative aspects, simulation still is an important analysis technique. However, caution must be exercised to ensure that the simulation results correctly model the design under consideration.

Analytic models have proven to be effective for modeling at the conceptual level. When developing such modeling techniques, it is necessary to determine the crucial portions of the design that must be modeled in order to understand its potential performance or cost. It is not alway obvious which aspects are most influential. In these cases, sensitivity analyses must be performed to determine which parameters are the most sensitive to change. Overall however, we have found one of the major difficulties in modeling to be the absence of satisfactory models. Therefore, the taxonomies we are developing are paired with the relevant analytic models.

IV. Verification

Provided that the appropriate tools and models are developed to aid the expert designer in the development of new designs, it is essential to provide verification techniques. As stated earlier, it is our goal to aid in formulating more accurate designs at an early stage. This goal relies on the ability to verify that a design meets the user-specified design requirements. A portion of the verification process can be accomplished using the performance and cost models discussed. However, to determine if power, delay, or other such specifications are satisfied, additional analysis and verification techniques are needed. The ability to provide accurate, comprehensive, verification tools is a fundamental part of our overall goal.

In no way is it our goal to replace the expert computer designer with an entirely automated design system. We feel that the expert should play an integral part in the design process. These designers possess important assets such as previous experience, extensive knowledge about current design goals, and information about implementation preferences (e.g. custom chips or off-the-shelf products), to name a few. At this point, Artificial Intelligence has not progressed far enough to allow a computer to reason as a human does. The cooperation of designer and design tools such as described should result in our goal of more accurate and complete, early designs.

V. Design Environment

We have found an interactive design environment to be the most productive. Allowing real-time interaction, as opposed to a design language approach, closely matches the desired design environment that we have been describing. We have also found that the use of menus is a good interface for presenting the taxonomy to the designer. At any time during the design process, the designer should be allowed to backtrack to an earlier design stage and explore other options. This should always be provided as a choice in each menu for maximal flexibility.

VI. Documentation

A significant source of error in the design process today is the translation of design ideas to the specification from which more detailed design proceeds. For digital systems, documentation generally consists of block diagrams, English prose, and, in some cases, flow charts. This informal documentation is prone to error and misinterpretation.

Our research has led us to the conclusion that documentation should be formal and automatic. That is, as the design is formulated from the pruning of taxonomies presented as menus, a LISP description is constructed by the system. Upon completion of the design, we have an executable specification in LISP that can be used to verify the design by the execution of a program on the specification.

VII. Additional Considerations

During our research we have learned much about the design process. In addition to the characteristics that have been outlined for a high-level design system, two additional characteristics are important. The flexibility to add "knowledge" to the design system is needed. We have pointed out that few truly new concepts are developed. However, when such an idea is formulated, a design system should be able to make that concept part of its repertoire, along with any new performance models that may be required. Also, for a design system to be fundamentally sound, it should be developed relatively independent of current technology. We are interested in determining the foundational theory underlying computer design. Ideally, it is desired that this theory is somewhat independent of technology when dealing at the architectural level. Of course, at the implementation level, as defined by Brooks [1], this probably would not be the case.

In the pursuit of the defined goals, several prototype design systems have been developed or are currently under development. These include, but are not limited to, instruction set design tools [6,7,9], an I/O system design tool [4], and a memory system design helper [8], as well as design space analysis techniques such as intraconnection minimization algorithms [2,3].

VIII. Conclusion

Our research has led us to the conclusion that a design system has the following components:

1. A design taxonomy
2. Analytic cost and performance models for each of the elements of the taxonomy
3. Automatic formal documentation of each of the elements of the taxonomy
4. A menu-based design environment that is host for the taxonomy

The research described in this paper has resulted in a number of M.S. theses and doctoral research topics. The following is a list of the work being done by this group:

Masters Theses Completed:

1. Natraj Arni, "Design of an Efficient Dynamic Memory Allocator," August 1988.
2. Jo Dale Carothers, "Adjacency-Matrix Reduction Algorithms," December 1986.
3. Jennifer L. Gill, "Processor Performance as a Function of Memory Bandwidth," May 1987.
4. Dehui Guo, "Binary Logarithm, Exponentiation Generators and Their Applicatons," May 1987.
5. Nadeem Malik, "Expert Aid for Execution Unit Design," December 1985.
6. Kaveh Massoudian, "Bit Traffic Analysis of Instruction Sets from Compiled Programs," May 1988.
7. Robert A. Mayer, "A New Algorithm for Evaluating Integrated Circuit Interconnections," December 1986.
8. Charles Edward Nuckolls, "An Alternative Approach to the Computer Design Process at the Instruction Set Architecture Level," May 1986.
9. Yongjae Rim, "Memory System Design Helper: Implementation and Analysis," December 1985.
10. Glenn D. Rosenberger, "Logic CALC: A Design Tool for Digital Systems," August 1987.
11. Robert Stepanian, "Performance Evaluation of Multiple and Overlapped Register Set Architectures," December 1987.
12. Michael W. Strevell, "A Taxonomy of Analytical Computer Performance Models for Computer Design," December 1986.
13. Ishaq H. Unwala, "A Novel Environment for Design and Simulation of Digital Systems Architecture," May 1986.

14. Krishnamurthy Venkatramani, "A Design Methodology for Object-Based Instruction Sets," December 1986.

15. Paul W. Vaughan,"A Visual Design Environment for Rapid Prototyping of Computer Systems," December 1987.

16. Zu-Yi Wang, "A Graphics Processor Design and Simulation in a Spreadsheet Environment," May 1987.

Masters Report Completed:

1. Waqas S. Khan, "Analytical Study of Garbage Collection Techniques for Linked Data Structures," December 1987.

Ph.D. in Progress:

1. Jo Dale Carothers (passed Q.E. September1987), "Computer Intraconnection Design System."

2. Kyungsan Cho (passed Q.E. September 1986), "Knowledge-Based Disk I/O Subsystem Design Aid and Performance Estimator."

3. Dehui Guo, "Verification of Computer Architecture."

4. William C. Hobart (passed Q.E. March 1988), "Locality of Reference of Memory Accesses During Symbolic Program Execution."

5. Nadeem Malik (passed Q.E. August 1987), "Performance Modeling in Use of Tagging Concepts in Computer Design."

6. Yongjae Rim (passed Q.E. November 1986), "Multi-Cache Coherence Problem Solving by Using Rule Based System."

7. Robert B. Rowen (passed Q.E. December 1986), "Manufacturing Software Requirements Analysis Using a Knowledge Guided Tool."

8. Michael W. Strevell (passed Q.E. November 1987), "Data Type Transformation and Coherence in Heterogeneous Parallel Computers."

9. Krishnamurthy Venkatramani (passed Q.E. March 1988), "A Rationale for Instruction Set Design."

10. Zu-Yi Wang, "Performance Modelling of Central Processor Unit."

References

[1] Blaauw, Gerrit A. and Frederick P. Brooks, Jr., *Computer Architecture*, Tecnische Hogeschool Twente, Enschede, The Netherlands, preprint.

[2] Carothers, Jo Dale, *Adjacency-Matrix Reduction Algorithms*, M.S. Thesis, Dept. of Electrical and Computer Engineering, Austin, Texas, University of Texas at Austin, 1986.

[3] Carothers, Jo Dale and Harvey G. Cragon, "Design and Minimization of Computer Intraconnections," *Proceedings of Midcon*, 1988.

[4] Cho, Kyungsan and Harvey G. Cragon, "Knowledge-Based Disk I/O System Design Aid," *Proceedings of CompEuro '88 System Design: Concept, Methods, and Tools*, Brussels, Belgium, April 11-14, 1988, pp. 127-133.

[5] Cragon, Harvey, "An Architecture Design System," *Computer Architecture News*, Vol. 13, No. 4, September 1985, pp. 15-21.

[6] Massoudian, Kaveh, *Bit Traffic Analysis of Instruction Sets from Compiled Programs*, M.S. Thesis, Dept. of Electrical and Computer Engineering, Austin, Texas, University of Texas at Austin, 1988.

[7] Nuckolls, Charles Edward, *An Alternative Approach to the Computer Design Process at the Instruction Set Architecture Level*, M.S. Thesis, Dept. of Electrical and Computer Engineering, Austin, Texas, University of Texas at Austin, 1986.

[8] Rim, Yongjae, *Memory System Design Helper: Implementation and Analysis*, M.S. Thesis, Dept. of Electrical and Computer Engineering, Austin, Texas, University of Texas at Austin, 1985.

[9] Venkatramani, Krishnamurthy, *A Design Methodology for Object-Based Instruction Sets*, M.S. Thesis, Dept. of Electrical and Computer Engineering, Austin, Texas, University of Texas at Austin, 1986.

OBJECT-ORIENTED DATA STRUCTURES FOR DESIGNING BY FEATURES: INTEGRATING FUNCTIONALITY AND GEOMETRY

Alice M. Agogino, P.E., Principal Investigator, Associate Professor,
Department of Mechanical Engineering, University of California at Berkeley

Ramanathan Guha, Member of Technical Staff, MCC Corporation, Austin, Texas

Abstract

It is our thesis that feature sets in design and manufacturing need to capture more than physical geometry in order to be of use in mechanical systems design and manufacturing. Object-oriented data structures that provide both features in a representation scheme and a calculus for reasoning about functionality, manufacturability and diagnosability in mechanical systems design are proposed. The computer-based methodology integrates multiple data structures with the thesis that the suitability of any particular knowledge representation scheme for a system depends on both the problem domain and the operations to be performed. Though at a first glance knowledge about mechanical systems seems to include a diverse range of static and dynamic phenomena, we propose that it can be represented by two major data structures: (1) physical objects and (2) functional interactions between these physical objects (processes). Physical objects are constructed of primitive geometric features that are connected to each other by means of connection objects. Connection objects can be viewed as virtual bodies that are attached to two or more physical bodies and contain information describing the connection, such as degrees of freedom of motion. Functional interactions between bodies or physical objects is represented by *process objects*. A process object is a unit that represents some dynamic interaction between bodies (or just some transformation process acting on a body). The parallel object-oriented data structures of geometry objects and functional objects, along with multiple inheritance subject to constraint satisfaction between them, provide the mechanism for the integration of geometry and function in computer-aided mechanical systems design. The OMDesign (Object-oriented Mechanical Design) prototype has been developed to test out features used in reasoning about functional limitations in preliminary designs and serves as a pre-processor for developing a formal model of mechanical design problems.

Acknowledgements: This research was partially sponsored by NSF grant #DMC-8451622 and IBM grant #705012. The associated paper is based on an invited presentation at the 1987 ASME Design Automation Conference, Boston, MA, August 29, 1987 and can be obtained by writing: Alice M. Agogino, Berkeley Expert Systems Technology Laboratory, 5136 Etcheverry Hall, University of California, Berkeley, CA 94720 (Working Paper 88-0303-1)

PART 5

ANALYSIS TO SUPPORT DESIGN

DESIGN FOR THE OPERABILITY AND CONTROLLABILITY OF CHEMICAL PROCESSES

Warren D. Seider
Department of Chemical Engineering
University of Pennsylvania
Philadelphia, PA 19104

ABSTRACT

Quantitative and systematic methods of design are evolving to improve upon the operability and controllability of chemical processes. Progress is described in the development of new analytical and algorithmic methods for the design of complex processes that are difficult to operate and control. The methods are being demonstrated for the design of four processes and a prototype computing system is being created as the designs proceed.

INTRODUCTION

Our research is concerned with the design of processes that are difficult to operate and control due to complex physical and chemical interactions. These complex effects are often the reason a new design has the potential of being more effective; for example, more energy efficient and more profitable. To achieve robust designs, more advanced methodologies are needed for the design of these processes and their control systems.

Process design usually begins with the analysis of simple structures using approximate models. Gradually, as the synthesis tree is pruned, more complex models can be justified to represent the real phenomena more accurately. However, the more complex models usually have a richer solution space and the number, type (steady, periodic, or chaotic) and stability of the solutions varies with the specifications and the parameters of the model. In design calculations, as in all mathematical modelling, the engineer must learn to beware of algorithms that converge to solutions that are not physically correct. Often, in the design stage, experimental data are not available and the design may continue based upon the wrong solution. For this reason, it is important to develop methods that can easily determine the number and kinds of solutions that exist. Then, if multiple solutions are expected, care can be exercised to find the correct one, or to avoid designing with specifications that lead to multiple solutions.

In many cases, the solutions correspond to real physical solutions. Here, it is important to recognize the existence of alternate attractors and to design control systems that can control around the desired steady-state. As parameters of the processes change, possibly due to disturbances, highly nonlinear processes can sharply alter their operating regimes. In cases where such swings are undesirable, the designer should be alerted to their existence and, consequently, given the opportunity to reject the designs as too difficult to operate. Alternatively, he/she may chose to design a robust control system that can sense disturbances and make adjustments to prevent the undesirable swings from occurring. In the event of failure, when a disturbance is so great that the process shifts to an undesired attractor, the control system should be designed to sound the appropriate alarms.

A prototype computing system is being developed as we experiment with analytical and algorithmic methods to more quantitatively and systematically rationalize the design process. The methods are expected to enable experimentation in all phases of engineering, but are being tested for complex chemical and biochemical processes.

Chemical Processes

In chemical processing, complex physical and chemical interactions arise in many designs. Three examples include separation trains with azeotropic distillation towers, natural gas separation plants with a high degree of heat and power integration, and catalytic reaction systems that operate periodically. Steady-state multiplicity and controllability are real problems, for example, in a gas expander plant where disturbances in the critical region can cause a process to shift from two to three or just one phase. In this paper, recent experiences in the design and control of these processes are presented, with emphasis on the problems that arise in the use of the more complex models necessary to adequately describe these processes.

Conventional computer systems for process design permit analysis of processes in the steady-state using an assortment of preprogrammed models. Most design engineers begin with the simplest models and add complexity as a particular configuration becomes more promising. More modern tools are being developed that assemble large sets of algebraic equations to be solved simultaneously (Perkins and Sargent, 1982; Perkins, 1983; Locke and Westerberg, 1983; Hutchison, et al., 1986a, b). These promise to be more efficient, although thus far they have not been applied to solve very complex systems.

In connection with our research on azeotropic distillation towers, we have developed an equation-oriented approach which utilizes a differential, arc-length, homotopy-continuation method with sparse matrix handling (Kovach and Seider, 1987b). This algorithm has successfully found solutions that other algorithms, such as the Newton-Raphson method, have not been able to locate. Furthermore, by tracing the continuation path in parameter space, we are able to locate multiple solutions when they exist.

Our research has also involved the development of nonlinear programming algorithms for optimizing the performance of azeotropic distillation towers (Prokopakis and Seider, 1983) and for the retrofitting of chemical plants to achieve a higher degree of heat and power integration (Colmenares and Seider, 1987). In the latter designs, the alternatives include heat engines and heat pumps, cascaded when desirable, with different working fluids. We have synthesized more efficient and less costly configurations in natural gas separation plants and similar energy-intensive processes. These near-optimal configurations, from an energy and cost perspective, are more integrated, difficult to start-up and control, and present more possibilities for an accident.

For these processes, it is important for the designer to consider their dynamics and controllability earlier in the examination of the design alternatives. Methods are being developed to project their controllability using the simplest models (e.g., steady-state gain analysis), which can be embellished as simulations show a satisfactory response to the projected disturbances.

With the increasing emphasis on operability, controllability, and safety, new computing systems for process design often permit dynamic simulation of chemical processes and their associated control systems. This usually involves structuring an equation-oriented simulator to integrate the differential equations, which are often stiff, with associated nonlinear algebraic equations.

This paper presents an overview of the methodologies being explored in our research work for the design of four complex chemical processes: (1) azeotropic distillation towers, (2) supercritical extraction processes, (3) utility systems to supply heating, refrigeration, and power demands, and (4) bioreactors. As we learn more about these and other processes, new design methodologies are evolving and the prototype computing system is taking shape.

HETEROGENEOUS AZEOTROPIC DISTILLATION

Heterogeneous azeotropic distillation towers are widely-used to separate close-boiling compounds by introducing a mass-separating agent that forms an azeotrope with one or both of the compounds and thereby exaggerates the differences in volatility between them. These towers are extremely sensitive to the amount of the mass separating agent. They exhibit concentration and temperature fronts that move through the stripping section in response to small changes in the circulation rate (reflux ratio).

Although mathematical models have been in existence since the late 1940s, the models for some azeotropic towers have been inadequate for design purposes until recently. For example, recently Kovach and Seider (l987a, b) have simulated an ARCO tower for the dehydration of sec-butanol with di-secbutylether and shown that the concentration and temperature fronts in the stripping section occur at the interface between trays having one and two liquid phases. This simulation reproduces the sensitive response of these fronts to the aqueous reflux ratio, which is adjusted frequently by the operator in response to disturbances.

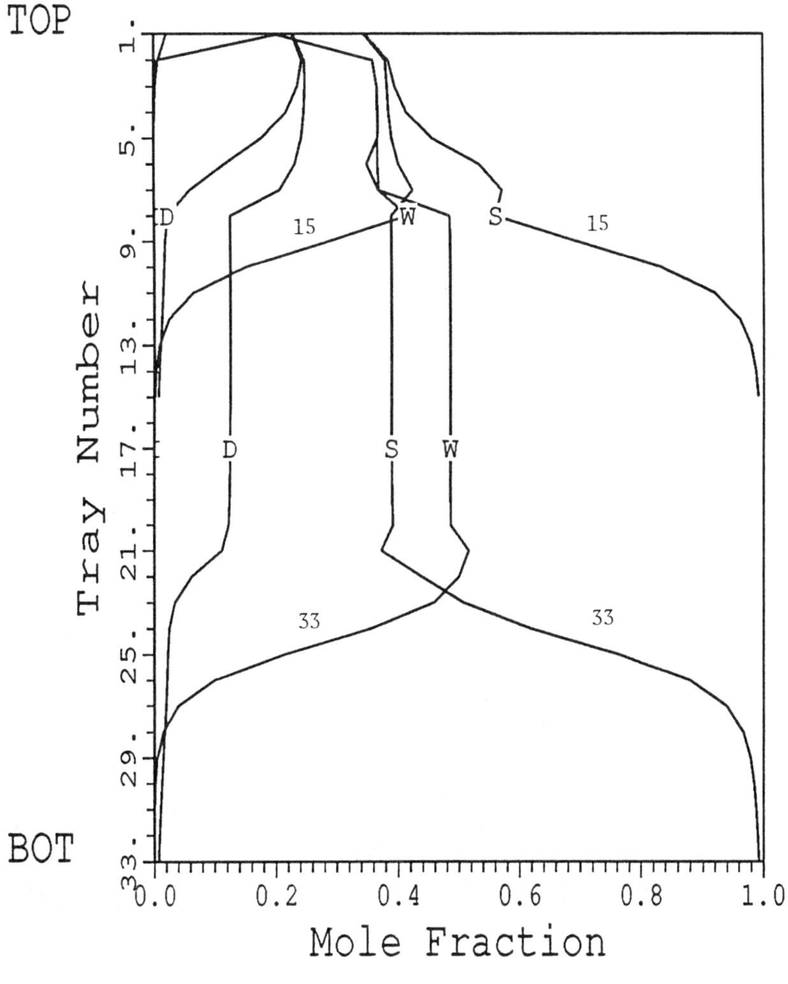

Figure 1 Azeotropic distillation tower to dehydrate sec-butanol. Liquid mole fractions on the trays in the steady-state for towers with 33 and 15 equilibrium stages; sec-butanol (S), water (W), di-secbutylether (D).

More significantly, recent design calculations have demonstrated that the tower can produce nearly the same high purity with less than half of its trays. Figure 1 shows the steady-state liquid mole fractions (average when two liquid phases occur on a tray) for the ARCO Tower (41 trays simulated assuming 80 percent efficiency, equivalent to 33 trays at equilibrium) and a tower with just 15 trays. In the smaller tower, the region of nearly constant liquid composition is virtually eliminated, leaving much less time for a manual control system to respond to disturbances that shift the concentration and temperature fronts up and down the stripping section.

To halve the tower, sharply reducing the cost of new designs, a robust control system seems necessary to respond to disturbances before off-spec product results. The designer must first develop a dynamic model of this nonlinear system and then experiment with alternative control strategies. Such a dynamic model is being developed, with a model predictive control algorithm that will be tuned to control a tray temperature in the stripping section. If successful, the tall tower will be replaced by a tower less than half its height and a microcomputer to implement the dynamic model and solve the NLP that determines the future control actions. In addition to sharply reduced capital costs, the operator will have more time to devote to other duties.

Similar designs should be possible for distillation towers with reactions carried-out on the trays; i.e., reactive-distillation towers. Often these towers involve mixtures with azeotropes and require detailed models to obtain efficient designs and implement robust control systems. One such tower has been simulated with simple PID controllers (Roat, et al., 1986) and is a candidate for tighter designs with model predictive control.

SUPERCRITICAL EXTRACTION PROCESSES

The potential for the extraction of foods, pharmaceuticals, insecticides, etc., using nontoxic supercritical solvents, like carbon dioxide, has intrigued chemical engineers for many years. It is the large variation of solute solubility in the solvent critical region that is so attractive. In principle, it should be possible to compress a solvent just beyond its critical pressure, extract the desired solute, reduce the pressure slightly (or make small adjustments in temperature) to release the solute, and recompress the solvent for recycle to the extractor.

However, high pressure processes are costly to build and must be highly-integrated to reduce the cost of heating, refrigeration, and power. These processes are characterized by narrow feasible regions in which two phases coexist and more confined regions in which the annual cost is sufficiently low to compete with alternative separation processes. Hence, the design problem is tightly constrained and the control strategies must be extremely robust to maintain optimal operating conditions.

One potential flowsheet for the dehydration of acetone with carbon dioxide is illustrated in Figure 2. Typical operating conditions are annotated, although these may not be optimal. Note that the elevated

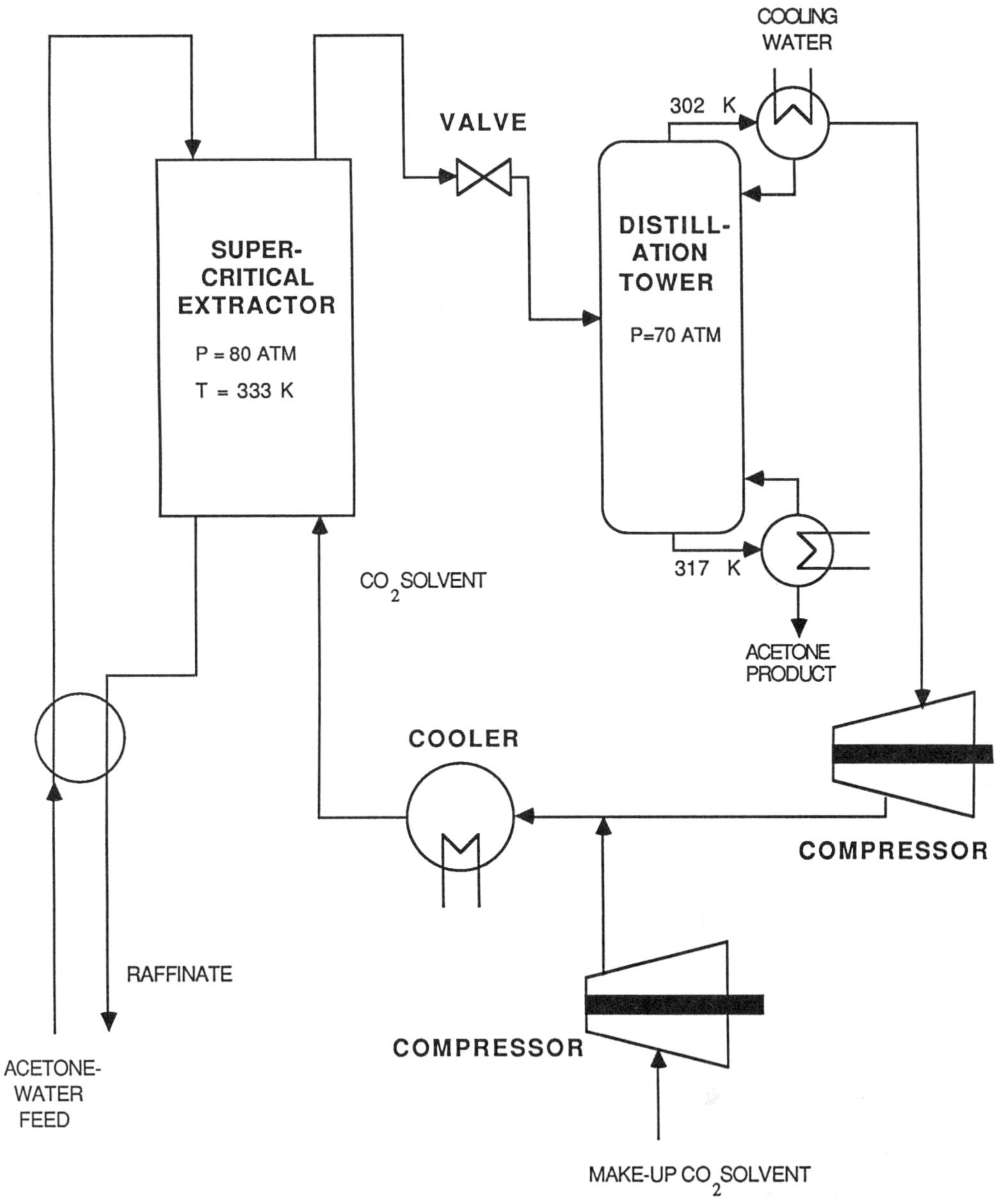

Figure 2 Dehydration of acetone with supercritical carbon dioxide. Supercritical extraction followed by distillation.

pressure in the distillation tower provides a small compression ratio and permits the use of cooling water in the condenser. At these elevated pressures, the data of Panagiotopoulos and Reid (1987) show that one, two or three phases exist. The two phase regions in the extractor and distillation tower are tightly constrained and, hence, the optimization of the temperatures, pressures, and carbon dioxide recirculation rate must be carried-out with robust algorithms.

At this point, the optimization strategy is nearly complete and dynamic models are being prepared to study how the potential designs will respond to typical disturbances. Simple PID controllers will be designed for operation about steady-state attractors and, very likely, will need to be replaced with more effective model predictive controllers.

These strategies are being perfected for the dehydration of acetone with carbon dioxide because extensive experimental data have been published and the Peng-Robinson equation has been modified to fit this data. We are optimistic about demonstrating the utility of this methodology for the design of processes to extract valuable foods and pharmaceuticals from their solutions and fermentation broths.

UTILITY SYSTEMS INTEGRATED WITH CHEMICAL PROCESSES

In the design of chemical processes, the performance of the utility system directly influences the operating cost and, hence, the optimal design of the utility system often leads to significant savings. For this reason, optimization strategies are being developed to design utility systems that "best" satisfy the process heating and power demands.

In one approach (Colmenares and Seider, 1988), the optimization algorithm considers the integration of the process heating and power demands with a cascade of steam engines, shown schematically in Figure 3. Note the potential for mass exchange between each of the steam engines. Energy from the condensers of the high pressure steam engines can be rejected to the process or to provide heat for a lower pressure steam engine. Power generated by the turbines can satisfy the power demands of the process. An NLP, with equality and inequality constraints, is solved to minimize an objective function, typically the annualized cost of the utility system. One optimal design, shown in Figure 4, involves three power cycles with heat and mass integration. However, several other designs are near-optimal and the choice should be based upon their operability and controllability.

In one approach, for each design, decentralization strategies are being considered to match manipulated and control variables based upon steady-state gains and the degree of interaction indicated by the relative gains. This approach has been applied with some success for the design of networks of heat exchangers (McAvoy, 1983) and is being extended to include turbines and pumps. The possibilities for calculating dynamic gains and examining the dynamic relative gains over a broad range of frequencies are being examined.

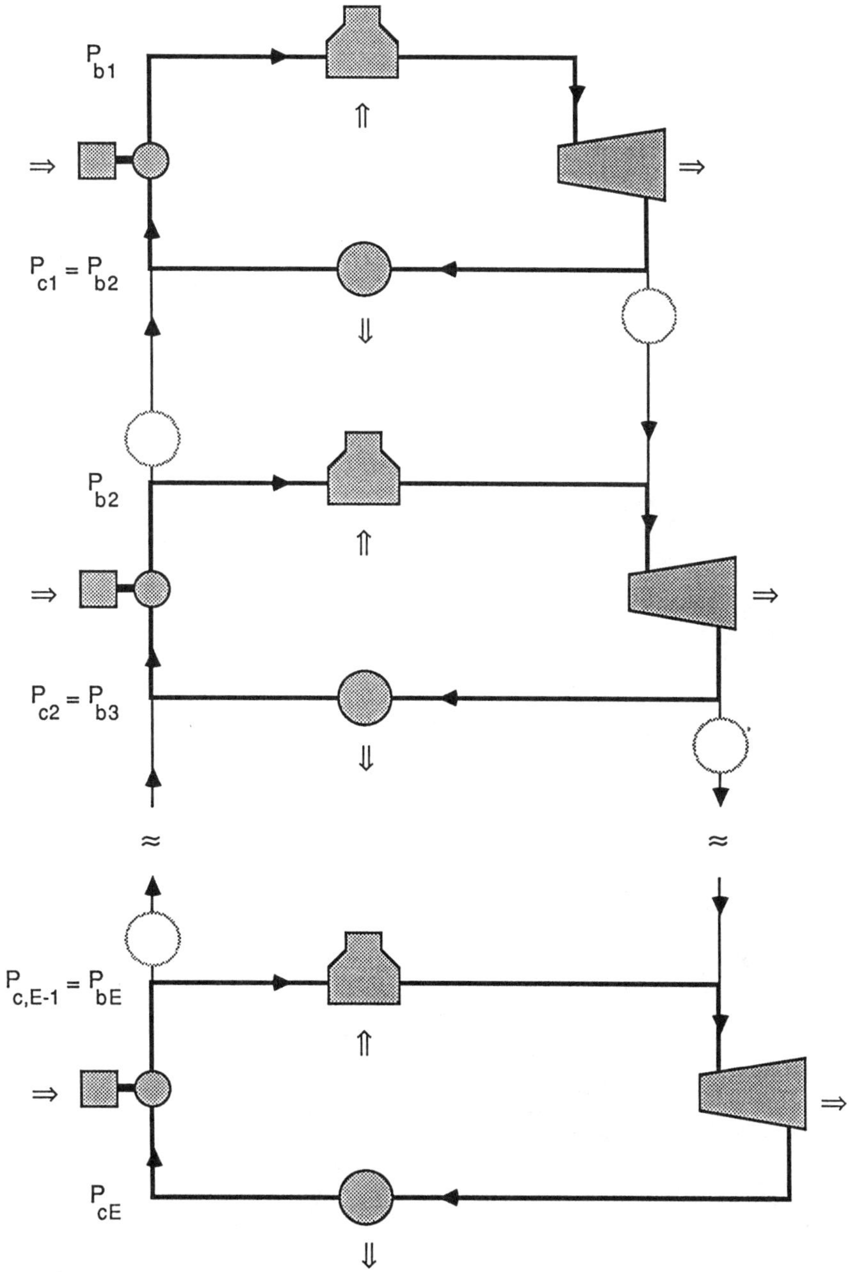

Figure 3 Cascade of Rankine cycles to potentially satisfy the heating and power demands of a chemical process.

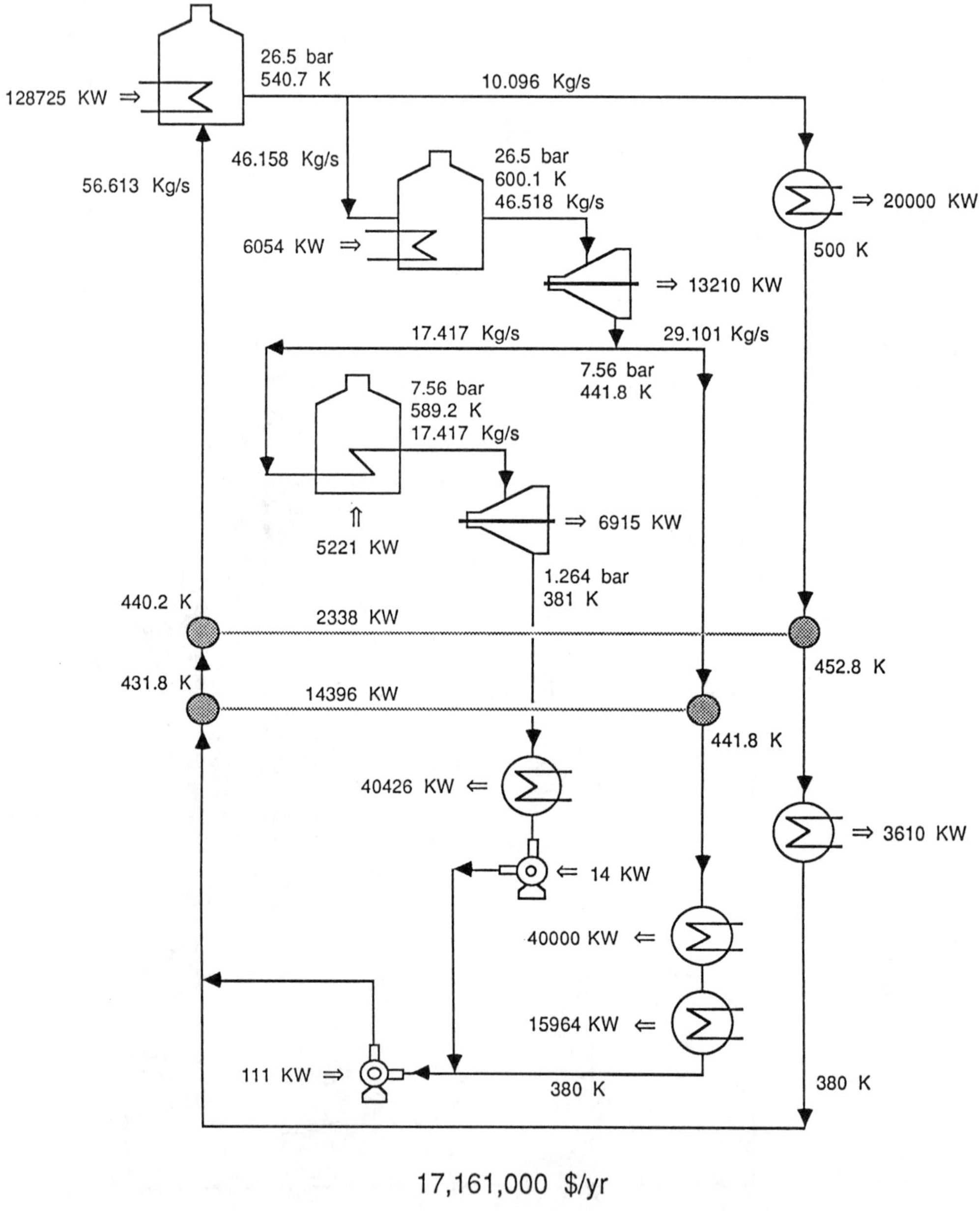

Figure 4 An optimal utility system that satisfies the heating and power demands of a chemical process.

Eventually, a potential design will be simulated dynamically and model predictive control strategies examined to permit robust control in the face of unmodelled disturbances.

BIOREACTORS

Bioreactors are difficult to design due to the complex mechanisms of cell growth and substrate consumption. In one model, introduced by Agrawal, et al. (1982), the "one hump function" accounts for the substrate inhibition of the growth rate and the yield coefficient is related linearly to the substrate concentration. The operation of its CSTR is typical of many fermentation systems in that its solution diagrams, shown in Figure 5, contain regions with three steady-state solutions as well as a Hopf bifurcation point and a branch of periodic solutions. At Da = 0.58, only the "washout" solution is stable. However, this corresponds to zero cell concentration and zero substrate consumption. Hence, at Da = 0.58 it is necessary to operate at one of the two other steady-states. But, both are unstable.

Chang and Chen (1984) have shown that, with conventional PI control, it is possible to stabilize the steady-states. However, the response to disturbances is uncertain, especially at low Da, near the Hopf bifurcation point, where the system is operated to achieve a high cell concentration. Furthermore, PI controllers introduce input multiplicity and regions of periodic oscillation over a broad range of controller gains.

Here, the designer must understand the nature of the attractors with and without control. And, furthermore, to achieve a satisfactory response to disturbances and set point changes, he/she must be prepared to consider the feasibility of more advanced control strategies, such as model predictive control. In recent work, we have been experimenting with new algorithms for the control of this highly nonlinear bioreactor.

PROTOTYPE COMPUTING SYSTEM

Each of the four processes is being analyzed using similar computing systems. These include:

1. DHTPGJ. This is our program for steady-state continuation, developed in connection with our studies on azeotropic distillation (Kovach and Seider, 1987b).

2. AUTO. This program, developed by Doedel (1986), performs steady-state continuation, locates real bifurcation points, identifies Hopf bifurcation points, and tracks the periodic branches that originate from them.

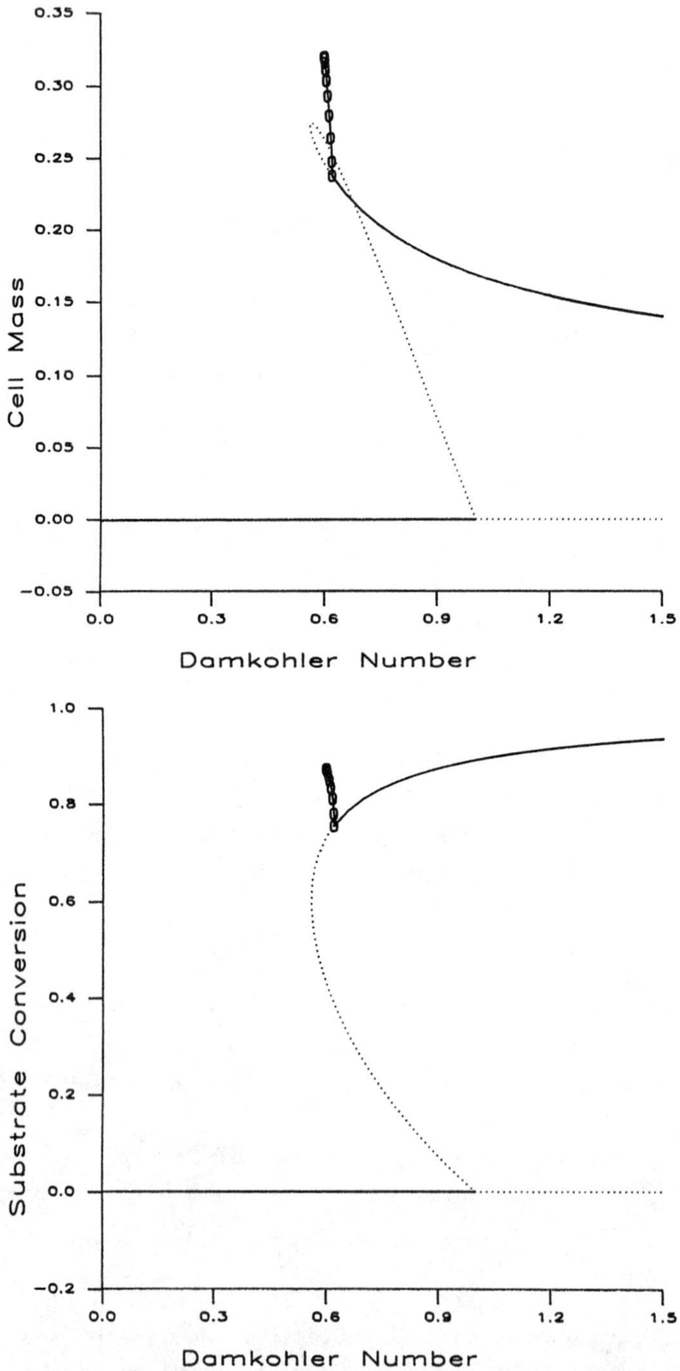

Figure 5 Solution diagrams for the bioreactor model of Agrawal, et al. (1982). Cell and substrate concentration as a function of the Damkohler number.

3. QPSOL. This program, developed at Stanford University (Gill, et al., 1982), performs successive quadratic programming to optimize an objective function subject to nonlinear equality and inequality constraints.

4. MINOS. This program, also written at Stanford (Murtagh and Saunders, 1983), implements a blend of nonlinear programming algorithms including quasi-Newton, reduced gradient, and augmented Lagrangian methods.

5. LSODE. This is the principal program in the ODEPACK (Hindmarsh, 1980) to integrate stiff ODEs.

6. DASSL. Also in ODEPACK, this program integrates stiff ODEs and nonlinear algebraic equations (Petzold, 1982a, b).

All of the programs require the values of function residuals and partial derivatives for the constraints and the objective function. Hence, it becomes very desirable to create interfaces to all of the programs that permit the routine that evaluates the residuals and partial derivatives to be accessed by all of the programs. We are beginning to define these interfaces.

REFERENCES

Agrawal, P., C. Lee, H.C. Lim, and D. Ramkrishna, "Theoretical Investigations of Dynamic Behavior of Isothermal Continuous Stirred Tank Biological Reactors," Chem. Eng. Sci., 37, 453 (1982).

Chang, H.-C., and L.-H. Chen, "Bifurcation Characteristics of Nonlinear Systems Under Conventional PI Control," Chem. Eng. Sci., 39, 7/8, 1127 (1984).

Colmenares, T.R., and W.D. Seider, "Heat and Power Integration of Chemical Processes," AIChEJ, 33, 6, 898 (1987).

Colmenares, T.R., and W.D. Seider, "Synthesis of Utility Systems Integrated with Chemical Processes," submitted to Comp. & Chem. Eng. (1988).

Doedel, E.J., AUTO: Software for Continuation and Bifurcation Problems in Ordinary Differential Equations, Comp. Sci. Dept., Concordia Univ., Montreal, 1986.

Gill, P., W. Murray, M. Saunders, and M. Wright, SOL/QPSOL User's Guide, Systems Opt. Lab., Stanford Univ., 1982.

Hindmarsh, A.C., "LSODE and LSODI, Two Initial Value Ordinary Differential Equation Solvers," ACM-Signum Newsletter, 15, 4, 10 (1980).

Hutchison, H.P., D.J. Jackson and W. Morton, "The Development of an Equation-oriented Flowsheet Simulation and Optimization Package - I. The Quasilin Program," Comp & Chem. Eng., 10, 1, 19 (1986a).

Ibid., "I I. Examples and Results," Comp. & Chem. Eng., 10, 1, 31 (1986b).

Kovach, III, J.W., and W.D. Seider, "Heterogeneous Azeotropic Distillation: Experimental and Simulation Results," AIChEJ, 33, 8, 1300 (1987a)

Kovach, III, J.W., and W.D. Seider, "Heterogeneous Azeotropic Distillation: Homotopy-continuation Methods," Comp. & Chem. Eng., 11, 6, 593 (1987b).

Locke, M.H., and A.W. Westerberg, "The Ascend-II System - A Flowsheeting Application of a Successive Quadratic Programming Methodology," Comp. & Chem. Eng., 7, 5, 615 (1983).

McAvoy, T.J. Interaction Analysis, Instrument Soc. of Amer., 1983.

Murtagh, B., and M. Saunders, MINOS 5.0 User's Guide, Systems Opt. Lab, Stanford Univ., 1983.

Panagiotopoulos, A.Z., and R.C. Reid, "High-Pressure Phase Equilibria in Ternary Fluid Mixtures with a Supercritical Component," in Supercritical Fluids: Chemical and Engineering Principles and Applications, ed. M.E. Paulaitis, ACS Press, 1987.

Perkins, J.D., and R.W.H. Sargent, "SPEEDUP: A Computer Program for Steady-state and Dynamic Simulation and Design of Chemical Processes," in R.S.H. Mah and G.V. Reklaitis, eds., Selected Topics in Computer-aided Process Design and Analysis, AIChE Symp. Ser., 78, 1 (1982).

Perkins, J.D., "Equation-oriented Flowsheeting," in A.W. Westerberg and H.H. Chien eds., Foundations of Computer-aided Process Design, CACHE, 1984

Petzold, L.R., "Differential/Algebraic Equations are not ODEs," SIAM J. Sci. Stat. Comput., 3, 3, 367 (1982a).

Petzold, L.R., "A Description of DASSL: A Differential /Algebraic System Solver," Sandia Report, SAND82-8637, 1982.

Prokopakis, G.J., and W.D. Seider, "Feasible Specifications in Azeotropic Distillation," AIChEJ, 29 , 1 (1983).

Roat, S. D., J.J. Downs, E.F. Vogel, and J.E. Doss, "The Integration of Rigorous Dynamic Modelling and Control System Synthesis for Distillation Columns: An Industrial Approach," in Chemical Process Control, CPC-III, eds. M. Morari and T.J. McAvoy, Elsevier, 1986.

STOCHASTIC MODELLING OF PHYSICAL PROCESSES AND OPTIMIZATION OF THE DOMAIN

Vadim Komkov
Air Force Institute of Technology
Wright-Patterson AFB OH 45433

Summary:
The present work introduces some basic techniques for both simulation of solutions to state equations directly from a physical description of a process and for computing the sensitivity of the numerical solutions with aspect to the perturbations in the shape of the domain. The main technique used for obtaining the numerical solutions is a version of Monte Carlo using a biased Bernoulli random walk. A cataloging of changes in the sensitivity of solution with respect to such changes is accomplished by means of R-function theory, that is the use of multi-valued logic, as pioneered by Rvachev and the Kiev group. Applications are offered to designs of domain for chemical reaction containers, for structural and mechanical applications. The numerical techniques introduced here are specifically suited to parallel processing and supercomputer applications.

This article is organized as follows:

1) Discussion of computational complexity and of available computational technology in optimal engineering design problems.

2) Classification of shapes and simplification of computer communications.

3) Bernoulli's random walks and stochastic solutions of deterministic processes.

4) Closing remarks.

5) References.

FOREWORD

Many design problems are ill-posed in the sense of Hadamard and present regularization difficulties (according to Tikhonov's difinition) because the ill posedness is physically, rather than mathematically or computationally motivated. This means that either the solutions are unstable, or on the verge of instability,or they do not depend continuously on boundary conditions,or they simply do not exist in the accepted sense. Some examples of solutions that are extremely sensitive to changes in the geometric domain were offered by Herschel Rabitz in his study of catalitic reactions. There may be problems with contradictions arising from constraints, or with poorly conceived or executed numerical techniques. However, more serious problems arise in cases when changes in the geometry of the domain destroy the validity of the mathematical model. For example, a change in thickness may invalidate the Lagrange-Sophie Germaine model of plate theory. Restriction on maximum and minimum allowable thickness will prevent this phenomenon, but it may also invalidate some physically meaningful improvements purely for the sake of mathematical convenience. Similarly, restrictions on rates of chemical reactions may save linearity in the mathematical model, but ruin meaningful sensitivity analysis. The trouble we encounter here was always with us in modelling of physical phenomena, but we were trained in our undergraduate courses to ignore it. Now we really have to face it. Changes in the shape of a domain will sooner or later introduce bifurcation phenomena. For an example of a simple mechanical system which illustrates this principle see the Irwin - Komkov article [43]. In cases where a simple check reveals collapse of a mathematical model, or for example a repeated bifurcations pattern leading to the initiation of chaos, the cure consists in a radical treatment. We have as alternative abandonement of the deterministic approach and proceeding with one of the known versions of Monte Carlo random walk solutions.

Advantages of the stochastic approach are many. Suppose that the system is close to an unstable point, and for the sake of simplicity we may assume that it represents the merging of two eigenvalues. The random walk solutions will converge to two eigenstates, say 70% of the time to state one and 30% of the time to state two. This information is extremely valuable and reveals the probabilistic nature of the output for small perturbations of the design (that is of the spatial domain), and probability of outcomes in a real physical system with a data close to the one accepted in our design. Moveover, the information provided by such random walk quickly reveals some "natural" decompositions of the domain into parts that only weakly interact with each other, and several partitioning schemes may naturally suggest themselves. There is, of course one serious drawback. Even for some very simple geometric shapes the number of computations can be frightening. Solving the heat equation for a rectangular domain with only 96 internal points takes about 20,000 separate walks on this grid to obtain a rough (three significant digits) approximation to a solution. Even on a hypercube with hundreds of

parallel computations such solutions present an intractable degree of complexity and require careful theoretical analysis.
The clever techniques of by-passing computation of the exact state of the system (such as the so called "Haug-Komkov trick") are not available, and one has to compute the state at each step. However, there exists a way of classifying shapes in a random walk,and elimination, whenever possible of this entire process, replacing it by a thermodynamic limit, that is by a partial differential equation that "works" in a given subregion. In that sense an efficient classification of regions and effects of small changes in their geometry is crucial in solving boundary value problems.
In this paper we introduce an idea of using a multivalued logic in the general style of Rvachev and the Kiev school, but with a new twist. We introduce also an idea borrowed from Kurt Godel (I refer to it as the poor man's Godel arithmetic) of encoding the geometric shapes as well as some mathematical properties of the state of some system by a single integer. Choosing primes either as our base shapes and base operators or as limits on our complexity, we can quickly produce very complex shapes. This is different from Rvachev's theory which is introduced first. The way we generate these Godel's-style numbers is simple-minded and no doubt it will be made more sophisticated in the future. We cannot follow his lead and introduce a number representation exactly in his style because large numbers overwhelm us. However, even this "naive" method generates a great variety of shapes and operators (both algebraic and differential operators), which suffices to test this technique on simple optimization schemes of intermediate complexity.

For the sake of illustration we shall propose a convenient arithmetic of shapes in which the most elementary shapes have in fact some intermediate values, then big numbers emerge by taking intersections and small numbers by taking unions,or vice versa. One (1) denotes the universe, or the space containing all admissible shapes. This is not the best possible solution for complex systems, but for the time being, it is convenient for introduction of some initiutive arguments.

This technique is combined with both analytical and probabilistic solutions of the state equations,permitting an application of domain optimization by either one of the well-developed analytical methods, or by a direct incorporation of the domain improvements into separate steps of a Monte Carlo process.

Some additional developments,borrowed from physics,which allow the use of Monte Carlo process have been suggested by Komkov and Dannon in the reference [40], permitting a random walk generation in some fairly novel applications. These techniques replace analytical methods only in situations where the mathematical modelling is uncertain,or experimental data indicates that the problems may be ill-posed. We stress that analytical methods for improvement of the domain are preferred to the massive computations required in random walk solutions,and in many cases the discretization into finite elements and by now standard approaches to optimization,such as given in chapter 1 of [2],produce excellent results,and the entire discussion that follows this introduction is superfluous. However in many cases of domain optimization solutions obtained by the "commonly used" algorithms are not reliable,and sometimes are obviously wrong. On other occasions such solutions really are optimal,but small changes in the enviroment may drastically negate such optimality. The example of Michell truss discussed by Prager is well known. It is in fact one of the favorite examples discussed in many structural design courses, which appeared in several engineering textbooks and monographs.

While stochastic techniques avoid many pitfalls of mathematical modelling, they suffer from what von Neumann called the curse of large numbers,that is from computational complexity.

1. Discussion of computational complexity.

In this article we concentrate on designing of mechanical, structural or chemical processing systems, concentrating on the design of large and complex systems. Designs which can be accomplished by a simple classification, and requiring only a well organized database, are of no interest in this article. It is assummed that appropriate selection of specified materials and sizes to comply with regulations or customer requirements can be accomplished by fairly routine means, and while the size of memory or the organization of database and spreadsheets may be complex, this is not the subject of our exposition and of proposed methodology. Our primary concern is a computer assisted analysis of design systems that are governed by partial differential equations, which are not necessarily linear or well posed. The mathematical model may have no solutions (that is with the prescribed data, and physically motivated constraints no exact solution may exist) or it may have more than one solution (due to insufficient data, ill-prosedness of the problem, or in particular, loss of stability and bifurcation). This could be further complicated by optimization procedures which may contribute to ill-posedness. In such complex situations it is hardly surprising that many algorithms which at first glance appear to follow sound numerical principles have produced completely ridiculous designs. This prompted W. Prager's advice to:"optimize for a while and then quit". While the reasons for a loss of stability within the mathematical model, or loss of computational stability are now much better understood, and various safeguards have been instituted in numerical algorithms, the degree of complexity of such computational schemes is very high. Let me offer an example from structural engineering. Let a complex structure be subdivided into ι substructures. The vectors $\boldsymbol{\mu}$ and $\mathbf{h}$ denote respectively the state vector and the vector of design variables for the entire system. Vector $\boldsymbol{\mu}$ is discretized. Here we assume that the knowledge of displacements at the joints is sufficient to describe both the state and the response to external loads of the entire system. The "joints" are either the boundary joints (which include interior and exterior boundary points) denoted by the subscript B, or interior joints denoted by the subscript I. Boundary joints are common to more than one substructure, while interior joint points belong to only one substructure. In the simplest case the optimization program can be restated in such manner that a single functional represents the quality (or the cost) for the entire system.

For complex systems with several quality-type functionals (for an aircraft frame we may wish to keep the weight down, produce a margin of stability, have a low drag, high maneuverability, good stall characteristics, low profile without cramped cockpit, etc.,) we may choose to apply Pareto's criteria for producing a joint cost functional: $J = J(\mu_B, \mu_I, h) \to \min_h$. A linear model which describes the response of this system can be written as: $L\boldsymbol{\mu} = \mathbf{q}$, where $\mathbf{q}$ is a system of discretized general loads acting on the structure. This is rewritten as a system of equations:

$$\begin{bmatrix} L_{BB} & L_{BI} \\ L_{IB} & L_{II} \end{bmatrix} \begin{bmatrix} \mu_B \\ \mu_I \end{bmatrix} = \begin{bmatrix} q_B \\ q_I \end{bmatrix}$$

with constraints:

$$\psi_i\ (\mu_B, \mu_I, \mathbf{h}) < 0,\ i = 1,2,\ldots m .$$

This is generally linearalized with respect to pertubations of the design: $\delta\mathbf{h}$ by introducing functions:

$$\psi_1 = (\partial q_b/\partial \mathbf{h}) - \partial/\partial \mathbf{h}\ \{L_{BB}\mu_B\} - \partial/\partial \mathbf{h}\ (\mathbf{L}_{BI}\mu_I)$$

$$\psi_2 = (\partial q_i/\partial \mathbf{h}) - \partial/\partial \mathbf{h}\ \{L_{IB}\mu_B\} - \partial/\partial \mathbf{h}\ (L_{II}\mu_I)$$

and corresponding linear forms can be derived:

$$\delta J = \{(\partial J/\partial \mathbf{h}) + (\partial J/\partial u_I)(L_{II})^{-1}\chi_2\}\delta\mathbf{h} -$$

$$\{[(\partial I/\partial u_I)(L_{II})^{-1}L_{IB}] - \partial J/\partial u_B\}\delta u_B ,$$

$$\delta\psi_i = \{\partial\psi_i/\partial \mathbf{h} + (\partial\psi_i/\partial u_I)(L_{II})^{-1}x_2\}\delta\mathbf{h}$$

$$\{\partial\psi_i/\partial u_I(L_{II})^{-1}L_{IB} - \partial\psi_i/\partial u_B\}\delta u_B ,$$

relating algebraically $\delta u_B, \delta u_I$ to $\delta\mathbf{h}$. Then the use of gradient projection, or one of the variants of the gradient approaches, an introduction of adjoint variables $\boldsymbol{\psi}, \boldsymbol{\nu}$ and assignment of

$$\delta J = \langle(-\text{grad}_h J)\ ,\ \delta\mathbf{h}\rangle$$

$$\delta\psi_i = \langle -\text{grad}_h \psi_i, \delta\mathbf{h}\rangle$$

produces a recognizable approach to optimization of such system based on solving of a large system of algebraic equations. Such an approach can be found in several monographs on structural optimization. See for example: Banichuk [1], Haftka [28], Haug, Choi and Komkov [2], or for historical interest, any of the early articles of Argyris, Zienkiewicz, and Gallagher

which were published in the early 1960s.
In fact, apart from speeding up convergence, and improving some fine details, such as using second variations to help identifying the crucial parts which need to be further subdivided, or adjustments in refining the finite elements and boundary finite elements, we roughly know how to proceed.
For a literature of modern computational ideas dealing with this problem see references [4] - [15]. I am not implying that there do not exist serious computational difficulties that need to be overcome, and that discrete optimization has not many challanges left. For example, reorganizing "local" rigidity matrix into a global system is not easy. Nor is the choice of a global cover. But, basically, we know what approach to take. However, in the modern theories dealing with very slender shapes and high accelerations, with possibilities of bifurcations, the linear approach obtained by perturbation analysis may not be applied routinely and in the structural case discussed here at least at the local level (that is inside each substructure) one has to consider distributed parameter approach to sensitivity analysis. Even if one uses one of the most efficient algorithms available, such as E.J. Haug's [16] or even more efficient proposal of Haftka of 1982 [17] the sensitivity analysis including a check on the second order sensitivity produces $n(K+1)$ computations at each iterative step, where n is the dimension of the design vector (generally in the thousands) and K the number of constraints (generally much smaller). Haug's computational proposal [16] requires $n + (n+1)n/2$ computations at each iterative step. Moreover, in these more traditional computational schemes the flow of information is basically sequential, thus not taking advantage of the latest developments in parallel processing techniques and in computer architecture. For a dynamic system with slender shapes the subdivision into subsystems becomes much more complex, because of the sensitivity of one subsystem to design changes in another one. Here the changes in the boundary conditions of a subsystem may have a crucial influence on its own sensitivity to design changes. Below we show optimal shapes of thickness for square panel plate subjected to exactly the same loads but with different boundary supports. Figure 1.1 indicates clearly the effects of boundary conditions on the shape of the plate panel. In the structural design case shown above of a substructure consisting of a single panel the equations of state are linear, the sensitivity relations have been linearalized, the loads do not depend on the design. Even in this simple case the volume of computations for such single element is very large. Almost any complications, such as possibility of buckling or appearance of very slender shape causing serious doubts about validity of Hooke's law, will produce a high degree of computational complexity, and we have to re-examine a sequential approach to optimization of structures modelled by nonlinear state equations and with a

demonstrated sensitivity to local design changes. Here we encounter serious computational difficulties. It should be clear that we cannot embark on design of a single substructure without much adaptive communication within the system and that parallel computation scheme must try to deal with a high degree of complexity. Sequential computational modes are much better understood than simultaneous computations in problems of design and optimization. However, the sequential mode is too restrictive for massive computations that accompany a practical design optimization for complex structural and mechanical systems. The theory of concurrent computation in distributed systems is only in the early stages of development. While many software manufacturers produced efficient schemes for multiplying large matrices there are still only skeletal constructs for a proper synchronization and decomposition algorithms. A large number of processors is essential to the efficient handling of large structural systems. The design of an entire project can be considered as an assembly of substructures (or sub-mechanical systems) which are joined at interfaces, or sets of boundary points, with the boundary consisting of a union of two sets: the outer (or true) boundary, defining the shape of the entire system and the inner or interface boundary consisting of interior points of the system which do not lie on the outer boundary, but are common points of at least two substructures (or subsystems). Each processor deals with the redesign of a specific subsystem. The key feature is communication between the subsystems. The processors communicate in an adaptive manner at intermediate stages of redesign. Such communication could be deliberately asynchronous, thus avoiding lengthy queues of information.

In the language of computer science the basic theoretical concept is of an actor. Such actors receive communications from individual processors and communicate the state of their subsystems to other actors. It resembles a theatrical play where actors walk onto the central stage to deliver their lines to other actors present on the stage, or maybe leave the stage for the remainder of the play. Thus, an actor may cease to perform, and a new group of actors may appear on the stage. Each actor has a "name", that is an address which is known to other actors. Here each actor knows only a part of the whole problem, performs some redesign actions within its domain, taking into account the actions of the other actors. The advantages contain our ability to produce multiple clones of the actors sharing the same codes. This type of architecture and software is being developed at MIT and is briefly described in the Agha monograph [18]. An approximate hardware for such applications exists, even though one would have to make some crucial choices. But the main challange lies in development of efficient software specifically oriented towards design applications for large structural and

mechanicalsystems. Perhaps the languages used for defining actors and effecting their communication need to undergo some serious study. Like Ada, one of the languages proposed for this application is loosely based on an Algol-type structure and syntax. A rapid development of computer architecture permitting large parallel computations has been unfortunately not utilized fully because of consistent lag in software development for "object oriented" purposes. (Again, a C.S. jargon does describe the main difficulty). Incorporating computer graphics for an adaptive input by the engineering community may be valuable in implementing some aesthetic features, prejudices of customer, or for implementing William Prager's warning about quitting in good time, even if some safeguards against loss of stability and bifurcation phenomena have been incorporated into the numerical algorithms.
In view of the complexity of such programs it is absolutely essential to reduce drastically the number of iterations and the computational complexity by first incorporating sound theoretical principles into design analysis, before attempting to write a computational algorithm. This analytical approach combined with large parallel computing facility is missing in many commercial C.A.D. applications where frequently "number crunching" is not an end result of careful mathematical analysis, but the beginning and the end of the entire process. The point of view is becoming recognized even among some "applied" numerical analysts and engineering experts working in C.A.D. and C.A.M., and hopefully it will be more widely adopted at all levels of engineering design practice.

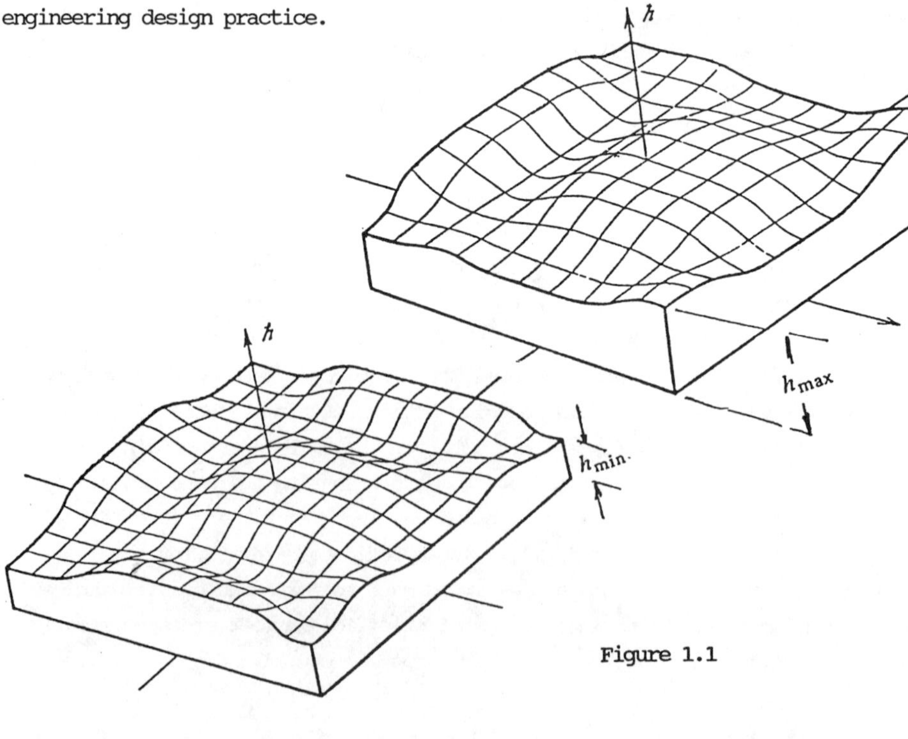

Figure 1.1

Chapter 2

Classification of Shapes and Simplication of Computer Communications

2. Some basic concepts of multivalued logic.

We assume some elementary knowledge of logic and set theory. A Bernays or Godel axiomatic system is presumed, permitting the existence of individuals. The primitive notions are: class, set, individual, and the membership relation of "being an element": ϵ. For the purposes of this presentation a "naive" set theory and corresponding intuitive concepts of predicate calculus suffices. Let χ denote some subset of real numbers, χ^n the n-tuple Cartesian product of χ with itself. A mapping f from χ^n into χ will be called a regular map, or function with alphabet χ. For example, a Boolean algebra is based on maps with the alphabet {0,1}. Here we can choose the alphabet to be a finite,or infinite set of real numbers. Let us suppose that the alphabet consists of the entire interval [0,1]. The values assigned to that alphabet may come from an algebraic structure such as a commutative ring,whose elements are sets,or functions,or operators. For example,we could consider a certain class of functions mapping n-copies of [0,1] into [0,1]. The algebraic structure would be a non-Abelian group with superposition of functions defining the group operation o. In this example we let n=2,and let a group generators to be the functions: $f_1 = \cos x$, $f_2 = 1/2x$, $f_3 = \pi/2$, $f_4 = -x+1$, $f_5 = x+y$. A variety of interesting functions can be generated by iterating superpositions of these "base" functions.

In general, we shall consider a finite number k of values in the alphabet χ, that is $\chi_k = \{0,1,\dots,k-1\}$. The values of " not χ" are as follows $\chi = k-i-1$ if $\chi = i$. Other, so called closure functions may be designed, according to the specific needs. Mapping of sets into n-values can be denoted by curley brackets. If **S** is a set we will denote by **S**{0}, **S**{1}, ..., **S**{k} subsets of **S** that are assigned logical values 0,1,...k. For example, we could consider a 4 - valued logic of a set S of C^2 functions (twice continuously differentiable) on Ω R^2. S{0} denotes subset of S satisfying $\Delta f(x,y)=0$ or Ω, S{1} : $\Delta f<0$ on Ω, S{2} : $\Delta f>0$, S{3} all other C^2 functions on Ω. In this case multiplication by a constant k is a regular operation. If k>0, it is the identity mapping, preserving logical values; if k<0, it permutes values 1 and 2; if k=0, it replaces all logical values by zero. Clearly, if $f\epsilon$**S**, $g\epsilon$**S** then the operation φ $(f;g) = f\int g\, dxdy$ assigns some logical values to Ω , but that value is not uniquely determined by (logical) values of f and g.

In that sense this is not a regular operation. However if we enlarge the values of our logic, it would become so. We need to introduce additional logical values for $g \in S$, denoting whether $\iint_\Omega g\, dxdy\}$ is positive, negative or zero. We replace the initial values (0,1,2,3) by say (1,2,3,5) and g has values (1,7,11), depending on $\{\iint_\Omega g\, dxdy\}$ having zero value, positive, or negative value. Then the values of $\varphi(f;g)$ are given by 1,14,21,35,22,33,55.

f	1	1	1	1	7	7	7	7	11	11	11
g	1	2	3	5	1	2	3	5	0	2	3
φ	1	1	1	1	1	14	21	35	0	22	33

If $f \in S\{7\}$ then $\varphi \in S\{35\}$, $S\{21\}$, or $S\{14\} \subset S\{2\}$.

In this outline we are not prepared to outline the so called Godel-type arithmetic based on identifying geometric shapes formed from unions and intersections of the type illustrated on figure 2.1 below.

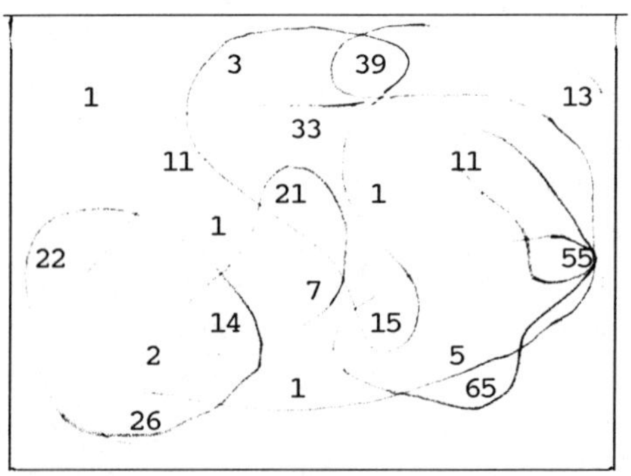

Figure 2.1: Both f and g have four values {1,2,3,5} and {1,7,11,13} respectively

2.1 Identification of shapes.

We shall suggest some simple identification schemes. Following the simplest possible idea (with some obvious shortcomings), we could consider the algebra of all continuous functions in $\mathbf{R}^2$, with an underlying commutative ring structure. The alphabet consists of three integers: {0,1,2}. The values are assigned to each function by considering a simple inequality relation: For each function $f_i(x,y)$ in $\mathbf{C}(\mathbf{R}^2)$ we assign value 2 to the set $I_i(f_i) = \{(x,y) | f_i(x,y) < 0\}$, the value 1 to the set $\mathbf{B}_i$ on which $f_i(x,y) = 0$, and value 0 to the set O_i on which $f_i(x,y) > 0$.

Such simple three-valued logic of sets with the set operations consisting of taking unions, intersection and complements allows us to identify quite complex shapes of regions.

As an example, we wish to identify the interior of a region shown on figure 2.2

below, bounded by a unit circle and some straight line segments:

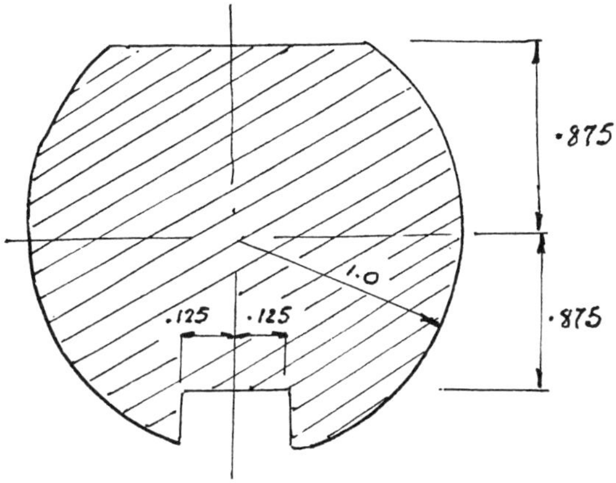

figure 2.2

Analytically the interior of this shape can be identified by logical relations between the sets $\{x,y\}$ satisfying statements:

$$f_1(x,y) = x^2 + y^2 - 1 < 0, \tag{1a}$$

$$f_2(x,y) = -.875 + y < 0, \tag{1b}$$

$$f_3(x,y) = -x - .125 < 0, \tag{1c}$$

$$f_4(x,y) = x - .125 < 0, \tag{1d}$$

$$f_5(x,y) = y + .875 < 0. \tag{1e}$$

It is easy to check that the interior of the shape on figure 2.2 is given by: $\{(1a) \wedge (1b) \wedge [(1c)\wedge(1d)\wedge(1e)]\}$, where, as before, bars denote the operation of complementation, and $\wedge$ stands for intersection.

In this example the three valued logic was unnecessary, except if we wish to identify boundary points. Using strict inequalities only, and assigning values 0, and 1, we identify a set of points $\{x,y\}$ with the interior of the region in figure 2.2 if:

$(1a)\wedge(1b)\wedge[\overline{(1c)\vee(1d)\vee(1e)}]$ has value 1.

The examples given above were extremely simple, and could be managed with the usual Boolean logic of sets.

2.2 The R-functions of Rvachev.

Various complex shapes, sets, or abstract properties may be characterized for purposes of identification, or for description of essential characteristics by a relatively small set of symbols. For example, a physicist would recognise

Einstein's metric by the symbol: {+,+,+,-}. The idea of R-functions is to reduce essential properties of a system to the minimal number of symbols represented by integers. Then properties of more complex systems can be derived by simple algebraic manipulation of these integers within the framework of k-valued logic. In engineering analysis and design even relatively simple designs usually contain almost unmanagable number of variations,arrangements, or changes. The so called "hard-nosed" search for improvements is almost impossible, except in classroom examples,which are deliberately concocted for that purpose.

Let us consider a set χ, and the Cartesian product χ^n. Set χ is decomposed into subsets χ_i, i=0,1,2,...,k-1. These sets may intersect each other,or they may be disjoint. We may identify position of points $\mathbf{x} = \{\mathbf{x_1}, x_2, \ldots x_n\}$ by finding to which subsets χ_i they belong. There exist k^n possible mutual intersections, which will be denoted by T_i. In the well-known Venn diagram for intersections of three sets we identify eight regions as shown on the figure 2.3 below:

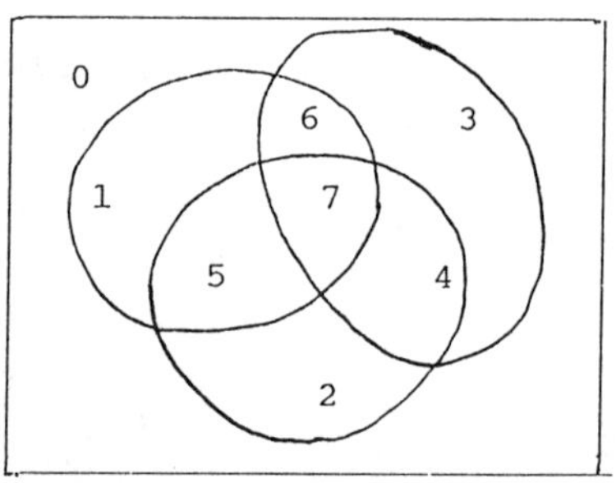

figure 2.3

In this case we introduce subsets T_i, i=1,...,8. Their union is χ^n. The mapping of this union of T_i into χ is called the R-function in the terminology of Rvachev. (We have followed without any modifications the ideas of Rvachev and his group in this exposition.)

The values of R-functions are determined by the logic assigned to a problem. For example,in a three-valued logic the value of X∧X is 0 if X is equal to either 0 or 2, and is equal to 1 if X is equal to 1. Thus,it is important to identify the type of logic,by denoting $\mathbf{R}_\alpha$, $\wedge_\alpha$, etc...to indicate α - valued logic

Moreover, the logic can be adjusted to the type of geometric, or physical problem. For example,our aim may be to arrange high dimensional objects in such manner that they only touch each other. Intersections are to be avoided between the interior regions. Each object may be originally coded by forming unions, intersections and complements of spheres, parallelopipeds, and tetrahedrons in such manner that the interior of each object is labelled by a numerical value that is a composite number, while boundaries are invariably labelled by one. Instead of the usual relation (min,max, respectively) we designate the value of an intersection by least common multiple and union by greatest common divisor. Interiors of different objects are labelled by different composite numbers. Choosing points on the boundaries of objects we can maximize the value of their sums, making certain that no boundary point of one object belongs to the interior region of another. If the shape is only approximately known we can perform some random sampling and then maximize the total by shifting the shapes according to some sequential algorithm. In the maximization process described above it is possible to assign some small prime to the "empty space" region, thus making certain that no damage is done when a point moves from the empty space to the interior of an object (that is we improve the quality indicator by filling the empty space by non-intersecting objects). Following these ideas we can originate a random walk over some region containing various arrangements of objects in a shape identification algorithms,when the boundaries are approximated by some a priori chosen analytical shapes (for example by planes). Space packing problems could now be assigned a fairly obvious algorithm.

2.3. Applications to solutions of partial differential equations of elliptic type.

The methodology indicated above is particularly suitable to Monte Carlo techniques used for solving partial differential equations of elliptic, or parabolic type. Suppose that we wish to compute the value of some integral functional over a fairly complex shape contained in the square region: $1 \leq x \leq 0$, $1 \leq y \leq 0$. Following the ideas of von Neumann and Ulam we generate pairs of random numbers : $1 \leq (x,y) \leq 0$, and find out the probability that a point with coordinate (x,y) will be located inside the region. This probability gives us the area of the shape contained in the unit square. To compute the value of the functional, we evaluate the corresponding functional density at each point that falls inside the prescribed shape, add these values and divide by the number of trials. The identification of "being inside the shape" can be carried out by the use of tri-valued logic. While one could argue that this process can be done more elegantly by defining appropriate Ito integral, we agree, but the purpose of this presentation is to produce simplest possible computational procedure with minimal use of advanced concepts. A well known technique of propagating boundary values in a random walk process can be used in solving Dirichlet problem for the Laplace equation in two or three dimensions, or for more general elliptic equation with variable coefficients. The very difficult problem of improving the shape of the domain can now be automated by ingenious use of a multivalued logic identifying the separate pieces of the domain, and permitting to compute the effects of removing or adding a piece which is given by one of the "standard shapes". Details of these types of manipulations may be found in Rvachev's monograph [20], and in related papers and monographs of the "Kiev group" (see for example [21]).

For our purposes the multivalued logic can be used to identify our position in a random walk procedure. This is also a more natural computer logic for identifying positions of holes from positions occupied by some lowly stressed material. We mention this problem because of the extensive research recently conducted in non-convex relaxation related to that class of problems (See, for example, articles of Strang and Kohn [22],and [23].) Related problems on positioning of holes and introduction of fibers made of varying materials were discussed at a more heuristic level by N.V.Banichuk in [24],[25],[26]. In addition, the multiple-valued logic may be used for data ordering (in order of importance) in a complex design optimization project, and for eliminating any conflicting instructions, or constraints by assigning numbers for such orders of importance. In this instance, the application of multi-valued logic competes with the usage of hierarchical arguments. At the present time I do not know of any efforts aimed at incorporating such multivalued logic into a multi-objective optimization schemes, where this approach appears to be "natural", or into replacing the popular Pareto cooperative optimization criteria. Such considerations deserve a separate discussion and should be pursued elsewhere.

However, it is worth mentioning that several problems where separating points representing different aspects of a problem is of great importance, could be handled by multivalued logic techniques. Applications to order statistics, to neural nets,and even to some apparently mundane problems in management and economics come to mind. However they will not be discussed in this paper.

It is a separate topic, and should be treated on its own, particularly, because of its potential importance.

Chapter 3
Bernoulli Random Walks and
Stochastic Solutions of Deterministic Systems

3.0 Introductory Comments

As has been frequently stated, the Monte Carlo approach is not a specific technique. Rather, it is a family of techniques or perhaps a philosophical approach to physics and other sciences. A modelling of thermodynamics or statistical mechanics processes can be attained by imitating the random behavior of molecules and defining probabilities according to the canonic or grand-canonic ensembles. The physical time may be ignored and some other sequential counts (generalized time) may be used to predict the occurence of certain physical states. For ergordic systems, where time averages and ensemble averages are the same, various imaginative averaging techniques could be introduced. For this reason theoretical basis for a specific Monte Carlo technique have to be stated before a computer simulation is undertaken for a specific physical process. In modelling of process where the random walk simulation follows the perceived random behaviour of the physical medium the averaging may be fairly straight forward and the statistical accuracy of the averaged quantities may be readily obtained, provided that no interaction or critical behavior is anticipated. A large body of literature has been compiled in recent years. For a sample of Monte Carlo techniques used in statistical mechanics of solid state and transition phenomena see for example the articles: [41] - [58].

For "classical" expositions of the Monte Carlo techniques see Spitzer [27], Hammersley and Handscomb [34]. These texts are now regarded as "classical", and are by no means that last word on the subject.

We assert here that many shortcomings of mathematical modelling that become apparent in optimal design process are by-passed if the physics of the process is directly mimicked by a Monte Carlo computational scheme. Let me offer an example. We wish to design a (hopefully 3-dimensional) vessel for a chemical process that is being commonly modelled by nonlinear partial differential equations somewhat resembling the Lottka-Volterra predator - prey systems. If we act like abstract mathematicians, we accept that Lottka-Voltera system as gospel truth, find out what happens at $t\to\infty$, obtain certain limit cycles and regard the problem as solved. The cycles seem to depend significantly on the shape of the boundary, so some design improvements in shape of the container are in order. We might even prove the existence (but not uniqueness) of a best shape if some smoothness conditions are assigned to the boundary surface and suitable a priori bounds are specified for the solutions. In fact, we have missed the most important features of this problem. We completely missed the fact that in some circumstances (which may depend on the domain of our equations) the mathematical model is not valid. These equations describe averages of some random collisions of molecules that produce all possible chemical bonds, formation and dissociation of such bonds. What the laws of large numbers predict, is exactly what will happen, except when probabilities of different outcomes are within our expectation for a managable number of trials. These terms are used loosely. But, suppose that in 20 years of operating a chemical process we expect one explosion. That is obviously unsatisfactory, even if the life-span of the installation is only 10 years. If we expect one catastrophe in 790 years according to the prediction of a random walk scheme for a plant lasting 10 years we may accept the design and concentrate on minor predicted variations. In a class of problems where shape of the domain affects "seriously" the accepted mathematical model, the Monte Carlo approach literally is simulating the behavior of a system. Averaging and thermodynamic limits may be suspect, and the averaging random walks must be examined and confirmed for theoretical validity. Various sampling and averaging schemes exist. Some simply work and consistently check with experimental findings. The names of Landau, Metropolis, Alexandrowicz are associated with some well-established schemes. In all these schemes the averages obtained by repetition of random walks are meaningful in the sense that accuracy of these averages is determined within a given expectation.

3.1. Technical aspects.

"Random" numbers determining the Monte Carlo runs are generated by the computer. Generally one does use a piece of software ("call Raaf (x)") without a serious investigation of the random number generating subroutine. Well known techniques, relying on simple multipying and truncation, or, for example, squaring and taking the middle, produce randomness of varying quality. Before a serious research into physical phenomena is concluded and estimated, it pays to investigate the randomness of "random numbers" generated in a subroutine that is designed by someone else. In a specific program imitating trimolecular reactions this author found that the random numbers subroutine appeared to have a periodic structure, the period being comparable in length to the typical length of the Markov Chain determining the random walk processes. On the other hand, several authors like Landau and Binder who investigated the effects of such non-randomness came to the conclusion that there were generally no noticable effects on the solutions caused by the pseudo-randomness of the supposedly random numbers.

Sources of errors caused by irregular boundary or by slow convergence were analyzed extensively in several numerical analysis articles. More subtle

difficulties may be traced to the logic used in interpreting stochastic theory of Kinetic processes. See, for example, the articles of Matheson, Walls, Gardner or of Gillespie [32], [33].
There is certainly no shortage of subtle difficulties accompaning the Monte Carlo numerical schemes. Despite it they give amazingly good predictions for many physical processes.

3.2. Direct solution of differential equations using a random walk or related techniques

In approaching the probabilistic techniques for solving differential equations we could use large volume of literature dealing with the inverse problem, i.e., with generating a system supposedly modelling an actual behavior of a physical system. Very few physical systems can be modelled by purely deterministic schemes. Say, the position of a planet in our solar system can be described with good or even excellent accuracy over a fairly reasonable time period (perhaps a century or so). Thermodynamic processes sometimes may be described by equations of the type pv=RT, and sometimes elude any deterministic predictions. Electrostatic potential, the deflection of soap bubbles and numerous other physical phenomena lead to the classical form of Laplace equation $\Delta^2\phi=0$, with appopriate boundary conditions. There are many numerical schemes based on probabilistic interpretation of Laplace equation. For example, one can generate a random walk of Bernoulli's type originating at each boundary point and redistribute the given boundary values along each non-self intersecting random path. The averages of such values approach the solution of the corresponding Dirichlet problem for a large number of trial walks. This scheme is not very practical because of large number of trials that is necessary for establishing high probability for an a-priori stated accuracy of the solution. Alternate schemes that start sequentially at each internal point and assign the boundary value at which the random walk terminates give both simpler and a more efficient numerical solution. A so called "explosive" scheme used by Japanese engineers in solving the Laplace and biharmonic equations will not be discussed here despite the claims for exceptional numerical efficiency. In all of these schemes inaccuracy is caused by positioning of the boundary points on a Cartesian grid, with only possible exceptions of domains containing only parallelopiped subregions.
Let us suppose that all such technical problems are quite satisfactorily handled. The most serious problem remains of choosing a numerical (discrete) scheme for a representation of a "thermodynamic limit" of such scheme, that is for a validation of the representation of our given differential equation.
The actual situation is much more complicated than the fairly superficial "thermodynamic limit", which represents in some sense a solution to a given differential system. Individual runs may exhibit a behavior that is in direct

contradiction of the expected properties of the averages as represented by a solution of the differential equation. Let us offer a specific example.

The predator-pray game discussed in this work, is described by a master equation of the general form

$$\frac{dp(x_i;n_i)}{dt} = \frac{1}{s} \{(x_{i+1}+1)p(x_i+1;n_i) - x_i p(x_i;n_i)\} .$$

This equation has an explicit solution indicating that as $t\to\infty$ the number n_i approaches a constant value and the oscilliations about an asymptotic value diminish. This is a very reassuring prediction until one examines the random walk solutions to this problem. There is a non-zero probability that all the prey will be devoured (i.e., that at some time we could have $n_i = 0$), thus terminating the cycle at a singular point $n_i = 0$; $i = 1,2,\ldots k,\ldots$. Thus, the law of large numbers and the "thermodynamic limit" predict a behavior at large values of t that contradicts the prediction based on more elementary probabilistic arguments. In this specific example the random walk technique models the actual prey-predator cycle much better than the prediction based on deterministic interpretation of the master equation for the probability of existence of a given number of species at a given time and then substituting a deterministic prediction based on the average of a large number of trial runs. In justifying the random walk simulation of processes caused by random phenomona we can quote Dansker's theories and the extensive results of R. M. Anderson published in his PhD thesis [87] and in subsequent articles [87] and [88], and related works such as [39], [41].

It is not the purpose of this work to extend these results. We quote them to justify the modelling of a random process by a random walk on a Cartesian lattice. In simulating a physical phenomenon we do not claim that our random walk model is the only valid model (such claim is ridiculously false) or even that it is an optimal random walk or a game whose outcome closely approximates the observed physical phenomena. It is essential that we compare the physical process with the equations modelling that process. Different techniques of modelling using the Monte Carlo approach may lead to the same thermodynamic limit which may be represented by a single partial differential equation.

For example, the predator - prey games may simulate a simple chemical reation $A \to B$. The large number limits that are predicted are identical, but there is a distinct patttern to each of these games, with different expectations of outcomes. The inverse process of generating a random walk or a game sequence that produces a solution to a system of differential equations has a great degree of arbitrariness. In this work we replace the differential equation by a discrete algebraic system. The trick introduced elsewhere by the author in collaboration with V.Dannon consists in introducing some scaling factors that permit us to interpret such algebraic system as a Markov process.

The idea is somewhat resembling the proposal of Herman Weyl of introducing a gauge factor into relativistic quantum mechanics. For example, the scale factor necessary to replace the finite difference scheme for a nonlinear diffussion equation by a Markov chain must obey a number of conditions described by a system of simultaneous inequalities. Quantities representing transition probabilities must be posisive and their sum must be equal to one. However, the resulting inequalities have a much deeper meaning. For example, the relation between the time increments and spatial step sizes for a two or three-dimensional linear diffusion equation turn out to be a well-known inequality quoted by numerical analysts as a necessary condition for stability of the numerical scheme. It is easy to show that no scale factor can be found for a backward heat equation or for some other ill posed physical problems. This relation between scaling factors for a random walk modelling of physical processes and stability of corresponding numerical solutions appears to be an entirely new and interesting development.

A series of articles utilizing this approach by this author in cooperation with V. Dannon is being presently submitted for publication at this time. Applications to design theory are in early stages and only very simple cases have been tested. Some design problems can be regarded as restatements of "inverse" problems in the theory of partial differential equations, and are notoriously ill-posed. The Monte Carlo techniques proposed here are perfectly suited to such problems. Specifically, different outcomes will be produced, but the probability of each outcome will be a part of such generalized solution. Let us offer additional comments regarding this aspect of Monte Carlo. It is an attitude towards modelling, rather than a specific procedure. Thus, the name Monte Carlo may be used for evaluation of definite multi-dimensional integrals by counting "hits" and "misses" after selecting some coordinates, or for direct processes modelled by simulating kinetic behaviour of molecular processes directly from the physical description. The possibility of relating a physical process to a stochastic process, or of determining numerical solutions of a deterministic process by generating a stochastic process was first pointed in the von Neumann-Ulam report (which was originally marked "secret" . Simulation methods that are using stochastic processes are now recognized as a powerful tool for modelling of processes that either fail to yield to analytic techniques, or are modelled with dubious simplifications. After relating a system under study to a stochastic process that is implied by the state equations, we "play dice", generating random steps, and average a large number of separate trial runs. The well-publicized old shortcomings of a random choice, or random walk, or other of the established Monte-Carlo simul-ation schemes which concern mainly the large amounts of repet-ative computations, are now becoming inconsequential in view of recent developments in computer technology, computer memories, speeds of computation, and the availability of parallel processing. Also, this approach may be describing a physical process more faithfully by directly modelling the underlying "small scale" phenomena. Studies of Oppenheimer and Kurtz show how some correlations and fluctuations vanish in the termodynamic limit, and in the steady state solutions. However, if the system is close to instability, or has more than one possible solution, such fluctuations may affect the loss of stability, cause bifurcations, or jumps between two admissible solutions, and occurence of singularities cannot be detected by studying a deterministic model of such process. Stochastic modelling may become the only choice, and the random walk procedure that mimics the turbulence or instability will exhibit these phenomena. Thus a random walk simulation will take account of the local fluctu-ations that disappear in the analytic formulation, and the predictions of the deterministic model may be checked out against the statistical estimates of appropriate stochastic simulation after a large number of trials. For instance, a classical analysis of the Lotka-Volterra prey-predator system shows that as time elapses the system approaches a limit cycle and the number and the amplitude of fluctuations decrease. However, the stochastic analysis reveals that (with an appropriately low probability) fluctuations persist, and that the extinction of a species not only may occur, but that it will occur in some, perhaps distant future, with non-zero probability. Therefore,"real-life" chemical kinetics that occur in many proceesses are, in general, beyond the scope of classical analytical techniques. In particular phenomena such as combustion may exhibit unstable behavior with possibility of local explosions, and the existence of unique solutions to the initial value problem is always in doubt, even for the simplest non-linear models. Moreover, stochastic simulation combines adventages of modelling weak solutions to nonlinear problems with a direct modelling of discretized schedmes. For example, we do not try to solve the Master equation that is simulated by a time evolution stochastic process, nor do we approximate infinitesimal steps by finite steps, derivatives by finite differences, or integrals by finite sums. We do not have to use a limiting process, or deal with averaged quantities in the termodynamic limit. In a Monte Carlo simulation of kinetic equations we do not have to assume molecular chaos. The key element in such kinetic equation simulation is the arbitrariness of the initial distribution of molecules in the container. They do not have to be randomly distributed. Also, many so called "paradoxes", like the celebrated Poincare's paradox, which appears to defy the second law of thermodynamics, have a simple interpretation in the stochastic simulation. For systems close to equilibrium the inevitable tendency towards greater randomness has been investigated by L. Landau, and has been re-examined in a series of remarkable papers by Prigogine and his collaborators (see, for example [65], [66], [67]) and in several "classical" works on statistical chemistry, such as [29], [59], [54], [61].

The effects of Prigogin's self-organization phenomena become apparent if one studies the computer output in a random walk simulation of a chemical process.

4. Closing Remarks

We hear frequently that computer aided design has revolutionized mechanical and structural design, and other statements concerning computerized design integration, incorporation of experimental facts and subdivision of complex systems into simpler subsystems. Complex systems, such as, envisioned in SDI proposals, present challenges that are absent in more traditional designs of mechanical and structural components. In systems containing thousands of subdesigns one has to decide on a basic organization of such subdesign tasks. Several hirarchical models for complex design with a basic root tree structure have been proposed by Sobieski, Rinderle and their coauthors. This author feels that full advantage has not been taken of some recent developments in parallel computing at the time of writing.

Because of the sheer magnitude of computations required in design changes of strongly interacting subsystems, utilization of most recent multiple processor technology is essential, and the flow of communications between the "actors" (in the C.S. jargon) makes or brakes "adaptive" attempts of that magnitude. ("Adaptive" is used loosely, implying only an adaptive nature of information from other subsystems influencing the redesign of a particular subsystem). There is no reason why some hiearchical features should not be included in any parallel scheme. The author also preposes a more efficient classification of complex domains and changes in the shape of a domain by using specifically designed methodology of multiple-valued logic. The combination of analytic and Monte Carlo method seems natural in projects where basic equations of state, or constituent equations are in doubt. The organization of such research is visualized as a massive feedback system (loops within loops), rather than a tree structure where designs are discarded once and for all. The delicate interplay between analytic and stochastic approaches has not been worked out for real-life engineering designs. The advantages of designing with intermediate state and sensitivity data are clear. This technique is an alternative to the "Polish School" probabilistic approach now pioneered in the United States by Arciszewski and his colleagues at Wayne State University.

The Monte Carlo had spectacular successes in many applications to chemical processes (references [41]-[58]) and is now developed in simulation of complex flows with exothermic reactions,and in reaction-diffusion simulation ([79]-[86],[63],[64], [70],[72]). It was developed for study of neutron transport by von Neumann and Ulam and is still an important technique in such studies (see [75],and [62]). Here we suggest its application to approximation of state of a complex mechanical or structural system. The advantage of obtaining a detailed description of the state are obvious in structural optimization. Even a rough computer graphics program offers an immidiate insight into potential changes in design, without studying optimization algorithms. As an example we mention a Monte Carlo printout for a solution of a nonlinear heat transfer in a vessel with conducting walls,and with a heat source located at the center of a rectangular domain. It is clear that the corners hardly participate in this heat exchange and cutting them off improves the process. Making the area too small would raise the average temperature,so some compromise will be reached. The actual near-optimal shapes can be quite surprising.

REFERENCES

[1] Banichuk, N. V., Problems and Methods of Optimal Structural Design, Plenum Press,New York and London,1986. (Translation by V.Komkov)

[2] Haug, E. J., Choi, K. K., Komkov, V., Design Sensitivity, Analysis of Structural System, 1986, Academic Press, New York.

[3] Arora, J. S., An Algorithm for Optimal Structural Design Without Line Search, In: New Directions in Optimal Structural Design, Editors E. Atrek et al., John Wiley, New York, 1984, pp. 429-441.

[4] Kiselev, V. G., An Algorithm for a level line search and its applications to optimization problems. In the volume: Applied Problems of strength and plasticity. All Union University Collection, Gorkii University Press, Gorkii, 1980, pp 81-87.

[5] Fleury, C., Ramanathan, R.K., Salama, M. and Schmit, L.A., Access Computer Program for Large Structural Systems, In: New Directions in Optimal Design; Editors E. Atrek et al,Chichester, Wiley & Sons, 1984,pp. 541-561.

[6] Hoerner, S., Design of Large Steerable Antennas, Astron, J., 1967, Vol. 72, #1, pp 35-47.

[7] Khot, N.S., Algorithms Based on Optimality Criteria to Design Minimum Weight Structures, Eng. Optim. 1981, Vol 5, pp. 73-90.

[8] Schmit, L.A., Structural Optimization-Some Key Ideas and Insights. In the Volume: New Directions in Optimal Design, editor E. Atrek et al, Chichester, Wiley & Sons, 1984, pp. 385-412.

[9] Diaz, A.R., Kikuchi, N., Taylor, J.E., A Method of Grid Optimization for Finite Element Methods, Comp. Meth. Appl. Mech. Eng., 37, 1983, pp 20-46.

[10] Kikuchi, N., Adaptive Grid Design for Finite Element Analysis in Optimization, in Computer Aided Design.; Editor C.A. Mota-Soares, 1986, Springer Verlag, Berlin and New York, pp. 544-562.

[11] Bensoe, M.P., Olhoff, N., Taylor, J.E., A Variational Formulation for Multicriteria Structural Optimization, J. Structural Optimization, 1984, Vol.11, #4, pp. 523-544.

[12] Kohn R., Strang, G., The Constrained Gradient Problem,Symposium on Nonclassical Continuum Mechanics, Durham U.K. Cambridge Univ. Press, 1987, pp 226-243.

[13] Thierauf, G., and Lawo, M., Optimal Limit Design-Comparative Studies for Large Scale - In the Volume: Optimization Methods in Structural Design., Editors: H. Eschenauer and N. Olhoff, Mannheim, F.R.G. Bibliogr. Institute, 1983, pp. 431-437.

[14] Eschenauer, H.A., Vector Optimization in Structural design and its Application to Antenna Structures, In: Optimization Methods in Structural Design, editors: H. Eschenauer and N. Olhoff, Mannheim, Bibliogr. Institute, 1983, pp 146-155.

[15] Fleury, C., Large Scale Structural Optimization by Finite Elements, In: New Directions in Optimal Design, editor E. Atrek etal, Chichester, Wiley & Sons, 1984, pp. 439-441.

[16] Haug, E.J., Second Order Sensitivity Analysis of Structural Systems, AIAA Journal, 1981, Vol 19, #8, pp. 1087-1088.

[17] Haftka, R.T., Second Order Sensitivity Deviations in Structural Analysis, AIAA Journ., 1982, Vol 10, pp. 323-330.

[18] Agha, G., Actors, Model of Concurrent Computation in Distributed Systems, M.I.T. Press, 1986, Cambridge, Mass.

[19] Rvachev, L.V., Elements of discrete analysis and of the theory of R-functions. Publ. Khakov Polytechnical Institute, 1972.

[20] Rvachev, V.L., Methods of Algebra of Logic in Mathematical Physics, Naukova Dumka, Kiev, 1983.

[21] Rvachev, V.L. and Slesarenko, A.P., solutions of mixed boundary value problems for different media in multiply connected complex regions using a structural technique; Diff. Urav., Vol.6, 1972.

[22] Kohn, R.V. and Strang, G., Optimal design and relaxation of variational problems, Comm. in Pure and Appled Math., Parts I, II, and III, Vol 39, 1986, pp. 113-137, 139-182, 357-377.

[23] Kohn, R.V. and Vogelius M., Homogenization theory for thin plates, Report of the Courant Institute, 1983, (also appeared in Int. J. Solids and Structures), Vol 20, 1984, pp. 333-350.

[24] Banichuk, N.V., Introduction to Optimization of Structures, Nanka, Moscow, 1986. (English translation to appear).

[25] Banichuk, N.V., Optimization of elastic bars in torsion, Int. J. Solids and Structures, Vol 12, 1986, pp. 275-286.

[26] Banichuk, N.V. Optimization of shapes of elastic bodies , Moscow, Nauka, 1980.

[27] Sptizer, F. Random Walk, van Nostrand, New York, 1960.

[28] Haftka, R.T. and Kamat M.P., Finite elements in optimal structural design; In: Computer Aided Optimal Design, C.A. Mota Soare, editor, Springer Verlag, Berlin and Heidelberg, 1986.

[29] Frank-Kamenetskii, D. A., Diffusion and Heat Transfer in Chemical Kinetics, Acad. of Sciences of U.S.S.R., Moscow, 1947.

[30] Gaydon, A.G., Wolfhard, H.G., Flames, Their Structure, Radiation and Temperature, Chapman & Hall, London, 1947.

[32] Gillespie, D.T., A General Method of Solving the Stochastic Time Evolution of Coupled Chemical Reaction, J. Comp. Physics, Vol. 22, 1976, pp 403-434.

[33] Gillespie, D.T., Exact Stochastic Simulation of Coupled Chemical Reactions, Symposium on Reaction Mechanisms, Models and Computers, Am. Chem. Society, New Orleans, March 1977.

[34] Hammersley, I.M., and Handscomb, D.C., Monte Carlo Methods, Methuen, New York, 1964.

[35] Ermakov, I.M., Monte Carlo Method and Related Questions, Nauka, Moscow, 1976.

[36] Halton, J.H., A Retrospective and Prospective Survey of Monte Carlo Methods, SIAM Review, Vol 12, #1, 1970.

[37] Rubinstein, Y.R., and Kreiner, J., About a Monte Carlo Method for Solving Lienar Equations, Mathematics and Computers in Simulation, XXV, (1983), PP 321-334.

[38] Havlin, S., and Ben Abraham, D., Corrections to Scaling in Self-Avoiding Walks, Phys. Review A, Vol. 27, #5, (1983), pp. 2759-2762.

[39] Hirsch, R., Brownian Motion and Non-Standard Analysis; Technical Report #277, (1973) University of New Mexico, Albuquerque, New Mexico.

[40] Komkov, V., and Dannon, V., Monte Carlo Simulation for the Solution of Partial Differential Equations, Part I (to appear).

[41] Keisler, H.Jerome, An infinitesimal approach to stochastic analysis, Memoirs of American Mathematical Society, #297, Providence,Rh.I. 1984.

[42] Kosciuk,S., Non-standard stochastic models in diffusion theory, Ph.D. Thesis Univ. of Wisconsin, Madison, Wi., 1982.

[43] Irwin,C. and Komkov,V., Sensitivity Analysis and Model Optimization for Reaction-Diffusion Systems, J. Optimization Theory & Applications, Vol. 44, #4, Dec 1984.

[44] Masur,E.F. and Mroz, Z., Non-stationary optimality conditions in structural design, Intern. J. Solids and Structures,vol.15, 1979, p.503-512.

[45] Demiralp,M and Rabitz,H. Chemical kinetic functional sensitivity analysis: Derived sensitivities and general applications, J. Chemical Physics, vol.75, 1981, pp.1810-1819.

[46] Koda,M., Dogru,A.H., and Seinfeld,J.H., Sensitivity of partial differential equations with applications reaction-diffusion processes, J.Computational Physics, vol.30, 1979, pp.259-282.

[46] Kuo,Y.S., and Korsin,S.J., J.Fluid Mechanics, vol.56, 1972, pp.447-479.

[47] Kuznetsov,V.R., Izv.Akad.Nauk SSSR, Mekh. Zhid.i Gaza, 5, 1977, p.3-15.

[48] Landsberg,P.T. The fourth law of thermodynamics, Nature, vol.238, 1972, pp. 229-231.

[49] Libby,P.A., Bray,K.N.C., and Moss,J.B., Combustion and Flame, vol.34, 1979, pp.285-301.

[50] Lavenda,B. Thermodynamics of Reversible Processes, Halstead Press, John Wiley & Sons, New York, 1978.

[51] MacKean, H.P., The Boltzman Equation, J. of Combustion Theory, Vol. 2, 1967, pp. 358.

[52] McQuarrie, D.A., Kinetics of Small Systems, Part I, J. Chem. Physics, Vol. 38, 1963, pp. 433-436.

[53] McQuarrie, D.A., Stochastic Approach to Chemical Kinetics, J. Applied Prob., Vol. 4, 1967, pp. 413-478.

[54] McQuarrie, D.A., Jackinnowski, C.J., and Russell, M.E. Kinetics of Small Systems, Part II, J. Chem. Physics, 1963, (the same issue).

[55] Mandelbroit, B.B., Fractals: Form, Chance, and Dimensions; Freeman and Company, San Francisco, 1977.

[56] Matsui, H., and Lee, J.H., Combustion & Flame, 27, (1976), pp. 217-220.

[57] Miller, I., On the Entropy Inequality, Arch. Rat. Mech. Analysis, Vol. 26, 1967, pp.118-141.

[58] Matkovsky, B.J., and Sivashinsky, G.I., Propagation of a Pulsating Reaction Front in Solid Fuel Combustion, S.I.A.M. J. Applied Math., Vol. 35, #3, 1978, pp. 465-478.

[59] Muller, J., A Thermodynamic Theory of Mixtures of Fluids, Arch. Rat. Mech. Analysis, Vol. 28, 1968, pp 1-39.

[60] Nicolis, G., and Prigogine, I., Self-Organization in Non-Equilibrium Systems, Wiley-Interscience, New York, 1977.

[61] Landau, L., and Lifshitz,E.M., Statistical Physics, Translation from Russian by Pegammon Press, New York and Oxford, 1959.

[62] Pasta, J.R., and Ulam, S., Heuristic Numerical Work in Some Problems of Hydrodynamics, Math. Tab. & Ads to Comput. 13., 1959, pp. 1-12.

[63] Perlee, H.E., Fuller, J.N., Saul, C.H., Bureau of Mines, Report RI17839, (1965).

[64] Povinelli, L.A., and Fuchs, L.E., Eighth Symposium on Combusion, Williams and Wilkins, Eds., Baltimore, Md. (1962), pp. 554-566.

[65] Prigogine, I., Nicolis, G., and Babloyantz, A., Physics Today, 25, #11 (1972), pp 23-37.

[66] Prigogine, I., and Lefever, J., J. Chem. Phys. 48 (1968), pp. 1956.

[67] Prigogine, I., Thermodynamics of Irriversible Processes, Wiley-Interscience, 3rd Ed., New York, 1967.

[68] Proceedings of the 15th Symposium on Combustion (in Pittsburgh), The Combustion Institute, New York, 1975.

[69] Saffman, P.G., Proc. Phys. Int. School Non-Linear Mat. Phys., Springer Verlag, Berlin, 1968, pp. 485-614.

[70] Sivashinsky, G.I., Acta Astonaut., Vol. 6, (1979), pp 631-645.

[71] Spanier, J., A Unified Approach to Monte Carlo Methods and an Application to a Multigroup Calculation of Absorption Rates, S.I.A.M. Review, Vol. 4, 1962, pp 115-134.

[72] Todes, O.M., The Theory of Thermal Explosions, Zhur. Fiz. Khimii, Vol. 13, #7 (1939).

[73] Tomita, K., Ohta, T. and Tomita, H., Fluctuations: in: Progress in Theoretical Physics, Vol. 52, 1974, pp 1744-17.

[74] Townes, H.W., Gow, J.L., and Webber, R.E., Trns. Am. Soc.Mech. Engnrs. 94, 1972, pp 353-362.

[75] Spanier, J., and Gelbard, E.M., Monte Carlo Principles and Neutron Transportation Problems, Addison-Wesley, Reading Mass., 1969.

[76] Truesdell, C., On the Foundations of Mechanics and Energetics in the Rational Mechanics of Materials - Continuum Mechanics, Vol. 2, Editor C. Truesdell, Gordon and Breach, New York, 1965.

[77] Valanis, K., A Theory of Viscoplasticity without Yield Surface, Archives of Mechanics, Vol. 23, 1971, pp. 517.

[78] Valanis, K.,and Komkov.V., Irreversible thermodynamics from the point of view of internal variable theory, Archives of Mechanics, (1980), Vol. 32, No. 1, pp. 33-58.

[79] Williams, F.A., Combustion & Flame, Vol. 26 (1976), pp 269-270.

[80] Wasow, W.R., A Note on the Inversion of Matrices by Random Walks, Mathematical Tables & Other Aids to Computation, Vol. 6, 1951, pp 78-81.

[81] Zaidel, R.M., Dokl. Akad. Nauk SSSR, Phys. Che. Section, #136, (1961), pp. 1142-1145.

[82] Zaikin, A.N. and Zhabotinski, A.M., Nature 225, 1970, pp.525-528.

[83] Zhabotinski, A.M., Self-Oscillating Concentrations, Nauka, Moscow, 1974.

[84] Zel'dovich, Ya. B., and Kompaneets, A.S., Detonation Theory (in Russian) Gostekhizdat, Moscow, 1955.

[85] Zel'dovich, Ya.B., and Frank-Kamenetskii, D.A., Zh. Fiz. Khimii, Vol. 12, No. 1, 1938.

[86] Sivashinsky, G.I., Instabilities, Pattern Formation and Turbulence in Flames, Ann. Rev. Fluid Mech., Vol. 15, (1963) pp. 179-199.

[87] Anserson, R.M., Nonstandard Representation of Brownian Motion and Integration, Israel J. Mathematics, Vol. 25, (1976) pp 15-46 (Also see PhD Thesis, Yale University, 1975.

[88] Anderson, R.M., Star-Finite Representation of Measure Spaces, Trans. Amer. Math. Soc. (1982) pp 667-687.

RETROFITTING HEAT EXCHANGER NETWORKS: A TWO STAGE APPROACH

Amy R. Ciric and Christodoulos A. Floudas
Department of Chemical Engineering
Princeton University
Princeton, NJ 08544

INTRODUCTION

A great deal of effort has been expended during the past two decades in the design of *grassroot* heat exchanger networks, to improve the heat recovery in chemical processes (e.g., Cerda and Westerberg (1983), Linnhoff and Hindmarsh (1983), Floudas et. al. (1986)). This research effort has resulted in over 200 publications, and a very good review is presented by Gundersen and Naess (1987).

It is interesting to note that, despite the fact that a lot of research has been focused on the synthesis problem of *grassroot* heat exchanger networks, there has been significantly less systematic study of the optimal redesign of existing heat recovery networks. Recently however, restricted profit margins and the high cost of building new chemical plants provide stong motivation to develop techniques for the optimal redesign of existing heat exchanger networks. The major objectives of these techniques are the reduction of utility use in the existing network, the full utilization of the existing heat exchangers, and the identification of the required structural modifications.

Jones et. al. (1985), (1986) presented a strategy for the retrofit of heat exchanger networks that is based upon the generation of a number of alternative designs and their evaluation using simulation runs. They selected the best design based merely upon the full utilitization of the existing equipment and the addition of area in some of the heat exchangers.

Tjoe and Linnhoff (1986) proposed a retrofit design philosophy for the retrofit of heat exchanger networks which consists of a) the identifiction of the cross-pinch exchangers; b) the elimination of the cross-pinch exchangers from the existing network; c) the positioning of new exchangers and, where possible, re-use exchangers removed in step b; and d) the evolution of improvements by considering heat-load loops between streams, process and utility heat exchangers. They stated, however, that such an approach can provide only bounds within which it is expected to find a good retrofit, and that the best retrofit design is difficult to identify due to the complication of process constraints and plant layout.

Saboo et. al. (1986) proposed an evolutionary strategy which is based upon nonlinear optimization, constrained MILP synthesis and feasiblity evaluation capabilities of RESHEX. Their procedure generates a number of successive retrofit design alternatives without the explicit consideration of economic data.

Yee and Grossmann (1987) imposed three targets: (a) maximum utilization of existing exchangers; (b) assignment of existing units to new required matches with minimum piping changes; and (c) minimum number of new stream matches that require the purchase and installation of new units. They developed a mixed-integer linear programming (MILP) assignment-transshipment model to determine the fewest modifications, at the level of matches. This model does not consider all the potential modification combinations, and does not explicitly take into account the use of existing area and its potential increase or decrease in the different heat exchangers. In an example problem, they demonstrated that the removal of cross-pinch heat exchangers, as suggested by Tjoe and Linnhoff (1986), is not always necessary.

In this paper, a systematic two-stage approach is proposed for the optimal redesign of heat exchanger networks. In the first stage, a mixed-integer linear programming (MILP) formulation is presented, at the level of matches, that incorporates explicitly the cost associated with each potential match of streams, and involves all possible options for modifications. The solution of this formulation provides information on which exchangers should be reassigned or newly installed, and whether there is a need to increase or decrease the area of the existing exchangers. The second stage takes advantage of this information, and a superstructure is postulated and formulated as a nonlinear programming (NLP) problem. The solution of the nonlinear programming problem provides the actual retrofitted network that minimizes the total modification cost.

PROBLEM STATEMENT

The problem to be addressed in this paper is stated as follows:

Given is an existing heat exchanger network, containing **K** exchangers of known area; a set **H** of hot process streams and a set **C** of cold process streams with fixed flowrate heat capacities, inlet and outlet temperatures; and a set of hot and cold utilities **HU** and **CU** respectively. Associated with each exchanger k is a set of matches $\mathbf{M(k)}$ containing matches (ij) that can be assigned to exchanger k. This set of matches may contain only the match originally housed in k, or matches with a particular type of duty, (e.g, condensing duties), or it may contain all possible matches. Associated with each existing and potential match (ij) is a set $\mathbf{Z(ij)}$ containing all exchangers that are allowed to be assigned to house it.

The objective of this problem is to determine an optimally redesigned heat exchanger network that features the minimum total modification cost.

This paper proposes a two stage approach for solving this problem. The first stage, process stream matches and match-exchanger assignments are made on the basis of estimates of the required heat exchanger area of each match and the total modification cost. In the second stage, a network structure is derived.

The first stage requires the following assumptions:

1. A heat recovery approach temperature, HRAT (Jones et al. (1986)) is specified.

2. The flowrate heat capacities of each of the process streams are constant.

3. The heat transfer coefficients for each match are known and fixed.

4. The cost per unit of heat transfer area in a new exchanger is equal to the cost per unit area in an existing exchanger.

The first two assumptions are requirements of the model: a single minimum temperature approach and constant flowrate heat capacities are required for modelling heat flow using the transshipment model. The assumption of fixed heat transfer requirements are required for estimating the heat transfer area. This assumption can be relaxed in the second stage of the procedure, when detailed information about the flowrates is available, and can be used to explicitly calculate variations in the heat transfer coefficients.

OUTLINE OF RETROFIT STRATEGY

The proposed strategy for the retrofit of heat exchanger networks is shown in Figure 1. It consists of the following five steps:

In the first step, a heat recovery approach temperature (HRAT) is selected. This selection of the HRAT can be performed either randomly within some reasonable range of temperatures, or using a targetting procedure that provides a good estimate of the optimum heat recovery approach temperature.

In the second step, a minimum utility cost calculation is performed which provides information on the hot and cold utilities, the yearly savings in utility cost, and the location of the pinch points that partition the temperature range into subnetworks.

In the third step, all potential pairings of matches and exchangers are considered, so that all decisions regarding reassigning heat exchangers, purchasing new exchangers, and repiping streams are incorporated in the proposed retrofit model at the level of matches. The solution of this model corresponds to the minimum modification cost predicted at the level of matches, and provides information on the reassignment of existing exchangers, stream repiping, and the purchase of new exchangers, and additional area for existing exchangers.

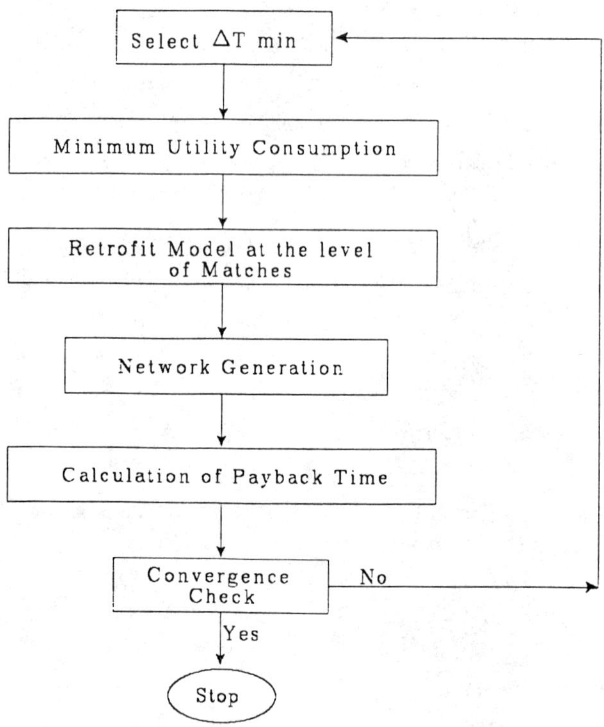

Figure 1: Outline of Retrofit Strategy

In the fourth step, a superstructure that has embedded all the alternative network structures is generated and formulated as a nonlinear programming problem, using the approach of Floudas et al (1986). The heat recovery approach temperature (HRAT) is relaxed for each match, and the temperature approach in each match is treated as a variable greater than a specified lower bound. The solution of this nonlinear programming problem provides the minimum investment cost heat exchanger network configuration.

In the fifth step, the total profit is calculated as the utility savings multiplied by the project lifetime, minus the investment cost. If the total profit and/or HRAT have converged, then the procedure is stopped. Otherwise, a new value of the HRAT is selected.

MATHEMATICAL FORMULATION AT THE LEVEL OF MATCHES

The purpose of the mathematical formulation at the level of matches is to select the combination of process matches and match-exchanger assignments that will minimize the cost of structural modifications, such as moving existing exchangers, purchasing new heat exchangers, adding area to existing exchangers and repiping streams. This is accomplished with an objective function that is based upon the cost of reassigning exchangers, purchasing new heat exchangers, increasing the areas of existing exchangers and repiping the apropriate streams, subject to a set of constraints that model the heat flow between streams, calculate the area requirements and new area associated with each match, and assign process matches to existing heat exchangers. To describe mathematically the contributions of each type of modification to the objective function and to the set of constraints, a classification of the structural modifications is required.

CATEGORIES OF STRUCTURAL MODIFICATIONS

The modifications of reassigning exchangers, purchasing new exchangers, and repiping streams arise from the potential pairing of matches and exchangers, and can be divided into six general categories. Each category is based upon characteristics of both the match and exchanger.

Category 1: *Existing match (ij) is housed in exchanger k in both the original network and the retrofitted network.* No structural modifications are required for this case, and consequently the piping costs of this category are zero.

Category 2: *The existing match (ij) was housed in exchanger k in the original network, and will be housed in exchanger k', in the retrofitted network, where k' is a member of* $\mathbf{Z(ij)}$. As the piping for this match is already in place, no repiping is required; the only required modification is to move the exchanger k' to the new match site. The costs of this category, C_p^2 reflect the labor costs of removing exchanger k from the match site, and installing exchanger k'.

Category 3: *The new match (ij) did not exist in the original network and will be housed in exchanger k, a member of the set* $\mathbf{Z(ij)}$, *in the retrofitted network; in addition, the original match housed in exchanger k contained either stream i or stream j.* This category has a cost C_p^3 that reflects the repiping of one stream.

Category 4: *The match (ij) did not exist in the original network, and will be housed in exchanger k, a member of the set* $\mathbf{Z(ij)}$, *in the retrofitted network; neither stream i nor stream j particicated in the match originally housed in k.* In this case two streams must be repiped. The cost of assignments in this category is C_p^4.

Category 5: *A new heat exchanger is purchased to house the existing match (ij).* This can occur when the original exchanger has been reassigned to a new match, whose area requirement is greater than that of match (ij) in the retrofitted network. In this case, no repiping is required, and there are no additional costs associated with this category. All existing matches may enter category 5.

Category 6: *A new heat exchanger is purchased to house a new match, that is, one that did not exist in the original network.* In this case, two streams need to be repiped, and a cost C_p^6, is assigned for the exchanger purchase and installation and stream repiping cost.

All six categories of structural modification can be accounted for in an explicit way, by introducing two sets of integer variables. The first set, z_{ij}^k, denote the assignment of a potential match (ij) to an existing exchanger k. The activation of a given z_{ij}^k variable represents the assignment of the match (ij) to an existing exchanger k. Therefore, the set of z_{ij}^k variables may be divided into four subsets, depending upon the category of the assignment:

$$z_{ij}^k \mapsto \begin{cases} C^1, & \text{if the assignment is in category 1;} \\ C^2, & \text{if the assignment is in category 2;} \\ C^3, & \text{if the assignment is in category 3;} \\ C^4, & \text{if the assignment is in category 4.} \end{cases}$$

The last two categories do not involve reassigning heat exchangers, but rather purchasing new exchangers. To model these categories, a second set of integer variables, denoted as m_{ij}, is introduced. The variable m_{ij} takes the value one if a new exchanger is to be purchased for the match (ij), and is zero otherwise. These variables can be partitioned into two subsets:

$$m_{ij} \mapsto \begin{cases} C^5, & \text{if match } (ij) \text{ exists in the original network;} \\ C^6, & \text{if match } (ij) \text{ is a new match.} \end{cases}$$

The introduction of these integer variables ensures that all possible combinatorial decisions at the level of matches are taken into account in the mathematical formulation of the optimal redesign problem of heat exchanger networks.

OBJECTIVE FUNCTION

The objective function consists of three main components: (a) the cost of purchasing new heat exchangers; (b) the cost of additional heat exchanger area; and (c) the piping cost resulting from structural modifications.

The following exponential form is a general expression that takes into account the cost contribution of the first two components of the objective function:

$$cA^b \tag{1}$$

where c is a cost constant and b a fixed exponent. In this paper, however, the exponential cost contributions are approximated with a piece-wise linear set of fixed charge cost expressions of the area of each exchanger. Therefore, for each potential match (ij), a cost consisting of piece-wise linear expressions of the following form is introduced:

$$C_{ij} = \alpha m_{ij} + \beta X_{ij} \tag{2}$$

$$X_{ij} = max(A_{ij} - A_{ij}^{ex}, 0) \tag{3}$$

where α is the installation cost of a new heat exchanger; m_{ij} represents the integer variable associated with the purchase of a new heat exchanger; β is the cost per unit of heat exchange area; X_{ij} is the additional area required for the match (ij), if any. Notice that X_{ij} is provided by the maximization problem of $A_{ij} - A_{ij}^{ex}$ and zero, as indicated by (3). If a new heat exchanger is purchased for the match (ij), then the existing area for match (ij), A_{ij}^{ex}, equals zero, $max(A_{ij} - A_{ij}^{ex}, 0)$ equals A_{ij}, and the additional area X_{ij} is equal to the area of the match. If a heat exchanger with an existing area A_{ij}^{ex} is reassigned to the match (ij), and the match requires a larger area, then the operator $\max(A_{ij} - A_{ij}^{ex}, 0)$ simply equals the difference between the total required area and the existing area of the heat exchanger. If however, A_{ij} is less than or equal to A_{ij}^{ex}, the cost contribution is driven to zero, since $max(A_{ij} - A_{ij}^{ex}, 0)$ is equal to zero.

It should be noted that for the first four categories of match-exchanger assignments the cost contribution comes from the expression

$$\beta X_{ij} = \beta max(A_{ij} - A_{ij}^{ex}, 0) \tag{4}$$

as only additional area needs to be purchased. Combining the expressions for the cost of purchasing new heat exchangers, additional area, and the cost of structural modifications results in the following objective function:

$$\begin{aligned} OBJ = min \sum_{(ij)} \beta X_{ij} + \sum_{(ijk) \in C^2} C_p^2 z_{ij}^k + \sum_{(ijk) \in C^3} C_p^3 z_{ij}^k + \sum_{(ijk) \in C^4} C_p^4 z_{ij}^k + \\ \sum_{(ij) \in C^5} C_p^5 m_{ij} + \sum_{(ij) \in C^6} C_p^6 m_{ij} \end{aligned} \tag{5}$$

CONSTRAINTS

The required set of constraints is classified into four main components: (a) the heat flow model;

(b) the area estimation at the level of matches; (c) the calculation of any additional required area; and (d) the match-exchanger assignments.

(a) Heat Flow Model

The heat flow model is represented by the transshipment model of Papoulias and Grossmann (1983), where the temperature range of the problem is partitioned into **T** temperature intervals by using the inlet temperature of each stream. Heat can flow in one of two ways: between hot stream i and cold stream j in temperature interval t, denoted by Q_{ijt}, or from one interval to another via a hot stream residual heat flow, denoted by R_{it}. A hot stream can either release heat to a cold stream in the same interval, or it may cascade the heat into a lower temperature interval through a heat residual. A cold stream may absorb heat from any hot stream in the same temperature interval, or from the residual heat of a hot stream in a higher temperature interval. Within each interval, hot stream i must release Q_{it}^H heat, and cold stream j must absorb Q_{jt}^C heat. The energy balances of the hot and cold streams can be represented with the set of the following linear equations:

$$\sum_{j=1}^{C} Q_{ijt} + R_{it-1} - R_{it} = Q_{it}^H \qquad i{=}1..H, \quad t{=}1..T \tag{6}$$

$$\sum_{i=1}^{H} Q_{ijt} = Q_{jt}^C \qquad j{=}1..C, \quad t{=}1..T \tag{7}$$

The existence of a match between hot stream i and cold stream j is designated with an integer variable, Y_{ij}. Y_{ij} takes the value one if streams i and j exchange heat; it is zero otherwise. If Y_{ij} takes the value zero, then the associated continuous variables Q_{ijt} should take zero values. This is represented by the equation

$$\sum_{t=1}^{T} Q_{ijt} - UY_{ij} \leq 0 \qquad i{=}1..H, \quad j{=}1..C \tag{8}$$

where U represents the minimium of the maximum heat loads of hot stream i and cold stream j.

(b) Area Estimation

At the level of matches, where no information is available for the actual structure of the heat exchanger network, it is not possible to incorporate a rigorous expression of the area of each match as a function of the inlet and outlet temperatures of the streams, unless a superstructure at the level of matches is introduced. Since such a superstructure will increase drastically the size of the problem, an estimation of the area expression of each match (ij) is introduced as follows:

$$A_{ij} = \sum_{t=1}^{T} \frac{Q_{ijt}}{\Delta T_{LM,ijt}\{\frac{1}{h_i} + \frac{1}{h_j}\}} \tag{9}$$

where $\Delta T_{LM,ijt}$ is the log-mean temperature difference of match (ij) in interval t, calculated based upon the inlet and outlet temperatures of streams i and j in interval t; h_i and h_j are the individual heat transfer coefficients of stream i and j , respectively; and Q_{ijt} is the heat exchanged between stream i and j in interval t.

Notice that the log mean temperature difference for each potential match (ij) and for each temperature interval t, $\Delta T_{LM,ijt}$ is fixed, and can be calculated a priori for each match (ij) for each interval t. Note that if h_i and h_j are specified, then the area of each match A_{ij} is a linear expression of Q_{ijt}, as indicated in equation (9). It should also be noted that the summation of equation (9) over all the matches results in an area target similar to the one proposed by Townsend and Linnhoff, (1984).

(c) Calculation of Additional Area

The cost expression of additional heat exchange area requires that the new area of each match be calculated. Let the variable S_{ij} represent the area of an exchanger assigned to match (ij), and variable X_{ij} represent the new area of match (ij). These variables can be calculated using the following two equations:

$$S_{ij} - \sum_{k \in \mathbf{Z(ij)}} A^k_{exist} z^k_{ij} \leq 0 \tag{10}$$

$$X_{ij} - A_{ij} + S_{ij} = 0 \tag{11}$$

The slack variable S_{ij} equals zero if the match (ij) is not assigned to an existing excanger k, as the z^k_{ij} for this match is equal to zero. If a match (ij) is assigned to an existing exchanger k, S_{ij} can take any value up to the area of exchanger k. Since the cost of the new area is minimized in the objective function, then X_{ij}, is minimized, and therefore the slack variable S_{ij} will be forced to its largest possible value. If the total required area for match (ij), A_{ij}, is less than the area of the assigned exchanger, then the slack variable will be equal to the required area, and X_{ij} will take the value zero. When the required area is greater than the existing area , the slack variable takes the value of the existing area, and X_{ij} equals the total area minus the existing area. Notice that as X_{ij} is minimized, these constraints become equivalent to the original definition of X_{ij}, namely, $X_{ij} = max(A_{ij} - A^{ex}_{ij}, 0)$.

(d) Match-Exchanger Assignments

An assignment model is required for selecting the appropriate pairs of matches and exchangers. The match-exchanger assignment subproblem can be modelled according to the following three conditions:

1. Each exchanger is assigned to only one match;
2. Each match is housed in only one exchanger;
3. No exchanger should be assigned or purchased for a match that will not occur in the retrofit network.

These conditions can be stated mathematically with two sets of equations:

$$\sum_{(ij) \in \mathbf{M(k)}} z^k_{ij} \leq 1 \qquad k=1..K \tag{12}$$

$$\sum_{k \in \mathbf{Z(ij)}} z_{ij}^k + m_{ij} - Y_{ij} = 0 \qquad i=1..H, \quad j=1..C \tag{13}$$

The first equation ensures that exchanger k is assigned to at most one match; it also allows for exchanger k to be removed from the network. The second equation requires that if match (i,j) occurs in the retrofit network, then either a new exchanger will be purchased for it, or an existing exchanger will be assigned to it. If, however, a match does not exist, then all the assignment variables associated with the match take the value zero.

Having presented the objective function (5), as well as the set of constraints (6) – (13), the overall mathematical formulation at the level of matches can be stated as follows:

$$min \sum_{ij} \beta X_{ij} + \sum_{(ijk) \in C^2} C_p^2 z_{ij}^k + \sum_{(ijk) \in C^3} C_p^3 z_{ij}^k + \sum_{(ijk) \in C^4} C_p^4 z_{ij}^k +$$

$$\sum_{(ij) \in C^5} C_p^5 m_{ij} + \sum_{(ij) \in C^6} C_p^6 m_{ij}$$

subject to

$$\sum_{j=1}^{C} Q_{ijt} + R_{it-1} - R_{it} = Q_{it}^H \qquad i=1..H, \quad t=1..T$$

$$\sum_{i=1}^{H} Q_{ijt} = Q_{jt}^C \qquad j=1..C, \quad t=1..T$$

$$\sum_{t=1}^{T} Q_{ijt} - UY_{ij} \le 0 \qquad i=1..H, \quad j=1..C$$

$$\sum_{(ij) \in \mathbf{M(k)}} z_{ij}^k \le 1 \qquad k=1..K \tag{P1}$$

$$\sum_{k \in \mathbf{Z(ij)}} z_{ij}^k + m_{ij} - Y_{ij} \le 0 \qquad i=1..H, \quad j=1..C$$

$$A_{ij} = \sum_{t=1}^{T} \frac{Q_{ijt}}{\Delta_{LM,ijt}\{\frac{1}{h_i} + \frac{1}{h_j}\}} \qquad i=1..H, \quad j=1..C$$

$$S_{ij} - \sum_{k \in \mathbf{Z(ij)}} A_{exist}^k z_{ij}^k \le 0 \qquad i=1..H, \quad j=1..C$$

$$X_{ij} - A_{ij} + S_{ij} = 0 \qquad i=1..H, \quad j=1..C$$

$$Q_{ijt}, R_{it}, A_{ij}, S_{ij}, X_{ij} \ge 0 \qquad i=1..H, \quad j=1..C$$

$$Y_{ij},\ z_{ij}^k,\ m_{ij} = [0,1]$$

This mathematical formulation corresponds to a mixed integer linear programming (MILP) problem, since it involves continuous variables, integer variables and a linear set of constraints. The solution of this mathematical model represents the minimum modification cost at the level of matches, and provides information on the reassignment of existing heat exchangers, the purchase of new exchangers and additional area, and the required structural modifications.

REMARKS ON THE MATHEMATICAL FORMULATION

(a) Restictions on Match-Exchanger Assignments

It should be noted that the summation in equation (12) is over the allowed set of matches $(ij) \in \mathbf{M(k)}$, as opposed to all combinations of (ij). Similarly, the summations in equations (10) and (13) are over the allowed set of exchangers $k \in \mathbf{Z(ij)}$, as opposed to all exchangers k. These restricted sets allow for cases where a given exchanger can only be reassigned to a subset of the existing and potential matches. For example, consider a retrofit problem with two hot streams (H1, H2), two cold streams (C1, C2), and three existing exchangers. Exchanger 1 can only be assigned to house match (H1 C1), exchanger 2 may only be assigned to house matches involving stream H1, and exchanger 3 may be assigned to any of the matches. In this case, equations (10), (12), and (13) become:

$$S_{11} - A^1_{ex} z^1_{11} - A^2_{ex} z^2_{11} - A^3_{ex} z^3_{11} \leq 0$$

$$S_{12} - A^2_{ex} z^2_{12} - A^3_{ex} z^3_{12} \leq 0$$

$$S_{21} - A^3_{ex} z^3_{21} \leq 0$$

$$S_{22} - A^3_{ex} z^3_{22} \leq 0$$

$$z^1_{11} \leq 1$$

$$z^2_{11} + z^2_{12} \leq 1$$

$$z^3_{11} + z^3_{12} + z^3_{21} + z^3_{22} \leq 1$$

$$z^1_{11} + z^2_{11} + z^3_{11} + m_{11} - Y_{11} = 0$$

$$z^2_{12} + z^3_{12} + m_{12} - Y_{12} = 0$$

$$z^3_{21} + m_{21} - Y_{21} = 0$$

$$z^3_{22} + m_{22} - Y_{22} = 0$$

Thus, the assignment variables representing undesirable match-exchanger assignments (z^1_{12}, z^1_{21}, z^1_{22}, z^2_{21}, and z^2_{22}) do not appear in the formulation. Consequently, the assignments will not be made.

(b) Decomposition of The Large Combinatorial Problem

The proposed mathematical model **(P1)** for the retrofit at the level of matches can be solved with standard branch and bound techniques, as it is classified as a mixed integer linear programming problem. It should be noted, however, that due to the explicit consideration of all potential matches in the assignment problem, there exists a large number of integer variables in the proposed mathematical formulation. This implies that if a standard branch and bound technique were to be used for the solution of **(P1)**, this could result in computational limitations because of the size of the retrofit problem.

This difficulty can be overcome by using an iterative solution method that isolates the assignment constraints in a separate subproblem, to create a linear programming (LP) assignment problem. Assignment problems have the desirable characteristic that the variables take integer values at the optimum. Thus, the z_{ij}^k and m_{ij} variables can be treated as continuous, greatly reducing the number of integer variables.

The MILP formulation **(P1)** cannot be efficiently decomposed into an assignment subproblem and another subproblem. It can be shown, however, that an equivalent mixed integer nonlinear programming (MINLP) formulation can be developed that is easily decomposed to give a pure assignment subproblem. The proposed MILP formulation **(P1)** can then be combined with elements of the equivalent MINLP formulation to provide an efficient iterative solution procedure.

ALTERNATIVE MINLP FORMULATION

An alternative form of the MILP formulation is a MINLP formulation, indicated as **(P2)**, which replaces the existing area variables, S_{ij}, and the new area variables, X_{ij}, with a set of variables X_{ij}^k representing the amount of new area associated with every possible assignment $z_{i,j}^k$. This equivalent formulation is as follows:

$$min \sum_{(ijk)\in C^1} \beta X_{i,j}^k z_{ij}^k + \sum_{(i,j,k)\in C^2} \{C_p^2 + \beta X_{i,j}^k\} z_{i,j}^k + \sum_{(i,j,k)\in C^3} \{C_p^3 + \beta X_{i,j}^k\} z_{i,j}^k +$$

$$\sum_{(i,j,k)\in C^4} \{C_p^4 + \beta X_{i,j}^k\} z_{i,j}^k + \sum_{(i,j)\in C^5} \{C_p^5 + \beta A_{i,j}\} m_{i,j} + \sum_{(i,j)\in C^6} \{C_p^6 + \beta A_{i,j}\} m_{i,j}$$

subject to

$$\sum_{j=1}^{C} Q_{i,j,t} + R_{i,t-1} - R_{i,t} = Q_{i,t}^H$$

$$\sum_{i=1}^{H} Q_{i,j,t} = Q_{j,t}^C$$

$$\sum_{t=1}^{T} Q_{i,j,t} - UY_{i,j} \leq 0$$

$$A_{i,j} - \sum_{t=1}^{T} \frac{Q_{i,j,t}}{\Delta T_{LM,t} U_{i,j}} = 0 \qquad (P2)$$

$$\sum_{(i,j)\in M(k)} z_{ij}^k \leq 1$$

$$\sum_{k\in Z(ij)} z_{ij}^k + m_{i,j} - Y_{i,j} \leq 0$$

$$X_{ij}^k \geq A_{ij} - A_{ex}^k$$

$$Q_{i,j,t}, R_{i,t}, A_{i,j}, X^k_{i,j} \geq 0$$

$$Y_{i,j}, z^k_{i,j}, m_{i,j} = [0,1]$$

The mathematical formulation **(P2)** corresponds to a mixed integer nonlinear programming (MINLP) problem since bilinearities of continuous and integer variables appear in its objective function. This equivalent formulation has the advantage that the assignment variables, z^k_{ij} and m_{ij}, appear only in the objective function and the assignment constraints. Notice that for fixed values of Q_{ijt}, R_{it}, Y_{ij}, A_{ij}, and X^k_{ij}, this MINLP reduces to a simple assignment problem:

$$min \sum_{(ijk)\in C^1} \beta X^k_{i,j} z^k_{ij} + \sum_{(i,j,k)\in C^2} \{C^2_p + \beta X^k_{i,j}\} z^k_{i,j} + \sum_{(i,j,k)\in C^3} \{C^3_p + \beta X^k_{i,j}\} z^k_{i,j} +$$

$$\sum_{(i,j,k)\in C^4} \{C^4_p + \beta X^k_{i,j}\} z^k_{i,j} + \sum_{(i,j)\in C^5} \{C^5_p + \beta A_{i,j}\} m_{i,j} + \sum_{(i,j)\in C^6} \{C^6_p + \beta A_{i,j}\} m_{i,j}$$

subject to

$$\sum_{(ij)\in \mathbf{M(k)}} z^k_{ij} \leq 1 \qquad (P3)$$

$$\sum_{k\in \mathbf{Z(ij)}} z^k_{ij} + m_{i,j} - Y_{i,j} \leq 0$$

$$0 \leq z^k_{i,j}, m_{i,j} \leq 1$$

ALGORITHMIC APPROACH

Any solution technique that implements the assignment problem formulation **(P3)** above will consist of two steps. The first step selects the matches, and calculates the heat load and area of each match. The second step assigns the matches to existing exchangers or to new exchangers and evaluates the total cost via the assignment problem.

The efficiency of the first step in generating good guesses of the matches and their areas largely determines the efficiency of the algorithm. Thus, a method which selects matches and areas that are near the optimum point of the MINLP formulation is desirable.

One such method involves relaxing the MILP formulation **(P1)**, by allowing the z and m variables to be continuous. This relaxed problem, designated as **(RP1)**, is a mixed integer linear programming problem, with integer variables $Y_{i,j}$ representing the potential process stream matches. The solution of this relaxed problem will provide process stream matches, their heat loads, and their estimated areas. The cost reported by this relaxed problem is a lower bound on both the final solution and the actual cost of this particular combination of process stream matches, heat loads, and estimated areas. The actual cost is evaluated by solving the the assignment problem **(P3)**, which also gives the match-exchanger assignments.

An iterative solution procedure can be built around **(RP1)** and **(P3)**. Solving **(RP1)** gives

TABLE 1: STREAM DATA

STREAM	T in(K)	T out(K)	FC_p (kW/K)	$ /KW yr
H 1	500	350	10	
H 2	450	350	12	
H 3	400	320	8	
C 1	300	480	9	
C 2	340	420	10	
C 3	340	400	8	
S 1	540	540		80
C W	300	320		20

$$U = 0.8 \; kW/m^2 \; ^\circ K$$

TABLE 2: COST FACTORS

$\beta = 171.4$	*Cost per square meter of heat transfer area*
$C_p^2 = 10$	*Relative cost for moving one exchanger*
$C_p^3 = 400$	*Relative cost of repiping one stream*
$C_p^4 = 800$	*Relative cost of repiping two streams*
$C_p^5 = 3460$	*Fixed charge cost of a new exchanger*
$C_p^6 = 4260$	*Fixed charge cost of a new exchanger and repiping two streams*

TABLE 3: AREAS OF EXISTING EXCHANGERS

Exchanger	Area (m^2)	Original Match
1	45.06	H 2-C 1
2	12.50	H 1-C 2
3	33.09	H 3-C 1
4	23.50	H 1-C 3
5	5.75	S 1-C 1
6	5.39	H 1-C W
7	11.49	H 2-C W

a lower bound on the final solution, while **(P3)** provides an upper bound. The procedure is driven forward by adding integer cuts to **(RP1)** at each iteration that will eliminate all combinations of process stream matches identified in previous iterations. Thus, at each new iteration, **(RP1)** will be forced to identify new combinations of process stream matches. The iterative procedure is terminated when the lower bound provided by **(RP1)** meets or exceeds the upper bound provided by **(P3)** .

Thus, a feasible solution algorithm is:

1. Solve the relaxed MILP **(RP1)** to obtain process stream matches, heat loads, and match areas. Set the lower bound on the optimum equal to the MILP objective value.

2. Determine the total cost and the exchanger-match pairings by solving the assignment problem **(P3)**. Set the upper bound of the optimum equal to the total cost.

3. If the lower bound from step one is greater than or equal to the upper bound from step 2, stop. Otherwise, add an integer cut to the MILP formulation and return to step 1.

EXAMPLE

This example, which is taken from Yee and Grossmann (1987), consists of three hot streams, three cold streams, one hot utility, and one cold utility. The data for this problem are given in Table 1. The relative cost factors for repiping streams and the cost factors for purchasing area are listed in Table 2.

The existing network, which is shown in Figure 2, involves 7 exchangers whose area are listed in Table 3. This network requires 360 kW of steam and 800 kW of cooling water, at a total cost of $44,800 /yr. The minimum utility cost of this network is $8800/yr, for a heat recovery approach temperature of 10° C. The use of steam is completely eliminated, but 440 kW of cooling water is required.

The proposed mixed integer linear programming problem (MILP) formulation of this retrofit problem involved 108 integer variables, (12 integer variables denoted the existence of a match, 84 were exchanger-match assignment variables, and 12 were new exchanger variables), 64 continuous variables associated with area calculations and the transshipment model, and 82 constraints.

The problem was solved on a Vax-Station II/GPX workstation computer, using a standard branch and bound technique (ZOOM/XMP). The branch and bound method took 1,970 CPU seconds to solve the full MILP.

The solution involved seven process stream matches, with an estimated total modification cost of $30,621. The heat loads and match-exchanger assignments for the seven matches are listed in Table 4. Four of these matches took place in the original network, and all of the existing heat exchangers were utilized. No new exchangers were selected to be purchased, and the area of each exchanger was estimated to increase. The assignment of exchangers to matches with larger required areas and the selection of no new exchangers reflect the maximal use of existing heat transfer area.

TABLE 4: RETROFIT NETWORK DATA

Match	Heat Load	Exchanger	Assignment Category	Existing Area	Estimated Area	Retrofitted Area
H 1 C 1	512	3	3	33.09	38.82	28.16
H 1 C 2	800	2	1	12.5	83.08	34.72
H 1 C 3	188	4	1	23.5	23.5	31.74
H 2 C 1	908	1	1	45.06	97.96	47.65
H 2 C 3	292	7	3	11.49	30.52	21.85
H 3 C 1	200	5	2	5.75	25.0	5.75
H 3 C W	440	6	3	5.39	15.84	12.84

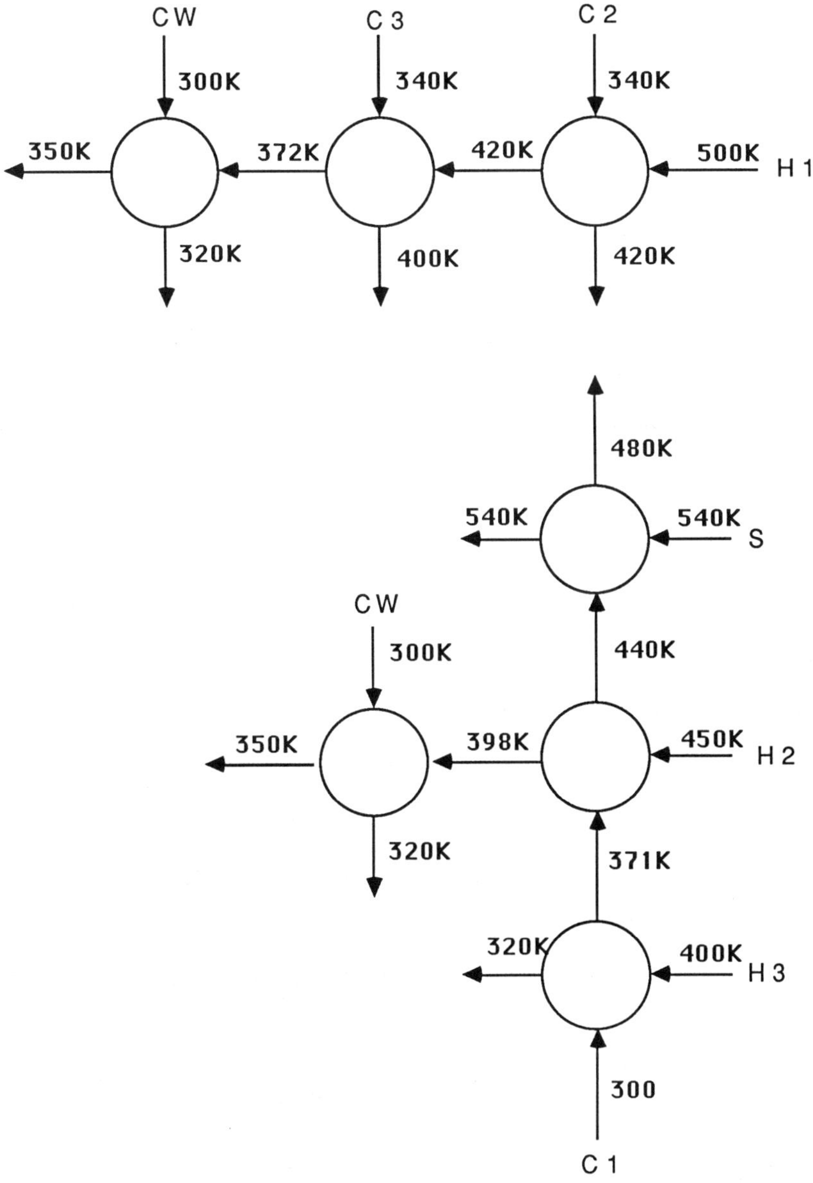

Figure 2: Existing Network in Example 1

Figure 3 shows a network configuration derived for these results and based upon a minimum temperature difference of 3°. This network requires an additional 50.9 m^2 of heat exchanger area. While the amount of additional area is less than the amount predicted by that MILP formulation, the amount of repiping is more than the predicted three streams. This extra piping arises from the two sets of parallel piped streams in the final network, a feature that cannot be predicted by the MILP formulation. Work is currently in progress to incorporate estimated repiping costs at the network superstructure level, and will be reported in a future publication (Ciric and Floudas, 1988).

The network derived from the proposed MILP formulation was compared with the network derived by Yee and Grossmann's (1987) technique that is based upon a minimum units approach. Their solution involved only six units, as opposed to seven in this paper, and featured 118.7 m^2 of additional heat exchange area. The reduction in the additional area occurs because the proposed method selects process stream matches and match-exchanger assignments based upon an estimate of the optimal use of existing heat exchange area.

CONCLUSIONS

In this paper, a two stage strategy is proposed for the retrofit problem of existing heat exchanger network configurations. Central to this strategy is a mathematical model for retrofit at the level of matches. This formulation features three main elements: a) *Every* possible match exchanger assignment is considered; b) The use of existing heat exchange area is optimized based upon an estimate of the required area of a match; and c)Repiping costs are explicitly modelled by dividing the match exchanger assignments into six structural modification categories. These elements allow the mathematical formulation to identify the process stream matches, heat loads, and match exchanger assignments of a retrofitted network at the level of matches, from the costs of stream repiping and additional heat exchange area.

As all possible match exchanger assignments are contained within the model, the use of existing heat exchange area can be optimized by moving an exchanger from one match to another. This was demonstrated in the example problem, where large exchangers were transferred from an existing match with a decreased area requirement to an existing or new match with a larger area requirement. Heat exchanger networks were derived in an example by using a superstructure methodology that minimized investment cost. The resulting heat exchanger networks featured low amounts of required additional area.

ACKNOWLEDGEMENT

The authors would like to greatfully acknowledge financial support from the National Science Foundation under Grant DMC-8617239.

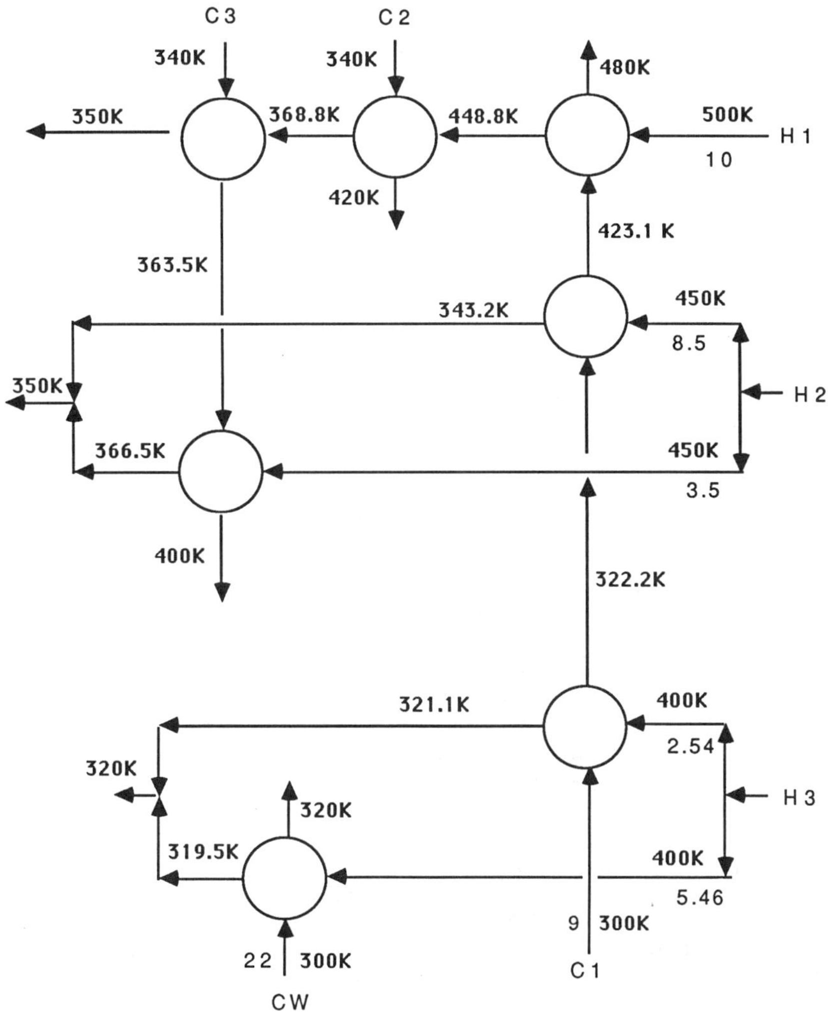

Figure 3: Retrofit Network for Example 1

REFERENCES

Cerda J. and A.W. Westerberg, "Synthesizing Heat Exchanger Networks Having Restricted Stream/ Matches Using Transportation Problem Formulations", Chemical Engineering Science, 38, pp 1723-1740, 1983.

Ciric A.R. and C.A. Floudas, "A Mixed-Integer Nonlinear Programming Formulation for the Retrofit of Heat Exchanger Networks", in preparation, 1988.

Floudas C.A., A.R. Ciric, and I.E. Grossmann, "Automatic Synthesis of Optimum Heat Exchanger Network Configurations", AIChE Journal, 32, pp 276-290, 1986.

Linnhoff B. and E. Hindmarsh, "The Pinch Design Method for Heat Exchanger Networks", Chemical Engineering Science, 38, pp 745-763, 1983.

Gundersen T. and L. Naess, "The Synthesis of Cost Optimal Heat Exchanger Networks, an industrial review of the state of the art", Proceedings of Chemical Engineering Fundamentals CEF '87: The Use of Computers in Chemcial Engineering, pp. 675-704, Taormina, Italy (1987).

Jones, D.A., A.N. Yilmaz, and B.E. Tilton, "Practical Synthesis Techniques for Retrofitting Heat Recovery Systems", paper 35c, AIChE Annual Meeting, Chicago (1985).

Jones D.A., A.N. Yilmaz, and B.E. Tilton, "Synthesis Techniques for Retrofitting Heat Recovery Systems", Chemical Engineering Progress, 82, pp 28-33,1986.

Kendrick D. and A. Meeraus, GAMS: An Introduction, Users Manual for GAMS, Development and Research Dept. of the World Bank (1987).

Marsten R., Users Manual for ZOOM/XMP, The Department of Management Information Systems, Univ. of Arizona. (1986).

Murtagh B.A. and M.A. Saunders, MINOS 5.0 User's Guide - Appendix A: MINOS 5.1, Technical Report SOL 83-20, Systems Optimization Laboratory, Dept. of Operations Research, Stanford University, CA (1986).

Papoulias S.A. and I.E. Grossmann, "A Structural Optimization Approach in Process Synthesis - II. Heat Recovery Networks", Computers and Chemical Engineering, 7, pp 707-721, 1983.

Paules IV, G.E. and C.A. Floudas, "APROS: A Discrete-Continuous Optimizer for Solution of Mixed-Integer Nonlinear Programming Problems," presented at the ORSA-TIMS Meeting, St. Louis MO (Oct. 1987).

Saboo A.N., M. Morari and R.D. Colberg, "RESHEX - An Interactive Package for the Synthesis and Analysis of Resilient Heat Exchanger Networks I: Program Description and Application", Computers and Chemical Engineering, 10, pp. 577-589 (1986).

Tjoe T.N. and B. Linnhoff, "Using Pinch Technology for Process Retrofit", Chemical Engineering, 93, pp 47-60, 1986.

Townsend, D.W. and B. Linnhoff, "Surface Area Targets for Heat Exchanger Networks", 11th

Annual Research Meeting on Heat Transfer, University of Bath, April (1984).

Yee T.F. and I.E. Grossmann, "Optimization Model for Structural Modifications in the Retrofit of Heat Exchanger Networks", Engineering Design Research Center, Report EDRC-06-25-87, Carnegie-Mellon University, 1987.

A FUZZY SET BASED ENVIRONMENT FOR DESIGN OPTIMIZATION

Alejandro R. Díaz
Department of Mechanical Engineering
Michigan State University
East Lansing, MI 48824

INTRODUCTION

An important aspect of design involves the unification of criteria for design evaluation into a global goal in which conflicting design objectives are balanced. This paper focuses on methods to represent this goal in an operational form, suitable for computations within the format of optimization algorithms. The methodology relies on fuzzy set models for design evaluation.

Standard optimization formulations do not model well the sequence of synthesis-evaluation-decision stages that characterize many design processes. In all forms of single objective optimization the decision stage occurs *before* the evaluation and is mostly limited to the selection of the performance measure to be used as the objective in the optimization problem. Multiobjective optimization goes one step further but is typically too rigid to deal effectively with more general decision strategies, involving for instance fully cooperative or non-cooperative criteria, subjective evaluations, or problems in which performance criteria are developed in a hierarchical fashion. Fuzzy optimization has been used to deal with some of these features but most applications have been limited to symmetrical aggregations of objectives and intersection operators. The present study of discusses how fuzzy optimization can be used to model more general problems.

BACKGROUND

Given a universe Y and a continuous function μ_G on Y with range [0,1], a fuzzy set G is an ordered pair $(a,\mu_G(a))$, $a \in Y$, in which transition between membership and nonmembership in G is gradual rather than abrupt and is measured by μ_G. The concept of fuzzy set can be associated easily with a performance criterion in design evaluation. For instance, one may informally make reference to the 'set of reliable structures' or the 'set of compact designs' when reliability and compactness are desirable attributes of a design. These loosely described sets are fuzzy sets in the universe of all possible designs. Associated with each set there may be one or several performance measures that can be used to quantify the membership of a particular design in the set. For instance, reliability of a structure may be measured by the maximum stress under a given loading condition and compactness of a design may be measured by some critical dimension or volume. The set

$$G \equiv \{ \text{ x is a reliable structure } \}$$

is associated with the proposition

$$p \equiv \text{x is a reliable structure}$$

which induces a possibility of degree $\mu_G(\sigma)$ that the maximum stress in the structure x is σ. μ_G also measures the agreement of $\sigma(x)$ with the proposition 'x is reliable'. It is worth noting that μ_G may exist even when an explicit performance measure such as stress or volume does not. This may be the case, for instance, in the sets

$$G \equiv \{ \text{ x is manufacturable } \} \text{ or } G \equiv \{ \text{ x is attractive } \}$$

Also, a performance criterion need not be a function of only one performance measure. Indeed, one may have

$$G \equiv H(G_1,G_2,\ldots,G_p)$$

where H indicates set operations on sets G_i that depend on different performance measures. These different ways to construct G are depicted in Figure 1.

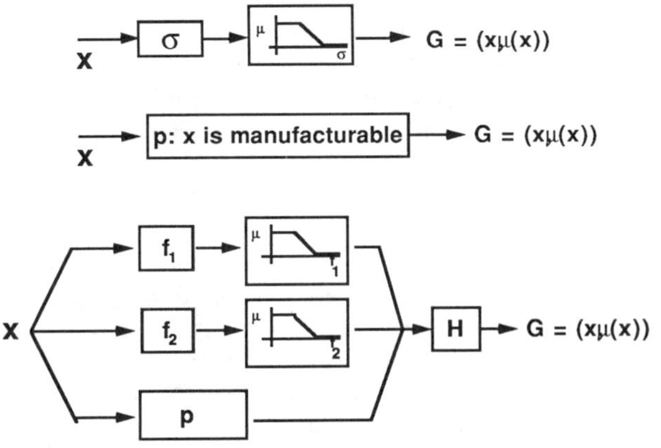

Fig 1. Construction of the goal set G

Changes in the relative importance of individual goals (performance criteria) can be introduced by means of *modifiers* . For instance, the goal

$G \equiv \{$ x is reliable $\}$

may be *strengthened* (*weakened*) into

$G' \equiv \{$ x is *very* (*somewhat*) reliable $\}$

using strengthening (weakening) modifiers. Modifiers affect the fuzzy sets G_i by changing membership functions μ_i. Different types of modifiers have been used in the literature (see, for instance, [1] and references therein). For instance *group 1* modifiers are strictly increasing mappings $m_1 : R \rightarrow R$ such that, if μ_G is the membership function of G,

$$\mu'(f(x))=\mu(m_1(f(x))), \quad f\in R$$

is the membership function of the *modified set* G' associated with G. Similarly, *group 2* modifiers are continuous, nondecreasing mappings $m_2:[0,1]\rightarrow[0,1]$ such that

$$\mu'(f(x))=m_2(\mu(f(x)))$$

Typical examples of modifiers are

$m_1(f)= a*f+b$, a,b scalars (threshold changes)
$m_2(u)= u^t, \in (0,1], t>0$ (linguistic hedge [2])
$m_2(u)= \min\{1,(u/\alpha)\}, \alpha \in (0,1]$ (weakening α-cut [3])
$m_2(u)= \max\{0,(u-\alpha)/(1-\alpha)\}, \alpha\in [0,1)$ (strengthening α-cut [3])

AGGREGATION MODELS IN FUZZY OPTIMIZATION

The fuzzy set representation of design performance is well suited to model more complex decision strategies than is currently possible using standard single or multiple objective optimization formulations. Examples of different complexity involve non-symmetrical aggregation of goals and hierarchical constructions of performance criteria.

A type of decision behavior that is not easily modeled with traditional single or multiple objective formulations involves fully cooperative and noncooperative and non-symmetrical decision strategies. In a cooperative strategy a design may be acceptable on the basis of one criterion alone, regardless of the level attained by other criteria (this is not single objective optimization because the relevant criterion is not known *a-priori*). In a noncooperative strategy a design alternative may be rejected on the basis of poor performance in one criterion alone, regardless of the levels attained by other criteria (this is not the same as a standard constraint, since no notion 'quality of design' is associated with a satisfied constraint). Models for cooperative (union '$\cup$') and noncooperative (intersection '$\cap$') aggregations of fuzzy sets have been extensively studied by Dubois and Prade [4].

By far, most fuzzy optimization problems found in the literature deal exclusively with symmetrical intersection operators, i.e.,

$$G=G_1 \cap G_2 \cap ,..,\cap G_p \quad \text{and} \quad \mu_G=\min(\mu_1,\mu_2,...,\mu_p)$$

It is possible, however, to build more complex models of an overall decision strategy. For example, the proposition

$$p \equiv \text{'x is reliable'}$$

may be found to be accurately modeled by some combination of the sets

$$G_1 \equiv \{ x_1 \text{ is reliable } \}, G_2 \equiv \{ x_2 \text{ is reliable } \}, \quad G_3 \equiv \{ x_3 \text{ is reliable } \}$$

where x_1, x_2, and x_3 are possibly three components of x (or perhaps the same x in three different environments). One may quite naturally associate p with any of the following sets G obtained from operations on G_1, G_2, and G_3:

$$G \equiv \{\text{either } x_1, x_2 \text{ or } x_3 \text{ is reliable}\} = G_1 \cup G_2 \cup G_3$$

$$G \equiv \{ \text{all } x_1, x_2 \text{ and } x_3 \text{ are reliable}\} = G_1 \cap G_2 \cap G_3$$

$$G \equiv \{\text{two of } x_1, x_2 \text{ and } x_3 \text{ are reliable}\} = (G_1 \cap G_2) \cup (G_1 \cap G_3) \cup (G_2 \cap G_3)$$

In the general case one may define the overall objective G as an arbitrary combination of individual objectives, e.g.,

$$G \equiv H(G_1,G_2,...,G_p).$$

μ_G is obtained from the functions μ_i associated with the G_i's by means of an *aggregation operator* h. h is a continuous operator from $[0,1]^p \rightarrow [0,1]$ that implements the set operations prescribed by H, i.e.,

$$\mu_G \equiv h(\mu_1,\mu_2,...,\mu_p),$$

A fuzzy set formulation can accommodate complex operations such as these in a natural and straight forward manner [4,5], once membership functions for the G_i's are known (Figure 2). It is noted here that no statement is made regarding which

aggregation of preformance criteria is the correct one and that the fuzzy set model does not provide an answer to this question. It does, however, provide powerful *tools* with which to study complex strategies in an organized and computationally efficient fashion and implement them in an optimization environment. Examples of such implementations are the single-level fuzzy optimization problem (SFOP) and the multi-level fuzzy optimization problem (MFOP) discussed in the following sections.

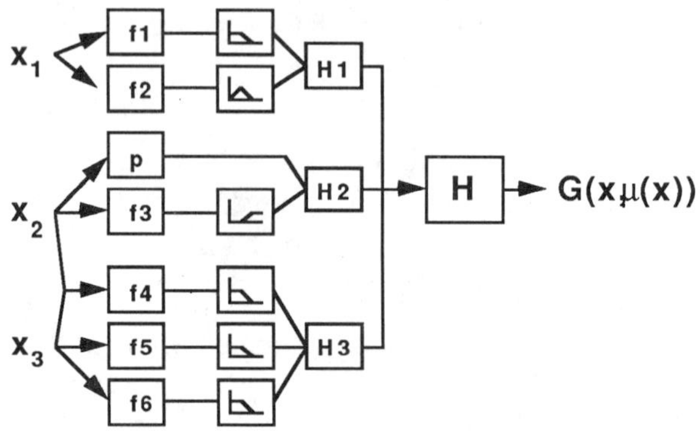

Fig 2. Construction of complex objectives

SINGLE-LEVEL FUZZY OPTIMIZATION PROBLEM (SFOP)

In the SFOP each goal G_i is associated with a *single* performance measure f_i. Also, if x represents a list of design variables with possible values in a universe U, all criteria associated with the evaluation of x are fuzzy sets $G_i=(x,\mu_i(x))$ with universe U. This requirement indicates that, whenever a design alternative a is described by design variables $(x^1,x^2)\in U^1 \times U^2$, the optimization of x^1 and x^2 can be carried out independently, i.e., the problem is separable.

The SFOP is a generalization of the standard multicriteria optimization problem with individual objectives equal to the performance measuring functions f_i. Operationally, the SFOP takes the following form: Find $x\in R^n$ that

$$\text{maximizes } \mu_G(x)=h(\mu_1,\mu_2,\ldots,\mu_p) \qquad x\in X\subseteq R^n$$

where h is the aggregation operator associated with $H(G_1,G_2,\ldots,G_p)$ and X denotes the usual set of *crisp* constraints on x. It is emphasized that, in contrast with typical fuzzy optimization problems, general, possibly non-symmetrical operators H are considered. A diagram illustrating this problem is shown in Figure 3 and a diagram illustrating the incorporation of modifiers in the SFOP is shown in Figure 4.

To make the decisions produced by the SFOP meaningful, some rules must be set for the construction of h [4,5]. The following set of rules is typical and can be traced back to what appears to be *reasonable* behavior:

1. A design that attains all individual goals must be fully acceptable
   $h(1,1,\ldots,1)=1$ and $h(0,0,\ldots,0)=0$

2. Small improvements in one goal do not produce large decision changes
   h is continuous

3. Improvement in *any* goal without without worsening of others does not produce worse solutions.

if $\mu_i(x^2) \geq \mu_i(x^1)$ for all i, $\mu_G(x^2) \geq \mu_G(x^1)$

4. Improvement in *some* goal without worsening of others improves partially acceptable solutions.

if $0<\mu_G(x^1)<1$, there exists a j such that, if $\mu_j(x^2)>\mu_j(x^1)$, $\mu_G(x^2)>\mu_G(x^1)$

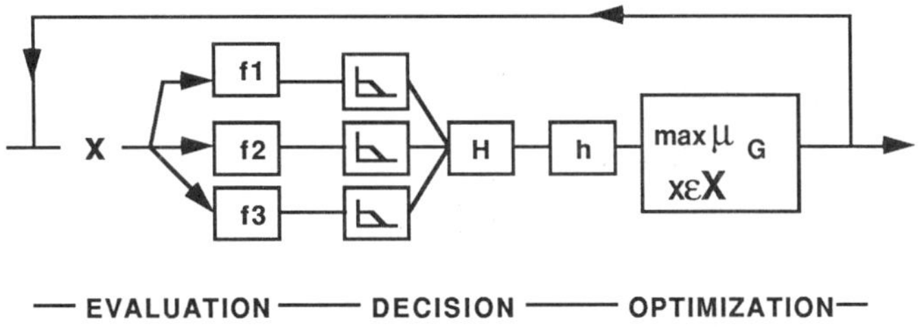

Fig 3. Single-level Fuzzy Optimization Problem

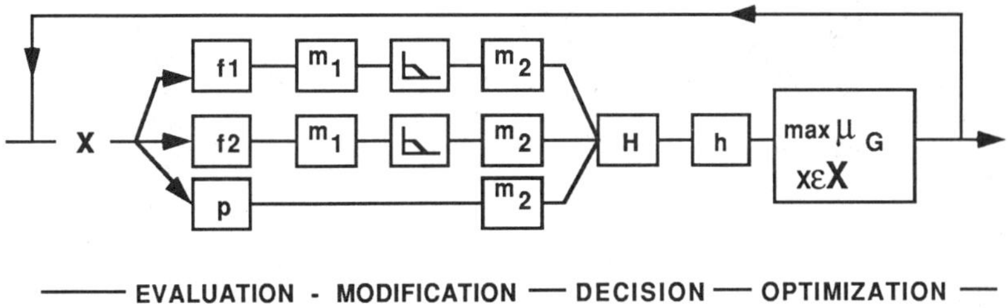

Fig 4. SFOP with modifiers

When these rules for the construction of SFOP's are followed it is possible to relate solutions of the (possibly non-symmetrical) SFOP to weakly efficient solutions arising from standard multiobjective problems. The basis for the comparison is the following theorem from [6] which applies to symmetrical or non-symmetrical operators h

Theorem: SFOP's and weakly efficient solutions. Let x^* be a solution to a SFOP and let $F(x)=\{f_1(x),f_2(x),...,f_p(x)\}$. Suppose that:

i. h satisfies rules 1-4

ii. The μ_i's are continuous and strictly decreasing over $\mu_i^{-1}(0,1)$

iii. $\mu_G(x^*)\in(0,1)$

Then x^* is a weakly efficient solution, i.e.,

$$x^*\in E^o=\{x\in X: \text{there is no } x^o\in X \text{ such that } F(x^o)<F(x)\}$$

As expected, performance criteria associated with measures f such that "less of f is always better" typically lead to efficient or at least weakly efficient solutions, i.e., solutions obtainable via traditional multiobjective optimization methods. However, thresholds of acceptability or rejection or fully cooperative decision strategies tend to make solutions to the SFOP not efficient and perhaps not even weakly efficient. A similar situation arises in problems with performance criteria of the type "*the closer f is to f^o, the better*" (μ not decreasing). These features arise in perfectly reasonable and well posed design problems . Application of traditional multiobjective or fuzzy optimization under these circumstances may not provide the best possible design solution.

COMPUTATIONAL IMPLEMENTATION

The SFOP is being implemented in a computational environment designed to provide support in the construction, interpretation, and modification of the model $M=\{\mu,H,h)$. The objective is to provide the tools necessary to construct a model M that reflects the user's (designer's) particular set of priorities and to help interpret how the results reflect these priorities. The tools being developed include the following features:

1. Membership functions can be selected from a library of standard functions to construct sub-goals G_i associated with *explicit* measures of performance. These may be modified to reflect agreement with a particular level of performance.

2. When no explicit function is available to measure performance, it may still be possible to construct μ_i *locally* , by sampling or by interfacing with a rule-based system that evaluates performance based on prescribed rules. The specific implementation of these features is currently under investigation.

3. System support is provided for standard set operations: union, intersection, addition, complementation. These are used to model cooperative, noncooperative and trade-off modes of aggregation.

4. An interpreter is provided to translate linguistic primitives (EITHER, AND, OR, BOTH, NEITHER, NONE, TWO OF THREE, SOME, SEVERAL, MOST) into aggregation rules.

5. Any goal at any level may be constructed from lower-level objectives by providing the appropriate aggregation rule. This feature is essential for future development of multi-level problems, outlined in the following section.

6. The level of interaction among goals is controlled by allowing different choices for the operator h. Options range from Zadeh's noninteractive "min" and "max" operators to controlled interaction operators such as Zimmermann's compensatory conjunction operator.

7. Strengthening and weakening modifiers can be used to change the decision model M. Simple linguistic primitives such as MORE, SOMEWHAT, MUCH, LESS can be used to affect the importance of goals at any level.

8. The interpretation of results associated with a particular model is enhanced by providing levels of satisfaction achieved by each goal and quantitative estimates of the effect of each goal on the decision (measures of dominance [6]).

This list of features is included here to sketch the kind of operations that are possible in a computational implementation of a fuzzy set based optimization problem. These operations are meaningful and useful in the construction of single-level fuzzy optimization problems but have a stronger impact in multi-level problems, outlined in the following section.

MULTI-LEVEL FUZZY OPTIMIZATION PROBLEMS (MFOP)

While standard fuzzy and multiobjective optimization (compromise programming, goal programming) problems can be viewed as special SFOP's, the potential of the fuzzy set based approach in design is more evident in multi-level problems.

In a multi-level fuzzy optimization problem (MFOP) optimization may occur at any level of the evaluation of the partial goals G^j. Also, the overall performance at each local level where the evaluation-decision-optimization occurs depends on design variables that are not in the local design space. This indicates that, if U is the design universe and G^j is a fuzzy set representing the local goal at level j, $G^j=(x^j,\mu(x^j))$, $x^j \in U^j$, the universe of G^j is a *strict subset* of U. This formulation corresponds to the notion of system decomposition in optimization.

An example of a MFOP is illustrated in Figure 5. In this example the design is evaluated by six performance criteria, each identified with a fuzzy set $G^1...G^6$. The design universe U is projected onto six subspaces $U^1,...,U^6$, each subspace becoming the universe of a corresponding fuzzy set, i.e.,

$$G^j=(x^j,\mu^j(x^j)),\ x^j=P^j(x)\in U^j,\ x\in U$$

Each P^j is a projection on U with range U^j. Let P^{jo} be the orthogonal projection such that, if $x^{jo}=P^{jo}(x)$,

$$x = x^j+x^{jo}$$

In the MFOP a single-level optimization problem is solved for each goal G^j keeping x^{jo} fixed (i.e., each circle in Figure 5 represents a SFOP). In general, the range of the P^{jo}'s is not null and some iterative scheme must be implemented to reach an optimum solution in U.

The criterion at the 'top' level, G^1, is a measure of the overall performance of the design x. It is an *explicit* function of only x^1, but it depends implicitly on x^{1o} via the relation between G^1 and the other goals. In the example, G^1 is connected (thick arrows) to G^2, G^3, and G^4, indicating the existence of an aggregation operator H such that

$$G^1 = H(G^2,G^3,G^4)$$

A similar connection exists between G^4 and G^5 and G^6. However, the explicit nature of this connection is not important in the computation of G^1. Furthermore, G^2, G^3, and G^4, may themselves be formed by the aggregation of several performance measures but the details of this internal aggregation of goals is not essential to the evaluation of G^1.

The MFOP has several attractive features. One important advantage is that it provides a framework for decomposition in *both* the design space (basis $\{x^j\}$) *and* the performance space (basis $\{\mu^j\}$), a natural result of the definition of fuzzy sets as ordered pairs of elements in both spaces. Also, in the MFOP formulation the relevance of each sub-problem in the evaluation of overall performance is explicit and quantifiable for any design alternative via measures of relevance [6]. This is invaluable information in computations since it suggests an ordering for the solution of the sub-problems that is not available in

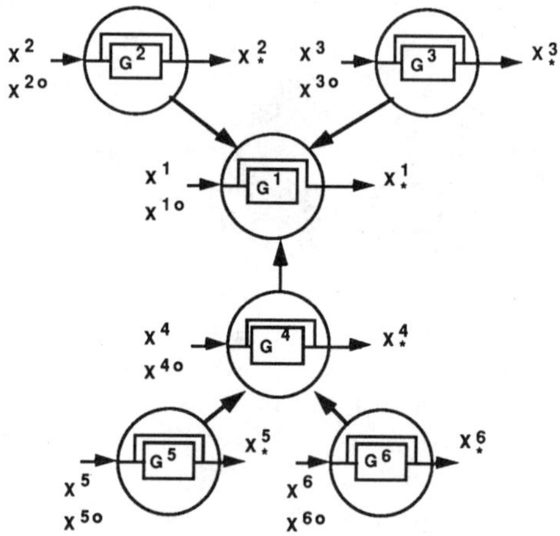

Fig 5. Example of MFOP

other decomposition schemes. This is not intended to play down the difficulties associated with the solution of strongly coupled problems but, instead, to add more solution tools to attack a difficult problem. Many difficulties are still unsolved and specific implementations of multi-level fuzzy optimization problems are still under investigation.

ACKNOWLEDGEMENT

This material is based upon work supported by the National Science Foundation under Grant No. DME 8706902. This support is gratefully acknowledged.

BIBLIOGRAPHY

[1] Dubois, D. and Prade, H., "Criteria Aggregation and Ranking of Alternatives in the Framework of Fuzzy Set Theory," in H. Zimmermann, L. Zadeh, and B. Gaines, Eds., Fuzzy Sets and Decision Analysis, TIMS Studies in the Management Sciences, Vol. 20, 1984, pp. 209-240.

[2] Zadeh,L., "PRUF - A Meaning Representation Language for Natural Languages," Int. J. Man-Machine Studies , Vol. 10, 1978, pp. 395-460.

[3] Negoitia C. and Sularia, M., "On Fuzzy Programming and Tolerance in Planning," Economic Computation and Economic Studies in Research, 1, 1976, pp. 3-14.

[4] Dubois, D. and Prade, H., "A Review of Fuzzy Set Aggregation Connectives," Information Sciences , Vol. 36, 1985, pp. 37-51.

[5] Díaz, A., "Goal Aggregation in Design Optimization," Engineering Optimization , Vol. 13, 1988, pp. 257-273.

[6] Díaz, A., "Fuzzy Set Based Models in Design Optimization," Proceedings of the ASME Design Automation Conference, Orlando, Florida, Sept. 1988.

CKAAD-EXPERT
A Computer and Knowledge Aided Architectural Design Expert *

Chris I. Yessios, Department of Architecture
Richard E. Parent, Department of Computer and Information Science
The Ohio State University **

An architectural problem solving and design oriented expert system has been designed and is being implemented as an integral part of an architectural modeling oriented CAD system. It is aimed at the enhancement of the design process in ways such that the individualistic and creative aspects of design are reinforced rather than restricted. The system does not dictate solutions but assists the architectural problem solver by offering advice and by evaluating candidate solutions generated by a user.

A knowledge base is built by a project team utilizing domain expertise from whatever sources may be available. In addition to entering knowledge through plain language and/or symbolic representations, the system supports the automatic extraction of knowledge from the graphic representations of prior architectural solutions. The knowledge derived from the variety of sources is merged into a single internal representation of a "model architectural program". This is externalized and communicated to the user in one of three formats: as a program listing, a functional diagram or an interaction matrix. These are equivalent and each format is supported by rather extensive editing capabilities which allow insertions, deletions and revisions. The manner in which the knowledge base is built allows the user to include individualistic preferences, which thus makes the generation of the knowledge base a part of the overall problem solving process.

Once a knowledge base has been constructed, it can be processed by an inference engine to perform the role of an advisor and the role of an appraiser. The user seeks advice during the tentative and exploratory stages of design as he/she uses the basic CAD system to generate a solution model for a building. Once a preliminary solution has been partially derived or completed the system may be requested to evaluate it. Such an appraisal consists of a grade depicting the degree of conformity relative to the requirements and rules of the knowledge base and the graphic identification of specific areas of the solution which contradict the programmatic requirements. The system does not by itself revise a solution but rather points out where corrections are needed.

The design of the CKAAD-Expert has been dictated by a recognition of the traditional methods of architectural design and its highly idiosyncratic and ill-defined nature. The system will initially be tested by solving two types of architectural problems: a campus library and a residential development.

---

* NSF Project # DMC-8609893, Knowledge Aided Architectural Problem Solving and Design, The Ohio State University.

** Graduate Research Associates to the project have been Wayne Brown, Department of Computer and Information Science, Constantine Terzidis and Sylvain Boulanger, Department of Architecture.

Computational Tools for Preliminary Engineering Design

Kristin L. Wood and Erik K. Antonsson
California Institute of Technology
Department of Mechanical Engineering

The primary objective of this work is to develop tools to aid the designer in developing designs; in making analyses; and in disqualifying unsatisfactory design configurations at the preliminary stages, where several different concepts (usually imprecisely defined) are simultaneously evaluated for feasibility and relative merit. Such tools are not "artificial intelligence", in that the designer is providing both the creative idea generation, and making the decisions. Instead, the idea is to provide computational aids which present the information to the designer in such a way that the behavior of the input parameters with respect to the functional requirements may be readily understood.

At this juncture in the research, the first class of such tools has been developed. The foundation of this class of tools is a technique to perform design calculations on imprecise parameters. By interpreting the input variables as fuzzy sets, the imprecision as well as the engineer's subjective knowledge may be encoded. Calculations may be performed with this technique, producing (imprecise) performance parameters from the imprecise input design parameters. Subsequent ranking of each input parameter's importance with respect to the design's functional requirements may be found by using a numerical measure developed by the authors.

Besides the first class of tools, current work is progressing in three other areas: (1) comparing the imprecise calculation technique mentioned above with results produced from applying probability calculus; (2) combining the classical methods of uncertainty analysis with the imprecision technique; and (3) developing a "data-flow" paradigm for design. The data-flow concept provides the means for schematically describing the design space, for integrating the computational techniques mentioned above into a usable computing environment, and for implementing parallel-computing techniques to the purely computational and simulation components of design.

For a more detailed description of the research and results, the reader is referred to [3], [2], and [1]. This material is based upon work supported by: The National Science Foundation (NSF) under a Presidential Young Investigator Award, Grant No. DMC-8552695; The Caltech Program in Advanced Technologies; and an IBM Faculty Development Award.

References

[1] K. L. Wood and E. K. Antonsson. Computations with imprecise parameters in engineering design: applications and examples. *ASME Journal of Mechanisms, Transmissions, and Automation in Design*, June 1988. Submitted for review.

[2] K. L. Wood and E. K. Antonsson. Computations with imprecise parameters in engineering design: background and theory. *ASME Journal of Mechanisms, Transmissions, and Automation in Design*, January 1988. Submitted for review.

[3] K. L. Wood and E. K. Antonsson. A fuzzy sets approach to computational tools for preliminary engineering design. In S. S. Rao, editor, *Advances in Design Automation*, pages 263–271, ASME, New York, September 1987.

Stochastic From Optimization

Abstract

Tomasz Arciszewski
Civil Engineering Department
Wayne State University, Detroit
ARCISZEWSKI%WAYNE_MTS@UM.CC.UMICH.EDU

Fax: (313) 577-3881

The objective of conceptual design is to develop a concept of a future engineering system. Very often, a standard, well-known concept is sought. Sometimes, however, an unknown but feasible concept is desired, one which is innovative and possibly patentable. In this case, conceptual design can be considered as a process of creative problem solving, and the principles of heuristics can be used.

The proposed conceptual design method, Stochastic Form Optimization, is based on the heuristic principle of delayed judgement, and on the morphological analysis developed by Zwicky. This method is proposed for engineering conceptual design when innovative concepts are desirable. It can be used by a single designer or by a group of designers. Domain and methodological experience improve the chances for the development of patentable concepts, but the method can be used even by novice designers, and is useful for teaching conceptual design.

A broad background on Stochastic Form Optimization is given, and the method is discussed in the context of engineering creativity. The method's basic assumptions are discussed, as well as its relationship to numerical and shape optimization methods. The procedure is given and its individual steps discussed. A computer program is described which was developed for the random generation of combinations of feasible states from a given morphological table. This program is based on the stochastic simulation of morphological analysis; a nonhomogeneous Markov chain is used as a mathematical model. The developed model can also be utilized for the development of a heuristic component in complex computer programs for conceptual design. This heuristic component can then be used for the random generation of concepts, when other, more traditional approaches to conceptual design fail, or when only innovative concepts are sought.

The developed method has been used to generate several inventions; two of these are mentioned here. The first is for a new type of connection between column and beams in a skeleton structure. This is now patented. The second invention, whose patent is still pending, is for a new type of spherical joint in steel space structures.

GLOBALLY OPTIMAL DESIGN USING SIMULATED ANNEALING

Alice M. Agogino, P.E., Principal Investigator
Pramod Jain, Graduate Research Assistant
Department of Mechanical Engineering
University of California, Berkeley, CA 94720

Sponsored by NSF PYI grant # DMC-8451622

Abstract

A new approach to optimization in nonlinear nonconvex mechanical design problems (e.g., kinematic synthesis of mechanisms, design of multi-speed gearboxes, scheduling and process planning in manufacturing processes, and shape optimization) is developed which is based on simulated annealing, a probabilistic combinatorial optimization technique [1,2]. Simulated annealing is a probabilistic *hill jumping combinatorial optimization* (CO) technique. A *hill jumping* technique is one which allows for an increase in the value of the objective function in a controlled manner, as opposed to a *hill climbing* interative technique, which only accepts moves which result in an immediate improvement in the objective function. Hill jumping implies the possibility of "jumping" out of a local optimum. The field of *combinatorial optimization* involves developing efficient techniques to find the minimum value of an objective function for problems with a large number of independent discrete variables. The advantage of this technique is that it converges to the global optimum asymptotically. Simulated annealing has been successfully applied to solve several combinatorial optimization problems in the fields of computer-aided design of VLSI chips, image processing and graph theory. However, it is not widely used as an optimization tool in the field of mechanical design. Applications of simulated annealing to several problems in mechanical design show that the method is fast and obtains globally optimal solutions.

References

[1] Jain, P. and A.M. Agogino (1988), "Optimal Design of Mechanisms using Simulated Annealing: Theory and Applications," to appear in: *Proceedings of the 1988 ASME Design Automation Conference*, (Working Paper 88-0102-P, Berkeley Expert Systems Technology Laboratory, 5136 Etcheverry Hall, University of California, Berkeley, CA 94720.)

[2] Agogino, A.M., J. Cagan, and P. Jain (1988), "Optimally-Directed Design: Innovation to Analysis," (Working Paper 88-0502-1, Berkeley Expert Systems Technology Laboratory, 5136 Etcheverry Hall, University of California, Berkeley, CA 94720.)

CONSTRAINT MANAGEMENT IN CONCEPTUAL DESIGN

by
David Gossard and **David Serrano**
Computer Aided Design Laboratory
Department of Mechanical Engineering
Massachusetts Institute of Technology
Cambridge, Massachusetts 02139

ABSTRACT

The management of constraints during conceptual design is a non-trivial task. Of particular interest are "mathematical" constraints that relate to a design's performance (i.e. function), physical laws it must obey (i.e. physics) and its geometrical and topological properties (i.e. form). Constraints are continually being added, deleted and modified as a design develops. The constraints are often numerous, complex and contradictory. The resulting constraint set may contain conflicting and/or unrealizable requirements.

Effective tools for managing large sets of constraints will play an important role in future conceptual design systems. This work develops basic methods for constraint management in such systems, specifically for representing, evaluating and maintaining consistency in sets of constraints.

A **constraint** relationship (or simply a constraint) is defined as a set of **parameters** which behave according to a specified mathematical relationship. The constraint relationships may be piecewise continuous equalities or inequalities. The parameters may take on continuous or discrete values. The set of parameters is partioned into two subsets: known parameters and unknown parameters. The **status** of a parameter is its classification as a (known) constant or a (unknown) variable. The **causality** of a constraint specifies dependency relationships between the constraint's parameters. Constraints may be **causal** or **acausal.** Individual constraints as well as systems of constraints (i.e. constraint networks) are modeled as a **constraint graph**: a directed graph (digraph) whose nodes denote parameters and whose arcs denote constraint relationships.

To evaluate a system of constraints, an assignment or **matching** of every unknown parameter to one constraint relationship is performed. The matching, represented by a bipartite graph, is used to convert the original constraint graph to a tree-like structure. When the resulting structure is a simple tree, the evaluation sequence is given by a **breadth-first traversal** of the constraint graph. The presence of a cycle identifies constraint subsets that must be solved simultaneously, In the experimental system developed in this work, constraint subsets are solved by a Newton-Raphson iterative procedure where the partial derivatives of the constraint equations are generated symbolically. Methods were developed to identify constraints which are redundant (i.e. inactive) and contradictory (i.e. in conflict with other constraints), and to eliminate conflicting constraints.

The constraint management methods discussed above were implemented in an experimental system for conceptual design. The system provides the user with a menu of predefined "concept-models" of common machine elements. Each machine element in the system contains a set of constraints which predict its performance, establish its physical limits, and define its topological (connectivity) restrictions. In addition, each object has an iconic representation for user interaction.

The user can interactively select individiual components from the menu, specify their connectivity and the causality of the resultant aggregate "concept model". The system creates the aggregate model by fusing the individual models, evaluates the aggregate model given the known parameters and identifies redundant or conflicting constraints. Concepts may be stored and retrieved and may be combined hierarchically into higher-level concepts.

PART 6

DESIGN FOR MANUFACTURING AND THE LIFE CYCLE

Design for Manufacturing and the Life Cycle

Howard M. Bloom
Factory Automation Systems Division
National Bureau of Standards

ABSTRACT

This paper describes "Design for Manufacturing and the Life-Cycle" and why such a research program is needed. The architecture of a design workstation that has the capability of performing concurrent design functions is described. The architecture of a larger system, a Manufacturing Commodity Exchange System, is also defined to illustrate the concept of unified life-cycle engineering. Various research thrusts into design technologies, manufacturing related applications, and information technologies are outlined. This topic is one of six topics that cover the Design program at the National Science Foundation. Comments are given on those NSF grants that fall into the area of design for manufacturing.

INTRODUCTION

The design of manufacturable products is an increasingly complex process for industry. Advances in technology have resulted in compartmentalized life-cycle product development activities. Isolated from the designer is (a) the sponsor who has a need for the product, (b) the process planner who understands how the product can be produced in the factory, (c) the manufacturing manager who understands the operation and performance of the shop floor equipment, (d) the quality assurance inspector who understands how to measure the functionality of the product, (e) the service engineer who is responsible for repair and preventive maintenance of the product, and In addition, it has been determined that approximately 80% of the cost impact is fixed by the time the design phase is completed. Another 10% is fixed by the end of the development phase, and the last 10%

during the production phase. It is unfortunate that designers are typically unaware of potential new materials, processes, or equipment, and develop the design based on the present state of the manufacturing environment. If this information were available, the development and production phases would have a much larger impact on the cost factor, and the designer would have more flexibility in creating the product design.

At each stage of the life-cycle, important knowledge about the product is acquired. Unfortunately, this information is seldom brought to the attention of staff working on other stages. Specifically, the designer typically has no understanding of the impact of his design on the creation of a true 'world-class' product - one built to minimum cost, highest quality, shortest time-to-market, and most functionality. More attention needs to be paid to the integration and information exchange that occurs during the product life-cycle.

"Design for Manufacturing and the Life-Cycle" can be defined as a product development process that is characterized by the cooperative participation of members of the management, research, marketing, design, production, distribution, and support divisions throughout a product's life. Cooperative participation implies the existence of an explicit shared goal toward which the members direct their efforts and an organization that encourages the free exchange of information in support of that goal. This process shares many labels such as "Concurrent Design", "Unified Life-Cycle Engineering", and "Total Quality Management."

As a system, it can be viewed as an integrated approach to product design, manufacture, and use. As a computer science application, it can be viewed as a computer-based architecture for managing the concurrent design, development, manufacturing, and servicing of products. In summary, all the views are concerned with the integration of people and information that will lead to the development of a world-class product.

ARCHITECTURE FOR A DESIGN FOR QUALITY SYSTEM

If a new approach for design is to be established, an architecture must be presented that illustrates what design systems will be in the future. The first step is to define the components of a "Design for Quality" system. The next step is to create a design workstation that can function in an environment that is using the unified life-cycle engineering approach. The final step is to create a "Manufacturing Commodity Exchange System" that can demonstrate the complete integration of all phases of the life-cycle.

If the designer is to learn from the results of other phases of the life-cycle, it is first necessary to define what is a good design. As stated earlier, a world-class product is one that has the highest quality, lowest production cost, shortest time response, and most modern features. The overriding issue is the design for quality, which can be subdivided into five categories: (a) manufacturability, (b) reliability, (c) functionality, (d) serviceability, and (e) reusability. For each of these categories, measurement techniques must be created that allow for the determination of a good product design. For manufacturability, the measure is how easy it is to make. For reliability, the measure is how often does it fail. For functionality, the measure is how good it is for its intended use. For serviceability, the measure is how easy it is to repair or maintain. Finally for reusability, the measure is how easy it is to use in another product or in recycling.

The next step is to define the system architecture for an ideal design workstation that allows the designer to create the product design through an interface to the other phases of the life-cycle. It also provides a mechanism for determining when the design has reached the desired level of quality. This system is shown in Figure 1. The system consists of four major elements: (a) the designer, (b) the decision databases, (c) the design system which itself has two elements, and (d) the product data specification.

The designer represents one or more people working together to produce the product design. Important issues to be addressed in describing this component include how to represent the product requirements as a design specification, how to maintain the link between a given product component and the overall product, and how to perform "cooperative design". These topics are covered in other parts of the Design Theory and Technology program.

The decision databases consists of four types of information: (a) performance measurement techniques, (b) quality feedback data, (c) design rules, and (d) life-cycle models. The performance measurements are definitions for determining how to measure the phases of the life-cycle in terms of a quality product. They address the quality subcomponents: manufacturability, reliability, serviceability, functionality, and reuseablity. They must include the ultimate cost, time, and features of the product. The quality feedback data represent all the information obtained from the other phases of the life-cycle (both real and simulated). The type of information includes such items as the number of days the product did not function, the cost of manufacturing one product component, the actual time to production, or the number of manufacturing rejects. The design rules represent how to apply the design to the product. For example, in a design for manufacturability, there would be rules as to the relationships between certain machinable features and where they can be positioned on the part. The fourth database is the set of life-cycle models which define the processes that occur during the various life-cycle phases. These models are used to simulate the real environment. For example, the designer, in interfacing to process planning, must have access to the possible processes that are allowable in the manufacturing phase.

The design system contains two types of internal structures. The first is the set of design technology features that can be used in optimizing the design process. The second is the set of functions for controlling the design process. The design technology features are discussed in other parts of the Design Theory and Technology Program, and include such topics as cognitive models of design processes, conceptual models of design processes, prescriptive models for design processes, and analysis to support design. The control function includes such concepts as feature-based design, integration with life-cycle activities, and configuration management. Feature-based design can be used to drive automatically a life-cycle system such as a machine tool from the design database with no intermediate generation of control code. The life-cycle integration involves defining and managing the flow of information between systems. The configuration management is mainly concerned with version control of the design database as the various life-cycle activities interact with the information.

The product data specification is the common repository of all relevant information about the product that needs to be maintained during the life-cycle. This information is represented as a part model that includes various views such as topology, geometry, features, and functionality. There is an effort in the United States to develop a product data exchange specification (PDES) that can serve as a standard for product information. There is a version 1.0 of PDES that describes a conceptual model of the data required by a range of applications such as product item, functional characteristic, configuration, form feature, topology, tolerance, geometry, manufacturing process, and logistics. The interesting aspect of PDES is that there are four levels of proposed implementation:

Level 1 - Passive File Exchange which is characterized by the definition of a neutral sequential file format, which can be written to and read by a stand-alone system through the use of translation software.

Level 2 - Active File Exchange which is characterized by the definition of a temporary working form used by access software to translate to and from a neutral sequential file format and to provide DBMS-like data structure navigational capabilities.

Level 3 - Shared Database Access which is characterized by the definition of a logical data structure for resident store of product data. Various application system views of the logical data structure may be defined to the database management system (DBMS) for purposes of data update and retrieval.

Level 4 - Integrated Product Knowledge base which is characterized by a data management system that can fully implement the product data standard directly, full enforcement of integrity constraints as an integral part of the data management system, and the ability to generate any required application view of product definitions.

Once the design workstation become operational, then it is possible to implement a Manufacturing Commodity Exchange System (MCES) (see Figure 2). The core of the system is the Manufacturing Commodity Exchange Manager (MCEM) which serves as a broker between the designer and the needed life-cycle systems (as well as other designers). The MCEM maintains a database of life-cycle models that represent the capabilities of the life-cycle systems and the product specifications (i.e. PDES files) that represent the designer requirements. In this manner the designer can be quickly matched with the necessary resource in order to create the final product design properly.

CRITICAL RESEARCH AREAS

This section addresses the key areas of research that are required if the Design Workstation and Manufacturing Commodity Exchange System are to become realities. There are three directions for the research. The first one addresses the application technologies needed to implement the five types of design for quality components.

For example, deterministic metrology (the technique of accurately measuring the process rather than the product) is important to design for quality. Group Technology should play an important role in design for reusability. The formalization of product functional specification is important to design for functionality. Emulation and Failure Mode Analysis are important to design for reliability. Finally, simulation and solid modeling are important to design for serviceability.

A second direction examines the research required in the area of information technology. There are appropriate topics for each of the four elements of the design workstation. The decision databases require the significant use of distributed database systems. The life-cycle models need support from information and data modeling techniques. The metric component requires the careful implementation of control systems that can direct the operations based on the measurement rules. Finally, artificial intelligence techniques must be applied to the successful generation and management of the knowledge rules.

A third direction is concerned with the generic nature of design as a separate technology, without the connection to a specific application area, such as manufacturing. Some of the important areas of research are assumption tracking and gaming, cooperative design, interaction among design constraints, knowledge database architecture, design rules, design models, design databases, and design analysis.

In summary, research in concurrent design is concerned with (a) the role of design in the life-cycle, (b) the interaction with new processes, materials, and equipment, (c) product definition, (d) information technology, (e) control systems, (f) metrology, and (g) materials research.

COMMENTS ON NSF PROGRESS REPORTS

There are three NSF grants that fall into the area of design for Manufacturing and the Life-Cycle. The first grant, "Computer-Aided Design of Flexible Assembly Systems", has Daniel Whitney of The Charles Stark Draper Laboratory, Inc. as the principal investigator working with the Massachusetts Institute of Technology. The second grant, "Research on Expert Systems in Automated Manufacturing, Knowledge Acquisition, and Design Methodologies", has Alice Agogino of the University of California, Berkeley as the principal investigator. The third grant, "Upgrading Design Systems", has Sarosh Talukdar of Carnegie Mellon University as the principal investigator. A brief description will be given of each of these grants in terms of the relationship of the intended research to the architecture and critical research issues described in the earlier sections.

The program on "Computer-Aided Design of Flexible Assembly Systems" has as its goal the development of a better understanding of the interrelationships of the life-cycle activities. The project uses the analysis of a flexible assembly system as the tool for developing the life-cycle interaction. There are three major thrusts to the research: (a) Strategic Product Design, (b) Analysis of life-cycle costs, and (c) Minimization of rework operations. Strategic Product Design is defined as the development of a systematic methodology for coordinated design of products and processes, and assembly systems. There is a clear implication that the assembly sequence analysis forces the concurrent design practice to be used. The analysis of life cycle costs looks at choosing assembly sequences in terms of both single products and multiple products that share resources. Single product costs consist of the resource and tool requirements, task time, operating costs for tools and resources, assembly tasks, and annual volume. The rework operations involve the analysis of the life-cycle maintenance and servicing activity. The analysis includes such items as the failure history and the cost of failures or repairs. Hence the service model attempts to minimize the testing, failures, and recirculation of parts in the assembly sequence.

The program on "Research on Expert Systems in Automated Manufacturing, Knowledge Acquisition, and Design Methodologies" has as its goal the development of an hierarchical architecture for integrating diagnostic reasoning in real-time in an automated manufacturing environment. There are two major thrusts: (a) reasoning about diagnostic testing in manufacturing, and (b) tieing product design, manufacturing and testing decisions to field quality, costs, and risks. The research is built upon the system IDES (Influence Diagram Based Expert System) which implements the concept of influence diagrams. This technique enables the modeler to capture the structure of design/manufacturing knowledge. It also provides a framework from which to access the available data on design and manufacturing. This system results in an intelligent expert system that can make high level inferences about the final design based on defined parameters. The uncertainty in the qualification of critical parameters for the product performance during its life time can be reduced. It works in three ways: (a) representation of the critical variables and their interrelationships, (b) minimization of costs, and (c) resolution of different views of design requirements.

The program on "Upgrading Design Systems" has as its goal the development of a formalized procedure for concurrent design through the use of TAO (test-aspect-operator) graphs. There are five major research thrusts: (a) decomposition of design tasks, (b) completeness of the product data model, (c) development of a design decision history database, (d) feedback from life-cycle models, and (e) flexibility for new processes and materials in design. The research experiment involves the upgrade of a window regulator design. The research can be divided into the design aspects and the simultaneous engineering activity. The design process is developed in an hierarchical fashion, first defining a stick model which outlines the major design decisions, and second defining a feature model that applies the manufacturing constraints to achieve the final design. The simultaneous engineering process is concerned with tolerance optimization that involves a life-cycle cost analysis. It also is concerned with the interference and clearance analysis which involves the use of a solid model for analyzing the maintenance and service activities in the life-cycle.

SUMMARY

This paper described the research thrusts needed for a program in Design for Manufacturing and the Life-Cycle. This research is unique from other aspects of the Design Theory and Technology topics in that there is an emphasis on integration. Standards play an important role with the Product Data Exchange Specification (PDES) being the core element in any concurrent design system implementation. There is a heavy reliance on other areas of design technology. The need to develop measures of quality is critical to any successful product design. Finally there is the need for a large involvement of many engineering disciplines to develop the ultimate design system that is capable of producing a world-class product.

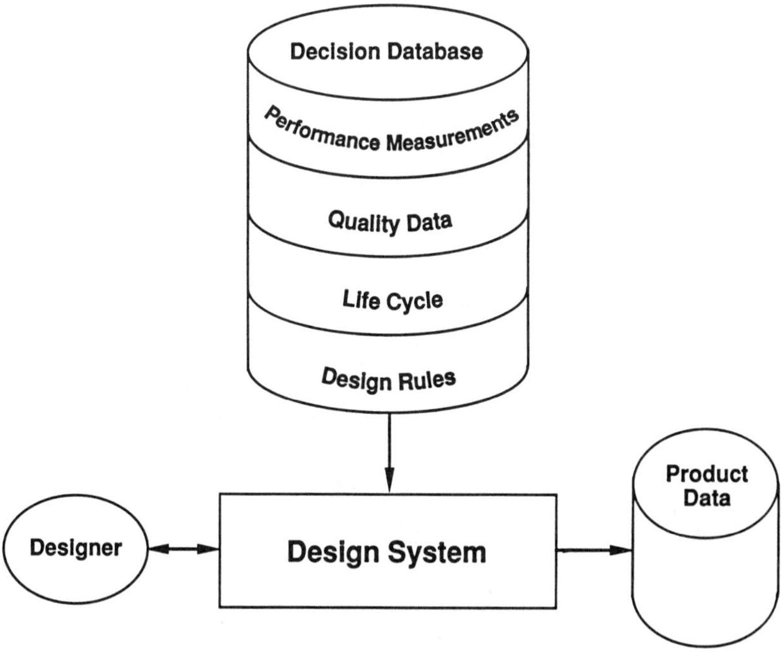

Figure 1: Architecture of a Design-in-Quality Workstation

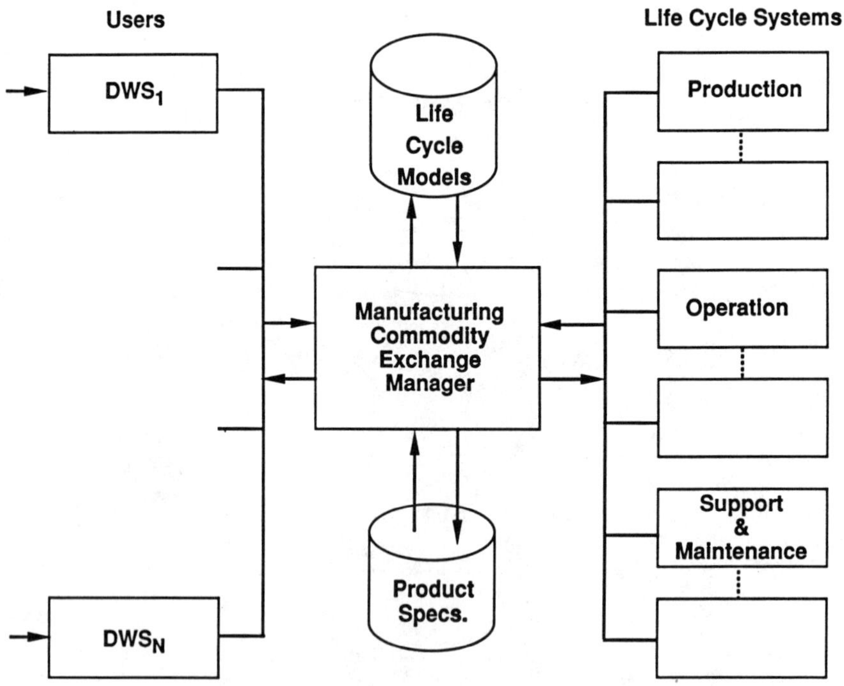

Figure 2: Manufacturing Commodity Exchange System

DECISION-ANALYTIC METHODOLOGY FOR LIFE CYCLE DESIGN

Alice M. Agogino, P.E., Principal Investigator
Department of Mechanical Engineering
University of California, Berkeley, CA 94720

Wayne Imaino and Sherman S. Wang
Mechanical Diagnostics Group, Almaden Research Center
IBM Corporation, San Jose, CA 95120

Sponsored by NSF PYI grant # DMC-8451622 and IBM grant # 705012

Abstract

A modular hierarchical influence diagram model is proposed as a decision-analytic framework for the life cycle design and manufacture of mechanical products. The influence diagram ties product design, manufacturing and testing decisions to field quality, costs and risks. With increasing international competition and concern for product performance and reliability in the marketplace, design and manufacturing groups must explicitly consider all life cycle aspects and uncertainties early in the product design cycle? We propose that design methodologies that enhance communication of experiential knowledge between design, manufacturing and marketing will be crucial in assisting in the early design phases. The focus of this research is in the use of influence diagrams for modeling and communicating critical interrelationships and quantifying their behavior. The IDES (Influence Diagram Based Expert System [1-4]) developed at the University of California at Berkeley is used to illustrate the potential of applying such a planning model to real-time diagnostic decisions in the manufacture and testing of mechanical components for high-speed printers [5].

[1] Agogino, A.M. and A. Rege (1987), "IDES: Influence Diagram Based Expert System," *Mathematical Modelling*, V. 8, pp. 227-233.
[2] Rege, A. and A.M. Agogino (1988), "Topological Framework for Representing and Solving Probabilistic Inference Problems in Expert Systems," in press: *IEEE Transactions, Systems, Man and Cybernetics*, V. 18 (3).
[3] Agogino, A.M., S. Srinivas and K. Schneider (1988), "Multiple Sensor Expert System for Diagnostic Reasoning, Monitoring and Control of Mechanical Systems," *Mechanical Systems and Signal Processing*, V. 2 (2), pp. 165-185.
[4] Jain, R. and A.M. Agogino (1988), "Stochastic Sensitivity Analysis using Fuzzy Influence Diagrams," *Proceedings of the Fourth AAAI Workshop on Uncertainty in Artificial Intelligence* (St. Paul, Minnesota, August 19-21, 1988).
[5] Agogino, A.M., O. Nour-Omid, W. Imaino and S.S. Wang (1988), "Decision-Analytic Methodology for Cost-Benefit Evaluation of Diagnostic Testers," in progress, *Transactions of the IIE* (Working Paper 87-0901-P, Expert Systems Technology Laboratory, Department of Mechanical Engineering, University of California, Berkeley, CA 94720.)

PART 7

RESEARCH IN DESIGN

This section addresses issues of research methodology in design. It includes a description of the National Science Foundation's program in Design Theory and Methodology.

the editors

ON RESEARCH METHODOLOGY TOWARDS A SCIENTIFIC THEORY OF ENGINEERING DESIGN*

John R. Dixon
Professor of Mechanical Engineering
University of Massachusetts
Amherst, Massachusetts 01003

ABSTRACT

The goal of this paper is to raise awareness and generate discussion about research methodology in engineering design. Design reseachers are viewed as a single communicating community searching for *scientific* theories of engineering design; that is, theories that can be tested by formal methods of hypothesis testing. In the paper, the scientific mathod for validating theories is reviewed, and the need for operational definitions and for experiments to identify variables and meaningful abstractions is stressed. The development of a design problem taxonomy is advocated. Generating theories is viewed as guided search. Three types of design theories are described: prescriptive, cognitive descriptive, and computational. It is argued that to seek prescriptions is premature and that, unless the human and institutional variables are reduced to knowledge and control, cognitive descriptive theories will be impossibly complex. A case is made for a computational approach, though it is also shown that computational and cognitive research approaches can be mutually supportive.

* Reprinted by permission from Artificial Intelligence for Engineering Design, Analysis, and Manufacturing (AI EDAM), Volume 1, Number 3, 1988

1.0 INTRODUCTION

It is widely argued that engineering design lacks sufficient scientific foundation, and that without an adequate base of scientific principles, engineering design education and practice are too much guided by specialized empiricism, intuition, and experience. In defense of this position, one observes that many traditional courses in engineering design are mostly collections of how to design this or that specific artifact type, and also that design education project experiences often lack critical and comparative feedback based on generally accepted principles. Such observations are the case in general despite the exceptional effort of a relatively small number of individual faculty who have worked long, hard, creatively, and well (and no doubt with too little recognition and reward) to bring structure and substance to the teaching of engineering design. These good efforts, however, are severely hampered by the absense of a solid science base for design. In contrast, the teaching of engineering subjects such as thermodynamics, mechanics, and materials are based on established principles like the first and second laws of thermodynamics, Newton's laws, and atomic and molecular theories of matter.

Can a scientific theory of engineering design be developed? Some work towards this goal has been done [Simon 1969, Yoshikawa 1983, Waldron 1987, Jacques 1980, Wallace 1983]. More work is in progress [Spillers 1988]. Nevertheless, there are those who question whether a *scientific* theory of design is possible. The argument is that there are too many fundamental differences among the various design fields, and too many people and organization variables, to make any sort of common theory possible at any useful level of abstraction. This argument may indeed turn out to be correct; however, this paper is based on the more optimistic view that useful theory or theories are possible, and that they will evolve from careful, creative research directed to discovering them.

To sponsor this research, the Engineering Directorate at National Science Foundation (NSF) in 1985 established the Design Theory and Methodology Program in the new Division of Design, Manufacturing, and Computer Integrated Engineering. The goal of the Program is to sponsor research leading to development of a scientific base of theory, principles, and generalized knowledge for engineering design. NSF is also sponsoring similar research via its Engineering Research Center Program with a grant to the Engineering Design Research Center at Carnegie-Mellon University.

Design is not an easy subject to research. It is quite different from domains like physics, chemistry, or biology where theories and hypotheses can be tested by laboratory or controlled field experiments. Because of the involvement of people and organizations, design is in part like cognitive psychology or sociology in terms of research. However, in addition to people and organizations, design also

to-be-manufactured-sold-and-used physical artifact or system). Thus design is extremely complex. Moreover, design is a process, and processes are not the usual subject of theoretical formulations.

How, then, should research in design be done so that the desired theoretical foundation will be the result? There has been considerable attention paid recently to the research *topics* that need to be studied in design theory and methodology [Rabins 1986]. In contrast, there has been relatively little attention paid so far to the research *methods* to be employed on these topics. Thus the goal of this paper is to raise awareness and generate discussion about research methodology as it relates to engineering design.

The remainder of this paper is organized as follows. In Section 2, several important assumptions and personal views that underly the paper are discussed. In Section 3, there is a definition of *theory,* and a discussion of the differences and relationships between generating and validating theories. In Section 4, the classical scientific method of validating theories is briefly reviewed. In Section 5, the not-so-scientific issues of generating theories are discussed. Sections 3, 4, and 5 say little about engineering design, but are no doubt limited by that context here. In Section 6, an attempt is made to relate the foregoing discussions more specifically to the current state and future directions of engineering design research. Section 7 is a summary and conclusion.

2.0 VIEWPOINTS

This paper is influenced by several of the author's viewpoints which should be made explicit at the outset. These are: (a) that researchers in design theory constitute a *single goal-directed research community;* (b) that the development of design theory is essentially still in a *pre-theory* stage; (c) that the goal of the design research community should be the development of a formal *scientific* theory or theories of design that will enable the generation of hypotheses for testing by traditional scientific methods, and (4) that the development of *design theory will improve design practice*. Each of these views is now discussed in more detail.

A Research Community. A viewpoint taken in this paper is that researchers in engineering design theory and methodology, though widely distributed geographically, constitute a single *goal-directed research community*. The common long range, top level goal of this community is (or, in my view, should be) to develop a scientific theory and foundation for engineering design.

It is my observation that at this point in time, though this community has some very fine individual research efforts, as a goal-oriented *community* we lack to a degree some important qualities necessary for the efficient attainment of the goal. These are: (1) a concensus on the nature of the common goal; (2) a recognized approach to the methods that will enable us to reach the goal; and (3) a high level

Item (3) above suggests we need to develop better means of communication and interaction among the researchers that make up the research community. We do need to read each other's papers. We need to be able to evaluate and constructively criticise work done by others without being offensive, and to be evaluated and criticized without being offended. We need a journal or journals dedicated to *engineering* design theory and methodology, including regular comprehensive review articles. We need fewer, but more concentrated, meetings. Efforts are in progress to help meet these communication objectives. This paper, however, is primarily about items (1) and (2): the nature of our common goal and the nature of our research methods.

As noted in [Kornfeld 1981], the problem solving successes of research communities are based on "concurrency" and "pluralism". That is, members of the community work in parallel on the different sub-goals each perceives to be the best way to acheive the common long range overall goal. However, what we have now is more like chaos than pluralism. This is quite natural for such a new research effort in such a complex field, but it is time now to begin the process of pulling towards the same goal -- albeit on different paths. We had better: the financial support for development of design theory will not last long without substantial visible progress soon.

Pre-Theory Stage. As noted above, there has been research done (by Simon and others) towards design theory, and more is in progress. However, despite the creative quality of some of these pioneering contributions, they are only a beginning. There is much yet to be learned and formalized before we may say that a scientific theory and foundation of principles exists for engineering design. Thus in this paper I will describe engineering design research as being in a pre-theory stage. In view of the work done, and the work in progress, this is an over-simplification. In view of the work still to be done, however, the over-simplification is reasonably accurate and, so long as we are aware of it, aids communication. In particular, it helps the subsequent discussion that distinguishes research methods for validating theories from research methods for generating them.

Scientific Methods. A key viewpoint is that the common goal of the design research community should be the development of formal theories of engineering design; that is, theories that generate hypotheses testable by use of traditional scientific methods.

It is important, however, to distinguish the validation of theories from their *generation*. To illustrate: if there is a single impression one gets from reading reviewer comments of past and present design research proposals, it is that many proposals lack what is variously called "sound research methodology". The criticisms of methodology are quite varied. One type was expressed by a reviewer this way: "Proposals rarely advance theories or hypotheses. When they do, it is rarely a testable theory or hypothesis. When it is testable, actual testing is

reviewer is expecting design research to begin with an hypothesis based on a theory and propose a test, or at least to generate an hypothesis and propose a test. The proposer, on the other hand, no doubt had in mind (but possibly failed to communicate) contributing only to some aspect of the the generation of a theory, and felt that testing was unnecessary, impossible, or perhaps a subject for a subsequent proposal. Research methods for generating theories and those for testing or validating theories are different, and should be judged by different criteria.

That said, however, I return to the implications of the view that the theories being sought should ultimately be formal and testable. There are, I believe, important consequences of this view that pertain to the research methodologies that will be most effective in generating such theories. At the very least, by examining, discussing, and debating the nature of our goal and our theory generating and testing research methods, I believe we will proceed both more quickly and more effectively. In any case, I can see no justification for not making the effort to understand our research methods better.

Why Theory? Finally, it is also my view that development of a scientific theory and foundation for engineering design will ultimately result in more efficient design practices and higher quality, lower cost products. The theory, and even some of the the steps on the way to the theory, will enable the development of new tools for improved design environments. There will also be pointers to or prescriptions for new organizational principles for design activities. All of this, I submit, is critical to our national economic well-being. Theory *will* help advance the practice of design, just as theory has advanced many other engineering fields.

3.0 THEORIES, MODELS, AND OPERATIONAL DEFINITIONS

3.1 Theories

Science seeks to develop theories that explain observed natural phenomena. Scientific theories consist of a statement or set of statements that define some kind of ideal or theoretical system [Giere 1979]. Newton's three laws of motion together with the law of universal gravitation constitute the set usually called [the theory of] Newtonian or classical particle mechanics. The four laws of thermodynamics (the so-called zeroth and the third laws are included to make four) constitute the set referred to as classical thermodynamics.

The statements that comprise a theory define relationships among the variables of a system. In the case of the Newtonian system, the variables include the mass, position, and velocities of particles making up the system. In classical thermodynamics, the variables include heat, work, temperature, and internal energy.

It is important to bear in mind the distinction between the theoretical system defined by the statements and the real world system that the theory purports to explain. The theoretical systems never match real systems perfectly, though it is

When considering theories, it is helpful to distinguish among: (1) data or observations from the real system; (2) generalizations of the data or observations; and (3) explanations of why the generalizations are what they are. Generalizations of data or observations are not theories; they are simply generalizations of data or observations. Theories emerge in the explanations of *why* the data and the generalizations of the data are the way they are.

Theories are arrived at inductively; they are not deduced from other statements. The inductive process may be very difficult, requiring great intuition or creativity; or, *once the right view or generalization of the data* is examined, the induction may be rather obvious. In any case, theories are inductive explanations from experience, observation, experimentation, and study.

To be scientifically valid, a theory must be objectively testable in a way that can determine whether it is true or false. A theory of such a nature that no test can be conceived which might refute it, is not a scientific theory. Sometimes the means for testing a proposed theory are not immediately obvious; in this case a test should at least be conceivable, and the method for testing should be actively sought. Until a valid test is performed, a theory is at best problematic.

An important point to note is that explanations must be at a level of abstraction higher than the data or generalizations they explain. However, tests for validating theories must be at the same level of abstraction as the data or generalizations.

Another distinction to keep in mind is the one between descriptive theories and prescriptions. A descriptive theory defines a system and explains why it behaves as it does. A prescription states how a system, or at least part of a system, *should* behave according to the author of the prescription.

3.2 Models

The term *model* is used in several different ways. One usage is to denote a physical model of another object. Another, as in mathematics, is to denote a system of things and relations governed by a set of axioms. Another, used in engineering analysis, is to denote an idealized system created by making assumptions (often known to be somewhat inaccurate) about some real system. A fourth usage, and the one that is prevalent in this paper, is to denote the representation of a process. Three types of processes of concern here for which models may be developed are prescriptive, cognitive and computational. A prescriptive model specifies how something should be done. A cognitive model is a representation of how people perform some mental task or activity. A computational model expresses a method by which a computer may perform a task or activity.

A model does not constitute a theory. Theory emerges when there is a testable explanation of *why* the model behaves as it does. This will be discussed in more detail in Section 6 in connection with engineering design.

3.3 Operational Definitions

characteristics of a system, those variables or characteristics must be well defined. Much science rests on the *operational definition* of terms [Bridgman 1927]. An operational definition is one which defines terms by the operations used to measure, observe, or compute them. For example, length is defined operationally by prescribing exactly how to measure the length of something. Also, in classical thermodynamics, the operational definitions of heat and work are essential to the usefullness of the theory. Operational definitions can exist for quite abstract terms, as is the case (say) for entropy in thermodynamics. However, in classical thermodynamics at least, such definitions are rooted in other operationally defined terms and mathematical operations. An operational definition always has a connection to some physical or computational operation(s) used to measure or quantify the defined entity. Note that there may be several operational definitions for the same term, even for physical concepts like length. There is no "right" operational definition; but when operational definitions are used, there is at least clear understanding of the meaning of terms, and meaningless terms are excluded.

Though the value of operational definitions is very well accepted in science, the degree to which strict operationalism should be adhered to is controversial. Opponents of strict operational definitions argue that strictness is unnecessary, that it can in fact inhibit the development of science, and that there must be room in science for terms that are plausible even if (at the moment) rigorous operational definitions are not possible. Less rigor is certainly more defensible in a pre-theory stage. However, it should always be borne in mind that operational definitions will ultimately be needed if a theory is to be validated scientifically. Moreover, progress can result from seeking operational definitions (as when Einstein, for example, considered how time is measured). The importance and value of working with operational definitions as closely as possible whenever possible is not disputed, though it is unfortunately sometimes ignored.

3.4 Validation and Generation

Philosophers of science have generally concentrated on the requirements and methods for validation of theories. (There are many books on the subject in every science library.) The issues treated are mostly those that relate to how to test whether, when, or to what extent theories that have already been stated are true or false. A brief description of the generally accepted scientific method for theory validation is in Section 4 below.

There is much less literature on the subject of how theories are arrived at in the first place, though see [Simon 1977, Hanson 1958, Hadamard 1949, Cohen 1988, Kuhn 1962]. One explanation for the origin of theories is that they are arrived at inexplicably - by intuitive induction, by the mysterious insights of genius' who are smarter, or luckier, or more blessed in some way than the rest of us. In the view of this paper, mystery and blessedness have nothing to do with it. The roles

persistence, hard work, and being well informed. The answer proposed in this paper, taken from [Simon 1977], is that theories are generated by guided search. This will be discussed in more detail in Section 6.

Theories about complex subjects are seldom generated solely by individuals or small isolated groups. This is true even though some individual or small group often makes the final step. Theories about complex systems are based on much data, on generalizations of data, on speculation and false starts, and on partial theories by many members of an entire research community focussed on the same theoretical goal. Is there a method in this process, or is theory generation essentially serendipitous? This question is discussed in more depth in Section 5.

4.0 THE CLASSICAL SCIENTIFIC METHOD FOR VALIDATING THEORIES

This Section presents a short summary review of the classical scientific method for validating stated theories. It is reviewed here for two reasons. One is my conviction that it should be the common goal of the engineering design research community to develop theories that can be scientifically validated; therefore we should be aware of the requirements the method will eventually impose. The other reason is that many of us (especially engineers) have had scant exposure to the traditional scientific method for many years, if very much at all. (One colleague confessed to not having heard it discussed since grade school!)

Readers will no doubt be well aware that what is summarized here in a page or so is the subject of entire books, and even entire careers. Thus the summary obviously omits much of interest and importance.

4.1 Hypotheses

As noted above, a theory defines a theoretical system, not a real system. The connections between theoretical and real systems are provided by hypotheses. An hypothesis asserts that some designated real system behaves the same as some defined theoretical system. For example, an hypothesis might be that "a pendulum on earth is a classical Newtonian mechanics system" [Giere 1979]. Others might be that a real gas turbine is a reversible adiabatic turbine, or that some real gas is an ideal gas under specified conditions.

Often, what we call physical "laws" can be deduced from hypotheses. For example, if a pendulum is a classical Newtonian mechanics system, then the familiar law (in this case, equation) relating the period of a pendulum to its length and to the acceleration of the earth's gravity can be deduced. Laws also result from considering reversible adiabatic processes of ideal gases.

To be useful, a theory must enable the generation of hypotheses. That is, if there are no real systems that can be purported to behave like the theoretical system, then the theory is likely to be of little value. Also, the hypotheses generated should be testable as discussed in the next Section.

4.2 Predictions and Hypothesis Testing

be generated that can be tested in an established way. The crucial element in testing an hypothesis is the existence of a *prediction* deducible from the hypothesis. A prediction is a statement that describes the occurence of a possible event or state of a real system, given a starting state and any relevant auxialiary assumptions. For a valid test, the condition must be met that if the hypothesis is true, then the prediction is correct.

However, the above condition, though necessary, is not sufficient for a valid test of an hypothesis. That condition does not account for the possibility that the prediction may turn out to be correct for reasons other than that the hypothesis is true. A proper hypothesis must also be refutable, and a proper test must also include the condition that if the hypothesis is *not* true, then the prediction is highly improbable (the more improbable, the better).

There are a number of pitfalls that commonly foil the proper testing of hypotheses. One is the use of a *vague prediction*. In this case, almost no matter what actually happens, the prediction is so vague that the result can be interpreted as successful. Predictions should be capable of failure, and if the hypothesis is false, the prediction should then also have a very good chance of failing. Another common pitfall is the use of *multiple predictions*. In this case, there are so many predictions that one or more is almost sure to succeed. Another error is to make *no prediction*, but instead to use logic or assumed examples to "prove" the hypothesis. Another mistake is to explain away a failed prediction, asserting that it would have succeeded if only.... Finally, an error can be made by believing or asserting that some phenomenon can only be explained by one of a finite set of causes, say A or B or C. Then when B and C fail to provide the needed explanation, it is concluded that A is the answer (falsely since it was never proved that the explanation had to be in the set A, B, or C.)

In the next section, the issues of theory generation are discussed. As will be seen, there is nowhere near the degree of formal methodolgy available for theory generation as there is for theory validation.

5.0 THE GENERATION OF THEORIES

In this section, the generation of theories is discussed in general without any special reference to engineering design. This is necessary in order to contrast theory generation with theory validation, and also to set the stage for the discussion in the next section related more specifically to engineering design.

Usually, theories are (finally) conceived and articulated by individuals or very small groups working hard on the explicit goal of explaining some set of observed phenomena. However, the final step is usually preceeded by a great deal of research over time - concurrent and pluralistic - by a larger, fully communicating research community. The more difficult and complex the domain, the greater is the need for this communicating community approach.

generating theories is *search*. The search is, first of all, for the variables (i.e., the meaningful abstractions) that describe the field. These are suggested by intuition or experience and/or by exploratory experiments. There is also search for the meaningful generalizations among the observations and experiments. As the variables and generalizations become clear, there is search for relationships that will constitute a theory. Thus, theory generation can be viewed as data gathering, data organizing, variable identification, data generalization, and pattern recognition.

Theory generation is not exhaustive search, especially in a large, complex domain. It also cannot be (at least, I assert, it should not be) random or serendipitous search. There will, of course, be dead ends, false starts, and fruitless projects, but even these should have been defensible in their beginnings as described below. Theory generation is goal-oriented, guided search by a community for variables and meaningful abstractions, for generalizations about these variables, and for testable relationships.

If the generation of theories is guided search, how can the search be guided? There seem to be two general categories: (1) by the experienced intuition of researchers familiar with the domain for which a theory is sought, and (2) by knowledge about previous research results, trial theories and hypotheses, failures, partial successes, etc. These two types of search guidance will be referred to here as (1) domain heuristics, and (2) research heuristics. These are imperfect labels at best, but it is still useful to distinguish between search guidance that derives from knowledge, experience, and intuition of design practice, and the guidance that derives from knowledge of, experience with, and intuition about design research.

The search for a theory is thus heuristically guided search, the heuristics originating in both domain experience and research experience. It is helpful then to define a *guiding dependency* as the relationship between some aspect of domain experience or some aspect of research experience and a proposed research question that will advance the community towards its goal. Though there is no way to be formal or rigorous about such dependencies, they should be made as explicitly as possible and be consciously articulated in decisions about research directions taken. That is, the generation and selection of the most effective pre-theory research projects should be based on (1) the proposed project's role in the community's total goal-directed effort to achieve a theory, and (2) explicitely stated rationale from either domain knowledge, experience, or intuition or from research knowledge, experience, or intuition, or preferably from both.

6.0 TOWARDS A SCIENTIFIC THEORY OF DESIGN

In this section, an attempt is made to relate the foregoing discussions more specifically to the development of a theory, or theories, of engineering design.

here: prescriptions, descriptive cognitive theories, and computational theories. Each of these is discussed below. First, however, a pre-requisite issue is that of defining the system to which the theories developed will apply.

6.1 Design System Issues and the Need for a Problem Taxonomy

When developing *any* kind of design theory, it is essential to identify the system or sub-system for which the theory is applicable. The system defined does not have to be an entirely general and complete design system; that is, it does not have to include all the people, all possible problem types, all possible organization and other design environment variables, etc. Incomplete and partial design systems can be studied so long as they are defined. Researchers are free to select any system or sub-system they wish, but they are not free to avoid focusing attention on *some* well defined system. This stricture is somewhat analogous to the requirement given to students of thermodynamics to "define a bounded system" and to students in mechanics to "define a free body". The theories involved in those subjects apply only to bounded systems and free bodies; that is, to the types of physical systems described by the creators of the theories. Since theories always relate to some limited system, the system must be described.

Every design system definition must indicate the type of problem under consideration. Thus, if we are to have design theories, we are first going to have to be able to classify the many and various types of design problems. In other words, we need a taxonomy of design problems. (Imagine a theory of botany without a taxonomy of plants.) We need a problem taxonomy not only so that we can define theoretical systems, but also so that we can better communicate with each other about the problems and especially about the *problem classes* on which we are respectively doing research.

A taxonomy of design problems must be independent of the other elements of a complete design system (i.e., the people and environment). The problem classification system must have to do with problem alone. A tentative taxonomy of this type for discreet part and assembly mechanical design problems is proposed in [Dixon 1988].

6.2. Prescriptions

Since we seek ultimately to improve the practice of engineering design, we want ultimately to have prescriptions that advocate how design *should* be done in particular circumstances. However, we cannot begin with prescriptions; there first must be valid descriptive theories on which prescriptions can be based.

Sometimes hypotheses are proposed for design research that take the form "If such-and-such is done, then the design produced will be *better* than otherwise" [Antonsson 1987]. This type of project is a search for a prescription rather than a search for descriptive theory. Providing the design system is defined, and there is a metric for *better*, and *such-and-such* is adequately defined, then such hypotheses (assuming they meet the general criteria for all hypotheses) are fine.

description of the resulting design than just *better* or *worse*. There should also be an explanation of *why* the result is what it is. I argue that it will advance the field faster to avoid concern for *better* at this time in favor of more fully describing the results and the reasons for them.

Examples of Prescriptions. There are at least four prescriptions that should be recognized here for their claimed relevance to a theory of engineering design [Nadler 1981, Christakis 1987, Pahl and Bietz 1976, and Suh 1978]. Both Nadler and Cristakis propose prescriptions for design in general, not just engineering design. Though these prescriptions are no doubt useful in certain design systems (i.e., combinations of problems, people, and environment), they lack operational or operational-like definitions for terms. Moreover, they cannot be validated as outlined above. Thus, they are subjective prescriptions, not scientific theories.

Pahl and Beitz present another prescription, this one more specifically directed to engineering design. They assert that design should be done in a prescribed sequence of phases, and they provide qualitative descriptions of these phases. They also provide exhaustive symbolic tables to assist designers in exploring alternatives in the various phases. This is a more operational-like approach, and there is an unstated but underlying taxonomy of problem types, but it is still a prescription, not a scientific theory.

Suh and his co-authors have taken yet a different approach. They propose a set of *axioms* for engineering design. The axioms are clearly prescriptive and, as axioms, are apparently not subject to verification. Many terms are as yet not operationally defined, though some work has been done to remedy this [Rinderle 1982] and more is in progress [Nakazawa 1984]. The axioms and their corollaries sound very much like statements of what might be called *good design practice*. However, it is not yet clear whether the axioms proposed apply equally well to design of parts as to design of processes, and to all engineering fields. For example, the stricture to avoid functional coupling in form may in some situations be in conflict with the usual mechanical design goal of reducing the number of parts. Also, it has not yet been shown that decoupling results in better parts or processes, even though it is intuitively satisfying that it will result in easier to control design processes. Continuing research may resolve such issues, however, and also tests for validation may also be found.

As noted above, it is my personal view that we should not yet be looking for prescriptions. First we should develop theories that describe design, and then base new prescriptions on these theories.

6.3 Descriptive Cognitive Theories

An essential ingredient in a descriptive cognitive theory is an active intelligent human designer. If the designer is present only as a user to provide initial input and receive some final output, then such a system is computational rather than cognitive. Combined systems (that is, people and computers interacting)

participant or decision maker.

A complete design system for a cognitive descriptive theory of design requires relating the meaningful operationally defined variables from each of the following:

- The person or persons;
- The problem;
- The organizational environment;
- The design environment, including (for example) computer tools, analytical and drafting tools, information resources, etc., and;
- Time.

Omitting time (to make things a little simpler) the basic structure of any descriptive cognitive theory will thus be of this form: When a person or persons with characteristics C, addresses a design problem of type P in an organizational and design environment E, then the result will be X, where X denotes the description of the resulting design and any other effects.

The immense complexity here is obvious. C, P, E, and X must all be described in terms of variables or meaningful abstractions, and the relations among these variables found. Assuming operational, or at least operational-like definitions for the variables, there are still a *very* large number of possible variables and combinations of variables for the theory to relate and explain. Moreover, at this time we simply do not know what distinguishing characteristics of people and environments should be included as variables, to say nothing of how to define them operationally.

Perhaps such theories can be developed for viably small partial design systems, but it seems out of the realm of practicality to attempt to develop completely general theories of this type.

Generating Descriptive Cognitive Theories. If descriptive cognitive theories are to be developed, even for small parts of design systems, the need for more formal experimental data about engineering design is obvious. See, for example, [Ullman 1987] and [Waldron 1987]

The need is for more data and generalizations (based on the data) that might reasonably be expected to form the basis for induction of a theory. At this time, therefore, the greatest need is for experiments that will help identify the variables, the important people and environmental variables that will be useful in a theory. The variables identified must have operational, or at least operational-like definitions. They must be detailed enough ultimately to enable useful hypotheses to be generated and tested, but also general enough that only the important complexities are included.

It is my opinion that one of the important variable types that we should be attempting to identify and describe is *knowledge*. We need to be able to develop theories that describe the effect of the designers' knowledge on the design process and on the design. The studies that divide designers into two classes, experienced

differences are in terms of knowledge that result in different designs and design processes. To get this data, we need to develop or select instruments for measuring designers' knowledge (and perhaps their skills and aptitudes, too). It might even be argued that certain affective issues are relevant. However, if we expect scientific theory to result, we will need to deal with the people characteristics objectively, not subjectively. Imperfect as they will be, measuring instruments for identifying the meaningful (to design) variables in $\underline{C}$ are essential. If the people variables can be reduced to knowledge, the situation is much less difficult, but by no means simple.

For the development of descriptive cognitive theories, we also need experiments to identify the critical variables in the environment (i.e., the organization, communication, tools, etc.) There must be sufficient detail here to serve as the basis for useful theories, but also sufficient generalization to keep the complexity to a manageable level. It may be that these variables can be reduced to those involving only issues of communication and decision making.

Those who argue that no *general, formal, testable* theory of design is possible usually base their position on the very large number of ill-defined variables that would have to be defined and related in any cognitive descriptive theory. I certainly agree. However, this is not to say that there can be no theory or theories at all. There can, I believe, be descriptive cognitive theories applicable to limited design sub-systems if human and social issues are reduced to knowledge and control, and there can be computational theories as described in the next section.

6.4 Computational Theories

A computational theory explains a computational model; that is, the theory explains why the model behaves as it does. The explanation is in the form of relationships among the variables present in the model. If we had computational models and associated theories for all the types of design problems in a taxonomy of design problems, that set of models and theories would constitute a computational theory of design.

A complete design system for a computational theory requires relating the meaningful operationally defined variables from each of the following:

- The problem;
- The knowledge;
- The control (i.e., communication and decision making); and
- Time.

It may be noted that if, in a cognitive descriptive system, people variables are reduced to knowledge, and environment variables are reduced to communication and decision making, then the resulting system is much like a computational system. The cognitive system will be concerned with the knowledge possessed by the people whereas the computational system will be concerned with the knowledge represented

decision making in the human organization whereas the computational system will be concerned with the control of knowledge, information, and decision making in the program. Thus an informal way to view these two types of systems is to consider that in the computational system as compared with the descriptive cognitive system, knowledge has replaced the people, control has replaced the organization, and the computer *is* the environment.

Except for time, each of the elements listed above that make up a complete computational design system has a large number of variables potentially relevant to a theory of design. Exactly what these variables are, or what abstractions of them are important, and how to represent them, has not yet been established. Thus even with the people and social variables reduced to knowledge and control in a program, there is an extemely large and complex set of variables to be related. Again, therefore, it seems reasonable to begin by looking for theories applicable to only partial design systems. We can establish the variables and relevant abstractions by looking at smaller systems, at partial systems or sub-systems, for which theories may be developed that can later on be integrated into larger, more comprehensive, or more abstract theories.

Generating Computational Theories. As outlined in Section 3, a theory is a testable relationship or relationships among operationally defined variables of some natural (real) system. The variables exist in and are measured in the natural system. The relationship(s) exist in the theoretical system. The hypothesis is that the relationship(s) can make accurate predictions about how the real system will behave given a starting place and any auxialary assumptions (constraints). Then experiments can be conducted in the real system to verify or refute the predictions, hence the hypothesis, hence the theory.

If we interpret the elements of a computational system in a certain way, the structure just described is reproduced. We view the computational model as embodied in a computer program. (Another way to say this is that the program expresses the model.) The program must have explicitely represented variables; otherwise it will not be useful in the development of a theory. The variables suggested here are knowledge and control of communication and decision making. A theory then expresses relationships among these variables, the program inputs (playing the role of starting place), and the program outputs (usually a final or intermediate design result). An hypothesis is that the theory can predict reasonably well the result when the program is given some input, specified knowledge, and specified control. The program can then be run as a kind of experiment, and the result compared with the predicted result in order to test the hypothesis. It must be noted that the inputs and the predicted results must be in real world terms, and any output of the program (such as a design) must therefore be evaluated or described as a real world entity. This means that if the output is a design, the design must be built or at least its performance simulated in order to make the connection with the real world

The structure described can be used to help generate computational theories as well as test them. Programs expressing computational models with knowledge and control as variables can be run many times as experiments to generate data on the effect of knowledge and/or control on the results in specified problems. From these data, the general relationships may be induced. Then, if suitable hypotheses and predictions are formed, the theories can be validated.

It cannot be emphasized too much that a program that simply runs is of little theoretical value. There must be variables that can be explicitely experimented with. Of course, the variables available for experimentation can be such that only relatively trivial theories are possible. For example, if the only controllable variable is the choice between two more or less equivalent anaysis procedures in the model, then not much of value will be learned. However, when the variables are raised to the level of more crucial knowledge, communication, and decision making issues, then much can be learned. This is especially true if these variables can be identified at domain independent levels. In any case, with these interpretations of computational models, the programs that express them, and the theories that explain them, we can have scientifically tested computational theories of design.

Of course, there is still the problem of identifying and defining operationally the variables, and of discovering the relationship(s) among them. What are the variables? I believe they are knowledge and control, or more accurately, operational attributes of knowledge, communication, and decision making. This can be illustrated with a simple example. We have been working with a simple computational model, the *redesign* model, that does parametric design of components or standard modules. Early in the work on this model, we developed a program for the design of extruded heat fins [Kulkarni 1985]. Everything was "hard-wired" into the program; all we could do was enter specifications and get out a good design. Suppose we arranged to have the analysis procedures used by this program be a variable. That is, suppose we provided a number of heat transfer correlations, of different accuracy and run time. Then we could have developed a small theory, perhaps relating the accuracy and run time of the analyses to the cost and efficiency of the designs produced. Running the program could then test the theory, assuming we could make valid predictions. (Results would show that the accuracy of the analyses have little effect; the important variables in this domain are manufacturing.) Other knowledge variables that could be added to this are, say, knowledge about other materials, or of how to find the properties of other materials. Different decision making criteria could also have been experimented with. All of this, of course, would have been relevant specifically only to a theory of heat fin design. Thus we chose instead to develop a more domain independent model for parametric component design so that we could experiment with more abstract and general types of design knowledge and control [Dixon 1987, Orelup 1988].

requires that great attention be paid to the issue of *representations*. (A computer representation of something is the computer code which contains the program's information about that something.) In fact, the issue of representations is a critical research issue in developing computational models. A number of *somethings* must be represented, including the problem itself, pointing again to the need for a problem taxonomy. The need for a problem representation also points out the need for each problem type to have a formal structure by which it can be entered into a program. Other somethings that need representing are function, form (topology, geometry, tolerances), standard devices (e.g., motors, clutches), manufacturing tools and processes, analysis procedures [Jones 1987], and more.

A most important representation issue is the designed artifact. How shall it be represented in the program so that the many logical and modification operations required can be performed on it? The answer seems to be: In terms of *features*. A feature, as of this writing, is an abstraction which has attributes of both form and function. However, research is still needed to understand exactly what this means, and how best to obtain the needed feature-based representation.

6.5 The Relationship Between Descriptive Cognitive and Computational Theories

Research on descriptive cognitive theories and research on computational models and theories have a great deal to contribute to each other. Either can help identify variables and meaningful abstractions useful to the other. One obvious example of this is knowledge. Computational models have the knowledge used represented explicitely in the computer code. A computational theory will have to explain why the knowledge used is important, a task not terribly different from that of identifing the important or meaningful knowledge possessed by people in a prescriptive theory. This argument can be extended to the case of group projects and assembly design computational models. Research results from the two approaches can be mutually reinforcing and useful.

Though it is by no means a necessity that computer programs do design in the same way that people do, there will also be times when the knowledge of how people do some task will be very useful to the development of a computational model or program based on a computational model. Thus protocol studies of how particular decisions are made, or how other specific tasks are done by people, may be useful to research directed primarily at computational models. This is not the same as development of a whole cognitive theory; it is simply using the way people do a task or sub-task as a way of finding some way to model and program the task.

Computational models can also illuminate issues helpful in the cognitive domain. For example, the nature and importance of redesign dependency knowledge is illustrated in the Dominic program, and the possibility that communications should be restricted to certain kinds is suggested in a computational model of larger decomposed systems [Meunier 1988].

6.6 Combined or Interactive Systems.

conjunction with computer programs that may, among other things, perform analysis procedures, produce graphics displays, or express computational models that perform design tasks. Of interest here is the combination of cognitive models and computational models in a single larger system. Such systems can be defined, and theories defined separately for the cognitive and computational parts as outlined above. Such systems will introduce a new complexity, however: the variables associated with the language L (primitives and operations) used for the human-computer interactions. Human-computer interactive systems are the most complex of all.

7.0 SUMMARY AND CONCLUSIONS

The basic thesis of this paper is that a scientific theory of design is a desireable and feasible goal, and that we as a research community should give more attention to the research methodologies that will achieve the goal.

We should be seeking theories that can ultimately be validated by traditional scientific methods involving hypotheses, predictions, and experiments. Generating theories is a less formal process, however. We can nevertheless be guided in our search by awareness that we are seeking to identify and operationally define the important variables and meaningful abstractions. We need a good taxonomy of design problems to help us communicate and to define the design sub-systems on which we are working. And we can be guided by explicit awareness of design practice and design research heuristics about where and how to search for varaibles and meaningful abstractions.

Probably we should not yet be concerned with prescriptions. First, we need valid cognitive and computational theories on which to base valid prescriptions.

Cognitive descriptive theories are going to be incredibly difficult if we do not reduce the people, institutional, and environmental complexity. We cannot hope to develop a theory of design that encompasses psychology and sociology as well as the natural and designed artifact worlds. It has been suggested here that the more important people variables may be reduceable to knowledge, and that the more important institutional and environment variables may be reduceable to control of communications and decision-making. Even if this is possible, however, the complexity is immense, and controlled experimentation with humans and real organizations is going to be virtually impossible.

The more fruitful approach, therefore, may be computational models and theories. A structure is presented in the paper that enables computer programs expressing computational models to play an important part of the role of experiments in conventional theory validation. If the variables in the computational model are representations of knowledge, communication, and decision-making, then it appears that scientific theories of design can be generated and validated using appropriately constructed computational models. Since

models, the two approaches can assist each other.

Conclusion. It is my observation and intuition that engineering design is knowledge based. Thus, I believe that the most important thing we as an engineering design research community can do is to discover and explicitely organize design knowledge, relating it the various classes and instances of design problems. In other words, it is my personal conviction at this time that the most important variables in the theory of design we are seeking have to do with knowledge and how knowledge is found, communicated, and used in design. Learning how to define and describe knowledge as a variable in a theory is a difficult task, but can be addressed either in the context of cognitive or computational studies.

It is *important* that we as a research community move as directly and as effectively as possible towards realization of the theory that is our goal. Our National economic well being is directly affected by how well our products and processes are designed. If we are to graduate better designers from engineering schools, then we must have faculty who are themselves specialists in design. If we are to have such faculty in appropriate numbers, then the academic respectabilty of design must be made equal with other fields. For this to happen, design must be based on a foundation of theory, principle, and organized generalized knowledge. Then there can be more graduate theses in design, significant numbers of new faculty in design, and new intellectual content for both design education and design practice. Research in design theory is the key.

The motivation behind this paper is to stimulate discussion, or even debate, on research methodology towards a theory of engineering design. It is not at all necessary that we all agree, only that we all contribute as best we can.

I believe that we will, in time, achieve the desired theoretical foundation for engineering design. To accomplish that goal we need only keep the goal in mind, keep informed about the work of others, thoughtfully use domain and research dependencies to guide our search, and employ research methodologies appropriate to the task of developing a science.

ACKNOWLEDGEMENTS

This work was sponsored in part by a grant from the National Science Foundation (DMC 8603076) to the University of Massachusetts.

I am also indebted to the following friends and colleagues who read and commented on an earlier version of the manuscript: Beth Adelson, Susan Finger, Mel Simmons, Steve Fenves, Herbert Simon, John Wesner, J. B. Jones, Chuck Eastman, John Goldak, Sandra Newsome, Bill Spillers, William Rasdorf, Rudy Motard, Clive Dym, Jim Douglas, and my students (Eugene Libardi, Eric Nielsen, Mark Orelup, Ken Meunier, Rohinton Irani, Rick Welch, Michael Guenette, and Michael Duffey.) To the extent that this is a useful and correct paper, these people helped significantly. To the extent that the paper is not helpful or contains errors, I take full

REFERENCES

Antonnson, E., "Development and Testing of Hyptheses in Engineering Design Research", Journak of Mechaniisms, Transmissions, and Automation in Design, 109/153, June 1987.

Bridgman, P. W., The Logic of Modern Physics, Macmillan, New York, 1927, 1955.

Christakis, A. N., Keever, D. B., and Warfield, J. N., "Development of a Generalized Design Theory and Methodology", in Proceedings from the NSF Workshop on the Design Process, M. B. Waldron, Ed., Oakland, February, 1987.

Cohen, P. R. and Howe, A. E., "The Invisible Hand: How Evaluation Guides Research", Department of Computer and Information Sciences, University of Massachusetts, Amherst, MA., 1988.

Dixon, J. R., Howe, A., Cohen, P.R., and Simmons, M. K., "Dominic I: Progress Toward Domain Independence in Design by Iterative Redesign", Engineering with Computers, 2, 137-145, 1987.

Dixon, J. R., Duffey, M., Irani, R., Meunier, K., Orelup, M., "A Proposed Taxonomy for Mechanical Design Problems", to be published in the Proceedings 1988 ASME Computers in Engineering Conference, San Francisco, July, 1988.

Giere, R. N., Understanding Scientific Reasoning, Holt, Reinhart, and Winston, New York, 1978.

Hadamard, J. S., The Psychology of Invention in the Mathematical Field, Princeton University Press, Princeton, N. J., 1949.

Hanson, N. R., Patterns of Discovery, Cambridge University Press, 1958.

Jacque, R. and Powell, J. A., Design: Science: Method, Proceedings of the 1980 Design Research Society Conference, Published by Westbury House, Guildford, England, 1981.

Jones, C., A Knowledge Representation Framework for Automated Engineering Analysis, Master of Science Thesis, Department of Mechanical Engineering, University of Massachusetts, February, 1987.

Kornfield, W. A. and Hewitt, C., "The Scientific Community Metsphor", IEEE

1981.

Kuhn, T., The Structure of Scientific Revolutions, The university of Chicago Press, chicago, 1962, 1970.

Kulkarni, V., Dixon, J. R., Sunderland, J. E., and Simmons, M. K., "Expert Systems for Design: thwe Design of Heat Fins as an Example of Conflicting Sub-goals and the Use of Dependencies", Proceedings of the 1985 ASME Conputers in Engineering Conference, Boston, MA., 1985.

Meunier, K., Dixon, J. R., and Simmons, M. K., "Iterative Respecification: A Computational Model for Hierarchical Mechanical System Design", to be published in Proceedings 1988 ASME Computers in Engineering Conference, San Francisco, July, 1988.

Nadler, G., The Planning and Design Approach, Wiley, New York, 1981.

Nakazawa, H. and Suh, N. P., "Process Planning Based on Information Concept", Robotics and Computer-Integrated Manufacturing, 1, 1, 115-123, 1984.

Orelup, M. F., Dixon, J. R., and Simmons, M. K., "Dominic II: More Progress Towards Domain Independent Design by Iterative Redesign", Proceedings ASME Winter Annual Meeting, December, 1987.

Pahl, G. and Beitz, W., Engineering Design, The Design Council, Springer-Verlag, New York, 1984.

Rabins, M., Ed., Goals and Priorities For Research in Engineering Design, An NSF Workshop, Published by ASME, July, 1986.

Rinderle, J. R. and Suh, N. P., "Measures of Functional Coupling in Design", ASME Journal of Engineering for Industry, 104:383-388, 1982.

Simon, H. A., "The Science of Design", Sciences of the Artificial, MIT Press, Cambridge, MA., 1969, 1981.

Simon, H. A., Models of Discovery, D. Reidel Publishers, Boston, MA., 1977.

Suh, N. P., Bell, A. C., and Gossard, D. C., "On an Axiomatic Approach to Manufacturing and Manufacturing Systems, ASME Journal of Engineering for Industry, 100(2):127-130, May, 1978.

Spillers, W., Newsome, S., and Finger, S., Proceedings of the 1988 NSF Design Theory and Methodology Grantees Worshop, Troy, N.Y., June 2-5, 1988, to be published by Springer International, New York.

Ullman, D. G., Stauffer, L. A., and Dietterich, T. G., "Preliminary Results of an Experimental Study of the Mechanical Design Process", Proceedings from the NSF Workshop on the Design Process, M. B. Waldron, Ed., Oakland, CA., 1987.

Waldron, M. B., Ed., Proceedings from the NSF Workshop on the Design Process, Oakland, CA., February, 1987.

Waldron, K. J. and Waldron, M. B., "A Retrospective Study of a Complex Mechanical System Design", Proceedings from the NSF Workshop on the Design Process., M. B. Waldron, Ed., Oakland, CA., February, 1987.

Wallace, K. N. and Hales, C., "The Application and Evaluation of Formal Design Engineering Methods, Proceedings of the International Conference on Engineering Design, ICED, Copenhagen, 1983.

Yoshikawa, H., "Automation in Thinking in Design", Computer Applications in Production and Engineering, North-Holland, Amsterdam, 1983.

program description

NATIONAL SCIENCE FOUNDATION

DESIGN THEORY AND METHODOLOGY

John R. Dixon, Program Director
(202) 357-7676

The long term goal of the Program is to establish a scientific foundation of theory, principles and knowledge for engineering design. The Program supports well conceived basic research in: (1) scientifically sound theories of design, (2) foundations for design environments, and (3) models of design processes.

SCIENTIFICALLY SOUND THEORIES OF DESIGN: Criteria for evaluating proposals stress the validity of research methods and the use of testable hypotheses. Results should be demonstrable in at least one engineering domain, and preferably extendable to other domains. Studies of large or multi-functional systems are preferred over small or single-function systems. Examples of research topics in design theory include, but are not limited to, *cognitive theories and computational models* that address:

- Large complex systems, including decomposition and synthesis;
- Function;
- Satisficing (in the presence of an evolving design and conflicting multiple goals, some qualitative, some quantitative);
- Redesign; and
- Preliminary, conceptual, or creative engineering design.

FOUNDATIONS FOR DESIGN ENVIRONMENTS: It is not appropriate for the Program to sponsor actual development of new design environments, but basic research is needed on which future developments depend. Criteria for evaluating proposals stress the advancement of understanding of fundamental principles. If results are specific to a domain, they should preferably be extendable to other domains. Examples of research topics include, but are not limited to:

- *Languages and representations* for function, topology, form, artifact classes, and design knowledge;
- Spatial reasoning, shape grammars, and geometric representations motivated by the needs of the design process;
- Designer-oriented representations for preliminary and conceptual design, including multiple views, multiple levels of abstraction, and partial designs; and
- Models for design data bases.

DESIGN PROCESSES: Research in design processes should seek to understand how design is done by people or computers, and why it is done that way. The emphasis should be on knowledge acquisition, organization, and generalization based on *experiments, software studies,* and *protocol studies.* If results are specific to one domain, they should preferably be extendable to other domains. Examples of research topics include, but are not limited to:

- Management of and communication in large design projects;
- Redesign, and preliminary, conceptual, or creative design;
- Design for the life cycle, including manufacturability, reliability, safety, serviceability, inspectability, disposability, etc; and
- Simultaneous or concurrent design (for function and manufacture).

APPENDIX

THE WORKSHOP

This Appendix describes the organization of the 1988 NSF Grantee Workshop on Design Theory and Methodology

the editors

Design Theory '88

1988 NSF Grantee Workshop on Design Theory and Methodology

Workshop Program

Thursday June 2

5:30 - 9 pm	REGISTRATION. Visitors Information Center
6:00 - 8:30	ICEBREAKER. 3rd Floor, Johnsson Engineering Center

Friday June 3

7:00 - 8:30	BREAKFAST
8:00 - 8:30	REGISTRATION AND COFFEE
8:30 - 8:45	WELCOME Dwight Sangrey, Dean of Engineering Rensselaer Polytechnic Institute
8:45 - 9:45	KEYNOTE ADDRESS Roland W. Schmitt, President Rensselaer Polytechnic Institute
9:45 - 10:00	COFFEE BREAK
10:10 - 10:20	THE NSF PERSPECTIVE Michael J. Wozny, Division Director, Design Manufacturing & Computer Integrated Engineering, NSF
10:20 - 10:30	WORKSHOP CHARGE John R. Dixon, Program Director, Design Theory and Methodology, NSF
10:30 - 12:00	SESSION 1. "Cognitive Models of Design Processes" Session Chair: B. Adelson, Tufts University Invited Speaker: Gary Perlman, Sloan School of Management, MIT
12:00 - 1:30	LUNCH (Tours of RPI's Center for Industrial Innovation will be available during this period.)
1:30 - 3:00	SESSION 2. "Prescriptive Models for Design Processes" Session Chair: L. Bucciarelli, MIT Invited Speaker: Charles Eastman, Graduate School of Architecture & Urban Planning, UCLA
3:00 - 3:30	COFFEE BREAK
3:30 - 5:00	SESSION 3. "Computational Models of Design Processes" Session Chair: S. Finger, Carnegie Mellon University Invited Speaker: Gale Nevill, Department of Engineering Science, University of Florida (Gainesville)
5:00 - 6:30	CASH BAR. Terrace, Russell Sage Dining Hall

(There are no planned Workshop events on Friday evening. Welcome to Troy, NY!)

Saturday June 4

7:00 - 8:30	BREAKFAST
8:30 - 10:00	SESSION 4. "Environments, Languages, Representations, and Data Bases for Design" Session Chair: James Rinderle, Carnegie-Mellon University Invited Speaker: George Stiny, Graduate School of Architecture and Urban Planning, UCLA
10:00 - 10:30	COFFEE BREAK
10:30 - 12:00	SESSION 5. "Analysis to Support Design" Session Chair: S. Newsome, Rensselaer Polytechnic Institute. Invited Speaker: Douglas Wilde, Department of Mechanical Engineering, Stanford University
12:00 - 1:30	LUNCH (Tours of RPI's Center for Industrial Innovation will be available during this period.)
1:30 - 3:00	SESSION 6. "Design for Manufacturing and the Life Cycle" Session Chair: David Ullman, Oregon State University (Corvallis) Invited Speaker: Howard Bloom, National Bureau Standards (Gaithersburg)
3:00 - 3:30	COFFEE BREAK
3:30 - 5:00	WORKING GROUPS' SESSIONS I (Workshop will split into small working sessions which will be organized around the earlier division of the Workshop into 6 sessions. Depending upon interest, additional sessions will be organized.)
5:30 - 7:00	CASH BAR. Terrace, Russell Sage Dining Hall
7:00 - 9:00	WORKSHOP DINNER - Main Dining Room, Russell Sage Dining Hall Speaker: Ray Lauzzana, Art Department, University of Massachusetts (Amherst)

Sunday June 5

7:00 - 8:30	BREAKFAST
8:30 - 10:00	WORKING GROUPS' SESSIONS II
10:00 - 10:30	COFFEE BREAK
10:30 - 12:00	CLOSING SESSION Session Chair: John Wesner, AT&T Workshop Summary: John R. Dixon, NSF Report from Working Groups' Sessions and Workshop Summary
12:00 - 1:00	BOX LUNCH

REPORTS FROM THE WORKING GROUPS

John R. Dixon, Program Director
Design Theory and Methodology
National Science Foundation
Washington, DC 20550

Five small working groups were formed at the Workshop around the topics of cognitive theory, computational theory, languages and representation, design for the life cycle, and analysis to support design. Each working group was asked to respond, if possible, to the following list of questions: (1) What is the state-of-the-art in their topic? (2) How will work-in-progress, when completed, contribute to development or understanding of a theory of design? (3) Are current research methods effective? (4) What is missing? What additional work should be initiated? (5) How can research and education be integrated? and (6) How can communication among the research community be improved? Summaries of the working groups follow here.

Cognitive Theory (Reported by David Ullman and Manjula Waldron)

The group consisted of David Brown, Mark Chignell, Ken Waldron, Mike Hagel, Manjula Waldron, David Ullman, Kevin Otto, Tom Kuffner, Mike Molezzi, and Art Erdman.

The goal of research into the cognitive theory design is to learn how to describe the process used by human designers. The motivation for this is partly scientific curiosity and partly to improve the design process by enabling improved design prescriptions, education, design aids, and automation.

The state-of-the-art is essentially embryonic. There are some descriptions based on conjecture, retrospective studies, and protocol studies - the latter two being recent and still evolutionary.

One thing missing is a taxonomy of design, an organizational structure. Such a taxonomy would include: (1) design types (problem characteristics and domain categories); (2) designer characteristics (the number and characteristics of participants); (3) cognitive methods (problem solving techniques and knowledge characterization); and (4) the level of description (abstraction and representation).

A major missing research tool is an effective research methodology. The contrasting approaches used by AI and behavioral psychology were noted. There is a need to learn how to establish scientific hypotheses and test them experimentally. This includes better methods for data collection, data analysis and verification of results. Methods seem to be lacking to move cognitive design research at a reasonable rate.

Computational Theory (Reported by Noshir Langrana)

Perhaps a distinction should be made between computational theory and computer models?

A sampling of research issues was discussed. In computer models, these included: progammable innovations; integration of innovation and analysis; innovation through interactive design; how to do domain specific capture and representation of knowledge; coupling of sub-problems; trade-off decision and management; dealing with multiple alternatives, redesign strategies; reasoning about numerical results; evaluating designs (including incomplete and imprecise designs).

In computational theory, the problems included: validation of design and design process systems; and design problem complexity.

Communication within the research community was considered, and the following suggestions made: (1) make more use of e-mail and fax; (2) have small focussed workshops; (3) encourage the new journal Research in Engineering Design; (4) publish compendium of individual research interests; (5) develop some on-line method of communicating abstracts of funded research proposals; (6) NSF funded inter-project collaboration and exchanges of intimate working visits.

Languages and Representation (Reported by Jim Rinderle)

The group consisted of Beth Adelson, Erik Antonsson, Leo Beltramini, Michael Besso, Susan Finger, David Gossard, Mike Gunderloy, Stephen Hoover, Helmuth Koller, Ray Lauzzana, Kincho Law, Steve Longnecker, R.L. Motard, Gale Nevill, Eric Nielson, Rick Parent, William Rasdorf, Jim Rinderle, Warren Seering, Vasanthan Sivagnanadasan, William Spillers, George Stiny, Dean Taylor, Karl Ulrich, Al Ward and John Watton.

Languages and representations in design include those for describing the design process as well as for describing the product as shown in Figure 1 below. Futhermore, the product description consists of a representation of the artifact per se and the transformations or operations which change the artifact description. The focus of discussion was the representation of the artifact. Data bases and environments were expected to follow from basic representational strategies and were not discussed extensively. Standards, such as IGES and PDES, were considered to be outside of the scope of the working group.

The state of the art in design languages and representations varies significantly from one design domain to another. Most of the representational strategies have been developed for specific problem domains. Each of the languages capture a variety of characteristics of products and are often layered or structured, (e.g. based on features or hierarchy). Many of the problem specific languages are imbedded in more formal languages based on well established algebras of grammars. The group felt strongly that the problem directed mode of design language development was appropriate and should be continued; however, it was also emphasized that it is important to identify the general characteristics of problems being addressed as well as the representational characteristics of the languages used. Excessive generality need not be forced at this time. It was emphasized that in pursuing application driven research, that scientific methods be employed to the greatest extent and that domains and languages are as completely characterized and reported as possible so as to provide a basis for an emerging science of design.

Design for the Life Cycle (Reported by Howard Bloom)

This group consisted of Alice Agogino, John Wesner, Douglas Allyn, Paul Sanchez, Mihendra Hundal, Keith Hall, Bill Cralley, John Goldak, Tony Woo, Farrokh Mistree, Can Dogan, Robert Davis, Stephen Graves, Herb Voelcker, Max Lui, Howard Bloom, and Vincent Vohnout.

Under the framework topic, the group felt that there were two modeling requirements: (a) a taxonomy that will serve as a planning mechanism for required research, and (b) a prototype that defines the life-cycle process. Under information requirements there were several issues such as (a) establishment of a standard for product data that will allow an efficient exchange of information between research projects, (b) formalization of design assumptions that are required throughout the life-cycle as well as input into future product development, (c) Conversion of information into procedural knowledge that can result in a better definition of the life-cycle process, (d) development of databases containing knowledge rules, performance measurements, quality feedback data, and process models, (e) studies into what types of information are needed to allow life-cycle processes to interact and be integrated, and (f) techniques for the formal procedures for data acquisition (e.g. sensor fusion technology) and compression as information travels through the life-cycle.

Under communications, there is a critical requirement to define an organizational network that describes how the product flows through the life-cycle. Specific issues include the specification of nodes (e.g. people involved in the decision making), security (who is permitted to access the information as well as maintain its integrity), and constraints (e.g. how is information updated through the system). This effort requires the development of representations and procedures for the establishment of such networks.

The group felt that the tools needed for life-cycle design would mainly come from research by other design areas and from other technical areas such as computer science. Some needed tools include (a) common workspace to do "team design", (b) simulation, (c) "How To" Guides, (d) graphic-oriented optimization tools such as IDES and TAO as developed by the grantees, (e) process planning, (f) geometric modelers, and (g) integration techniques. In this area of dissemination, the group felt that the life-cycle view was being communicated mainly through trade journals rather than technical journals.

In conclusion, the group felt that the state-of-the-art in life-cycle design is being practiced by many industries and very little is being performed by universities. The National Science Foundation should begin to place an emphasis on the development of the tools that will be needed to do life-

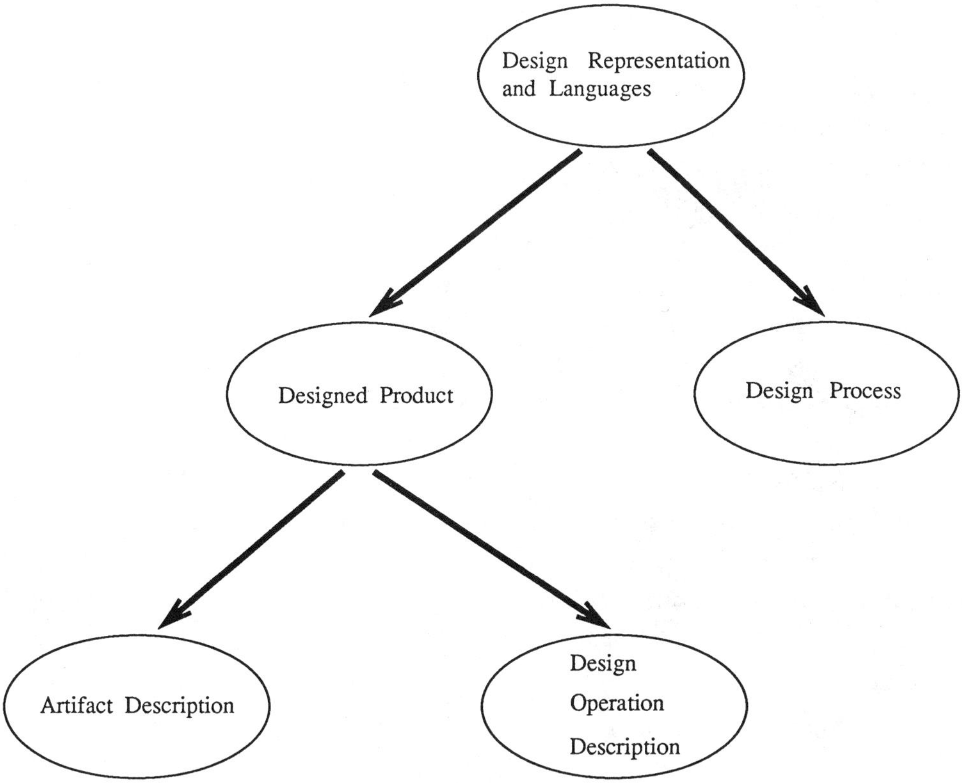

Figure 1 - Types of languages and representations for design

cycle design as feasible research projects for university investigators. Finally, there is the belief that any useful research in life-cycle design will require the implementation of two investigator grants, a designer and a life-cycle process expert, because the chief concern is to study the interaction in the life-cycle process.

Analysis to Support Design (Reported by Warren Seider)

The members of the working group were Shapour Azarm, David Brengel, Jo Dale Carothers, Sheldon Chang, Alejandro Diaz, Chris Floudas, Ignacio Grossmann, Charles Reif Hammond, Jean Hou, Pramod Jain, Vadim Komkov, Prakash Krishnaswami, Wei-Chu Li, Niki Mehta, Panos Papalambros, Warren Seider, H.S. Teoh, and Douglass Wilde.

What makes "analysis" important in support of design?

Some answers: Analysis is used to model designs, to provide evaluation and predictions of performance, and as an organizing tool (an abstraction of the design reality). Optimization relates to all design phases, especially conceptual design. Key considerations are (1) problem representations (model definition); (2) screening of alternatives; and (3) level of complexity. Analyses are part of a large feedback loop where possible use could be made of "relaxed optimization" or "sloptimization".

It was noted that the integration of optimization and knowledge based systems needs more emphasis. The use of heuristics might help prune the synthesis tree. Herb Simon's article was referred to: "Two Heads Think Better Than One - The Collaboration Between AI and OR", Interfaces, 1987.

The effectiveness of research meetings was discussed, and this meeting's format was considered an improvement over individual research presentations. However, speakers to plenary sessions should place less emphasis on their own work in favor of more coverage of the contribution of grantees. Finding ways to stimulate cross group interaction should get high priority.

In education, there is need to focus more on conceptual design - on the conceptual aspects of formulating designs. Less emphasis should be placed on the formalisms of optimization, especially at the undergraduate level. There is a need for new texts like Douglas, Conceptual Design of Chemical Processes, and the new Papalambros/Wilde book. Also, case studies are needed.

To improve communication among us, more conferences on design theory and methodology are needed. Also, we should try to cooperate with societies like ASME (CIME, DAC) and AIChE (CAST). The new design research journal will help. Also, perhaps an NSF e-mail network could be useful.

Concluding Remarks (John R. Dixon)

To conclude the Workshop, I first want to thank all the people who contributed: Dr. Roland Schmitt for an outstanding Keynote address; the invited speakers (especially Gail Nevill, whose well-done report was noted by all, and Gary Perlman, who had the toughest task and did it admirably); the grantees, who were patient while their work was discussed too much - or not enough; and our special friends not in any of the above categories - people like Herb Voelcker, Susan Finger, John Wesner, John Goldak, and J.B. Johns, who came just because they care about design, design research, and the Design Theory and Methodology Program. I also especially want to thank Dean Sangrey for the hospitality of RPI (and for the dinner wine) and Bill Spillers and Sandra Newsome for their very competent and hard work in planning the Workshop, making all the millions of arrangements, putting out all the little "fires", and keeping a calm pleasant exterior through it all. They did a magnificent job.

A reformation in engineering design is critical to the well being of U.S. industry. Our role in this reformation is to develop a theory of engineering design that is scientifically sound (i.e., "hard"), that leads ultimately to new tools, new knowledge, and new procedures for high quality, low cost, short time-to-market designs, and is also teachable. To do this, we will have to apply energy and take a scientific approach to our research methodology.

This is a community effort - as is all search for new theory in a complex field. NSF will continue to do its best, but funds are limited, processes and decision-making are imperfect, and Program Directors can do only a little by themselves. The research community must do the job.

Before I close, I want to tell you about a Workshop held recently (May 25-26) in Washington. The participants were H. Barry Bebb, Vice-President, Polaroid; Harry Cotrill, Mgr. Mechanical and Chemical Systems, JPL; Paul DeRusso, Assoc. Dean of Engineering, RPI; George Dieter, Dean of Engineering, Maryland; Jack Dinsdale, Professor of Machine Systems, Cranford, England; Steven Fenves, Professor of Civil Engineering, Carnegie-Mellon; Lee Harrisberger, Professor of Mechanical Engineering, Cincinnati; William G. Howard, Vice-President, Motorola (NAE); David G. Jansson, Assoc. Professor of Mechanical Engineering, Texas A and M; Charles H. Kruger, Chairman of Mechanical Engineering, Stanford; Hugh R. MacKenzie, Vice-President, Polaroid; Joe Mullins, Director of Manufacturing Engineering, University of New Mexico; Nam P. Suh, Professor of Mechanical Engineering, MIT; Robert Templin, Retired Chief Engineer, Cadillac Div. GM; Bruno Weinschel, President, Weinschel Engineering; George Kuper (Chair), Director of Manufacturing Studies Board, NAS. The Executive Summary of the Workshop's Final Report reads as follows:

Sixteen nationally recognized experts met on May 25-26 in a focused discussion on the critical need to bring new national attention to problems of advancing and integrating engineering design research, education (especially graduate education), and practice. Throughout the discussion, there was a clear underlying consensus that engineering design is not taught, researched, or practiced as effectively as needed by American industry. Much of the world's best existing design knowledge is not available to U.S. designers. Moreover, it was stressed emphatically that there is a need to convert the present "art form" of design into a theory-based and repeatable professional discipline. Perpetuation of the art form leaves major issues of competitiveness, education, and utilization of national resources to chance.

A consensus emerged that the National Science Foundation should, in conjunction with industry, states, and other Federal agencies, mount an aggressive program to bolster graduate-level engineering design research and education activities. Because a recognized body of principles and pedagogical material does not yet exist in this field, a unique combination of research, teaching, and experience is needed to develop new faculty and to revitalize mid-career faculty. Specific recommendations include: (1) establishment of an NSF Design Scholar Program to generate new faculty for design teaching and research; (2) support for projects to acquire, distill, and disseminate currently successful design practices and new research results to American practitioners; (3) development of imaginative, high-quality new graduate design programs; (4) research into the efficient teaching of design; and (5) support for projects that integrate engineering design research, teaching, and practice.

I am proposing an Initiative to NSF for new support of certain engineering design activities based on recommendations of the Workshop. The Initiative is in three parts as follows:

(1) NSF Engineering Design Scholar Program. This program will be available to faculty at any age and to doctoral candidates working on design theses. Design scholars will work one to two years in carefully selected participating industries in positions where they will get high quality, modern design experience. Upon return to their Universities, scholars will receive three year NSF research support, a portion of which will be contingent on matching funds from industry;

(2) Engineering Design Education Research and Development. This program will include support for: (a) establishment of three new experimental graduate engineering design education programs for demonstration and test purposes, the programs to be of very high academic quality, to include on-campus industrial projects, to be transferable to other Universities, and to be viable long term without NSF support; (b) establishment of three experimental undergraduate design education programs for demonstration and test purposes, the programs to explore ways to improve significantly the efficiency and effectiveness of design education; and (c) establishment of a program for updating previously published engineering design case studies and to develop a continuous supply of new case studies for use in education and research;

(3) Engineering Design Knowledge Acquisition and Dissemination Program. This program will support projects to acquire, distill, evaluate, organize, and distribute the world's best design knowledge, possibly in the form of a set of general purpose design manuals directed to the needs of practicing designers, design teachers and students.

This three part Initiative is a broad-based program involving research, education, and practice designed to address the critical National problems described and defined by the Workshop. NSF's leadership role is crucial in this regard. The proposed components to the Initiative will stimulate others in government, industry, and education also to address the problems of engineering design that are so important to our economy and security. I hope that NSF accepts the challenge and responsibility to provide National leadership in engineering design at this opportune time, but the task of obtaining the needed funding will be extremely difficult and is by no means certain.

I am very proud of the work being done by the engineering design research community. It is important work that you are doing. Let us continue to create a foundation of theory and principles on which a reformation in design practices can be based.

DESIGN THEORY '88

1988 NSF Grantee Workshop on Design Theory and Methodology

LIST OF PARTICIPANTS

Beth Adelson
Department of Computer Science
Tufts University
Medford, Massachusetts 02155
(617)-381-3214

Alice Agogino
Department of Mechanical Engineering
University of California
Berkeley, California 94720
(415)-642-6450

Douglas Allyn
SIE Department
University of Arizona
Tucson, Arizona 85721

Erik Antonsson
Department of Mechanical Engineering
California Institute of Technology
Mailcode: 104-44
Pasadena, California 91125
(818)-356-3790

Tomasz Arciszewski
Department of Civil Engineering
Wayne State University
Detroit, Michigan 48202
(313)-577-3789

Shapour Azarm
Department of Mechanical Engineering
University of Maryland
College Park, Maryland 20742
(301)-454-8851

Leo Beltramini
Department of Chemical Engineering
Campus Box 1198
Washington University
St. Louis, Missouri 63130

Michael Besso
Department of Civil Engineering
Virginia Polytechnic Institute & State University
Blacksburg, Virginia 24061

Howard Bloom
National Bureau of Standards
Building 220 - Room A127
Gaithersburg, Maryland 20899

David Brengel
Department of Chemical Engineering
University of Pennsylvania
220 South 33rd Street
Philadelphia, Pennsylvania 19104

David Brown
Computer Science Department
Worcester Polytechnic Institute
Worcester, Massachusetts 01609
(617)-793-5618

Megan Brown
Department of Psychology
Rensselaer Polytechnic Institute
Troy, New York 12180

Wayne Brown
Department of Architecture
Ohio State University
190 West 17th Avenue
Columbus, Ohio 43210

Louis Bucciarelli
E51-201B
Massachusetts Institute of Technology
Cambridge, Massachusetts 02139
(617)-253-4061

Jon Cagan
Department of Mechanical Engineering
University of California
Berkeley, California 94720

Jo Dale Carothers
Department of Electrical & Computer Engineering
University of Texas at Austin
Austin, Texas 78712-1084

Sheldon Chang
Department of Electrical Engineering
State University of New York
Stony Brook, New York 11794
(516)-632-8420

Mark Chignell
Department of Industrial & Systems Engineering
University of Southern California
GER 240, MC-0193
Los Angeles, California 90089-0193
(213)-743-2705

Ken Chin-Purcell
Mechanical Engineering Department
University of Minnesota
111 Church Street SE
Minneapolis, Minnesota 55455

Harvey Cragon
Department of Electrical & Computer Engineering
University of Texas at Austin
Austin, Texas 78712-1084
(512)-471-5368

Bill Cralley
Institute for Defense Analysis
1801 N. Beauregard Street
Alexandria, Virginia 22311
(703)-845-2168

Michael Danchak
School of Engineering & Science
Hartford Graduate Center
275 Windsor Street
Hartford, Connecticut 06120
(203)-549-3600

Robert Davis
Department of Industrial Engineering
104 Freeman Hall
Clemson University
Clemson, South Carolina 29634-0920

Alejandro Diaz
Mechanical Engineering Department
Michigan State University
East Lansing, Michigan 48824-1226
(517)-353-0825

John Dixon
Department of Mechanical Engineering
Engineering Laboratory Building
University of Massachusetts
Amherst, Massachusetts 01003
(413)-545-4242

Can Dogan
Department of Industrial Engineering
104 Freeman Hall
Clemson University
Clemson, South Carolina 29634-0920

Michael Duffy
Department of Mechanical Engineering
Engineering Laboratory Building
University of Massachusetts
Amherst, Massachusetts 01003

Clive Dym
Department of Civil Engineering
University of Massachusetts
Amherst, Massachusetts 01003

Charles Eastman
Graduate School of Architecture & Urban Planning
405 Hilgard Avenue
2217C Perloff Hall
University of California
Los Angeles, California 90024-1467
(213)-825-1174

Art Erdman
Department of Mechanical Engineering
111 Church Street SE
University of Minnesota
Minneapolis, Minnesota 55455
(612)-625-8580

Steve Ervin
E51-201B
Massachusetts Institute of Technology
Cambridge, Massachusetts 02139

Susan Finger
Robotics Institute
Carnegie-Mellon University
Pittsburgh, Pennsylvania 15213
(412)-268-8828

Greg Fisher
Department of Mechanical & Aerospace Engineering
Rutgers University
P.O. Box 909
Piscataway, New Jersey 08855-0909

Patrick Fitzhorn
Department of Mechanical Engineering
Colorado State University
Fort Collins, Colorado 80523
(303)-491-5505

Christodoulos Floudas
Department of Chemical Engineering
Princeton University
Princeton, New Jersey 08544
(609)-452-4595

John Goldak
Department of Mechanical & Aeronautical Engineering
Carleton University
Ottawa, Ontario, Canada K1S 5B6
(613)-564-3627

Allen Goldberg
Keitrel Institute
1801 Page Mill Road
Palo Alto, California 94304
(415)-493-6871

David Gossard
Department of Mechanical Engineering
Massachusetts Institute of Technology
Room 3-449
Cambridge, Massachusetts 02139
(617)-253-4465

Stephen Graves
A.P. Sloan School of MIT
E53-390
Massachussetts Institute of Technology
Cambridge, Massachusetts 02139
(617)-253-6602

Ian Grosse
Department of Mechanical Engineering
University of Massachusetts
Amherst, Massachusetts 01003
(413)-545-1350

Ignacio Grossmann
Department of Chemical Engineering
Carnegie-Mellon University
Pittsburgh, Pennsylvania 15213
(412)-268-2228

Michael Guenette
Department of Mechanical Engineering
Engineering Laboratory Building
University of Massachusetts
Amherst, Massachusetts 01003

Peter Hadjo
Department of Civil Engineering
Wayne State University
Detroit, Michigan 48202

Mike Hagel
Department of Industrial & Systems Engineering
University of Southern California
GER 240, MC-0193
Los Angeles, California 90089-0193

Keith Hall
Medicine and Computer Science
Stanford University
Stanford, California 94305

Charles Reif Hammond
Department of Mechanical Engineering
University of Pittsburgh
Pittsburgh, Pennsylvania 15261
(412)-624-9787

William Ho
Department of Electrical Engineering Systems
Sal-300 University Park
University of Southern California
Los Angeles, California 90089-0781

Steven Hoover
Mechanical Engineering Department
Carnegie-Mellon University
Pittsburgh, Pennsylvania 15213

Jean Hou
Department of Mechanical Engineering & Mechanics
College of Engineering
Old Dominion University
Norfolk, Virginia 23529-0247
(804)-440-3720

Kou-Cheng Hsu
Department of Electrical Engineering
State University of New York
Stony Brook, New York 11794

Mahendra S. Hundal
Department of Mechanical Engineering
Votey Bldg.
University of Vermont
Burlington, Vermont 05405
(802) 656-1930

Rohinton Irani
Department of Mechanical Engineering
Engineering Laboratory Building
University of Massachusetts
Amherst, Massachusetts 01003

Pramod Jain
Department of Mechanical Engineering
University of California
Berkeley, California 94720

Linda Jaynes
Department of Psychology
Rensselaer Polytechnic Institute
Troy, New York 12180

J. B. Jones
Mechanical Engineering Department
Virginia Polytechnic Institute & State University
Blacksburg, Virginia 24061
(703)-961-7186

Helmut Koller
Department of Civil Engineering
Wayne State University
Detroit, Michigan 48202

Vadim Komkov
Department of Mathematics
College of Engineering
Air Force Institute of Technology
Wright Patterson AFB
Dayton, Ohio 45433

Prakash Krishnaswami
Department of Mechanical Engineering
Kansas State University
Manhattan, Kansas 66506
(913)-532-5610

Tom Kufner
Department of Mechanical
Engineering
Oregon State University
Corvallis, Oregon 97331

Tai Hun Kwon
Department of Mechanical &
Aerospace Engineering
Rutgers University
P.O. Box 909
Piscataway, New Jersey 08855-0909
(201)-932-3282

Sivand Lakmazaheri
Department of Civil Engineering
North Carolina State University
Box 7908
Raleigh, North Carolina 27695

Steve Longenecker
Department of Mechanical
Engineering
Colorado State University
Fort Collins, Colorado 80523

Noshir Langrana
Department of Mechanical &
Aerospace Engineering
Rutgers University
P.O. Box 909
Piscataway, New Jersey 08855-0909
(201)-932-3618

Ray Lauzzana
Art Department
University of Massachusetts
Amherst, Massachusetts 01003
(413)-549-4646

Kincho Law
Department of Civil Engineering
Rensselaer Polytechnic Institute
Troy, New York 12180-3590
(518)-276-6937

H.H. Lee
Department of Mechanical &
Aerospace Engineering
Rutgers University
P.O. Box 909
Piscataway, New Jersey 08855-0909

Wei-Chu Li
Department of Mechanical
Engineering
University of Maryland
College Park, Maryland 20742

Eugene Libardi
Department of Mechanical
Engineering
Engineering Laboratory Building
University of Massachusetts
Amherst, Massachusetts 01003

Man-Cheung Max Lui
A.P. Sloan School of MIT
E53-390
Massachusetts Institute of
Technology
Cambridge, Massachusetts 02139

Niki Mehta
Department of Mechanical
Engineering & Mechanics
College of Engineering
Old Dominion University
Norfolk, Virginia 23529-0247

Farrokh Mistree
Department of Mechanical
Engineering
University of Houston
Houston, Texas 77004
(713)-748-4044

Mike Molezzi
Department of Mechanical
Engineering
University of California
Berkeley, California 94720

Rodolphe Motard
Department of Chemical Engineering
Campus Box 1198
Washington University
St. Louis, Missouri 63130
(314)-889-6072

Gale Nevill, Jr.
Department of Aerospace Engineering
Mechanics and Engineering Science
University of Florida
Gainesville, Florida 32611
(904)-392-0961

Sandra Newsome
Department of Psychology
Rensselaer Polytechnic Institute
Troy, New York 12180
(518)-276-6430

Eric Nielsen
Department of Mechanical
Engineering
Engineering Laboratory Building
University of Massachusetts
Amherst, Massachusetts 01003

Mark Orelup
Department of Mechanical
Engineering
Engineering Laboratory Building
University of Massachusetts
Amherst, Massachusetts 01003

Kevin Otto
Department of Mechanical
Engineering
111 Church Street SE
University of Minnesota
Minneapolis, Minnesota 55455

Panos Papalambros
Department of Mechanical Engineering
University of Michigan
Ann Arbor, Michigan 48109
(313)-936-0354

Rich Parent
Department of Architecture
Ohio State University
190 W. 17th Street
Columbus, Ohio 43210

Gary Perlman
Sloan School of Management
Massachusetts Institute of Technology
E40-136
50 Memorial Drive
Cambridge, Massachusetts 02139
(617)-566-6858

Kerstein Pollack
IRM MH270
Manufacturing Studies Board
National Academy of Science
2101 Wisconsin Avenue NW
Washington, D.C. 20007
(202)-334-2572

N. Ramachandran
Department of Mechanical & Aerospace Engineering
Rutgers University
P.O. Box 909
Piscataway, New Jersey 08855-0909

William Rasdorf
Department of Civil Engineering
North Carolina State University
Box 7908
Raleigh, North Carolina 27695
(919)-737-2331

Donald Riley
Mechanical Engineering Department
University of Minnesota
111 Church Street SE
Minneapolis, Minnesota 55455
(612)-625-0591

James Rinderle
Mechanical Engineering Department
Carnegie-Mellon University
Pittsburgh, Pennsylvania 15213
(412)-268-3677

Paul Sanchez
SIE Department
University of Arizona
Tucson, Arizona 85721
(602)-621-6558

Dwight Sangrey
School of Engineering
Rensselaer Polytechnic Institute
3006 JEC
Troy, New York 12180
(518)-276-6203

Roland Schmitt
President's Office
Rensselaer Polytechnic Institute
204 Pittsburgh Building
Troy, New York 12180
(518)-276-6211

Warren Seering
Massachusetts Institute of Technology
Room 3-461
Cambridge, Massachusetts 02139
(617)-253-2045

Warren Seider
Department of Chemical Engineering
University of Pennsylvania
220 South 33rd Street
Philadelphia, Pennsylvania 19104
(215)-898-7953

Vadim Shapiro
MAE Upson Hall
Cornell University
Ithaca, New York 14853

Mark Shephard
Department of Civil Engineering
Rensselaer Polytechnic Institute
7017 CII
Troy, New York 12180
(518)-276-6751

Charles Sheppard
Department of Chemical Engineering
Campus Box 1198
Washington University
St. Louis, Missouri 63130

Vasathan Siragnanadasan
Department of Mechanical Engineering
Colorado State University
Fort Collins, Colorado 80523

William Spillers
Department of Civil Engineering
Rensselaer Polytechnic Institute
4048 JEC
Troy, New York
(518)-276-6363

George Stiny
Graduate School of Architecture & Urban Planning
University of California
Los Angeles, California 90024
(213)-825-5751

Kwa-Sur Tam
Department of Electrical Engineering
Virginia Polytechnic Institute & State University
Blacksburg, Virginia 24061
(703) 961-4448

Marty Tannenbaum
Schlumberger
3340 Hillow Avenue
Palo Alto, California 94304
(415)-496-4600

Dean Taylor
Department of Mechanical Engineering
Cornell University
Ithaca, New York 14853
(607)-255-3575

Yahia Teboub
Department of Civil Engineering
Rensselaer Polytechnic Institute
7017 CII
Troy, New York 12180

Hok Sin Teoh
Department of Mechanical Engineering
Kansas State University
Manhattan, Kansas 66506

David Ullman
Department of Mechanical Engineering
Oregon State University
Corvallis, Oregon 97331

Karl Ulrich
Massachusetts Institute of Technology
Room 3-461
Cambridge, Massachusetts 02139

Herbert Voelcker
MAE, Upson Hall
Cornell University
Ithaca, New York 14853
(607)-255-9654

Vince Vohnout
Department of Engineering Graphics
Ohio State University
Columbus, Ohio 43210

Kenneth Waldron
Department of Mechanical Engineering
Ohio State University
Columbus, Ohio 43210

Manjula Waldron
Department of Engineering Graphics
Ohio State University
Columbus, Ohio 43210

Allen Ward
Massachusetts Institute of Technology
Room 3-461
Cambridge, Massachusetts 02139

John Watton
Mechanical Engineering Department
Carnegie-Mellon University
Pittsburgh, Pennsylvania 15213

Richard Welch
Department of Mechanical Engineering
University of Massachusetts
Amherst, Massachusetts 01003

John Wesner
AT&T
200 Laurel Avenue
Middletown, New Jersey 07748

Gio Wiederhold
Medicine and Computer Science
Stanford University
Stanford, California 94305

Douglass Wilde
Mechanical Engineering Design
Stanford, California 94305
(415)-725-3805

Tony Woo
Computer Integrated Engineering
National Science Foundation
Washington, D.C. 20550
(202)-357-7676

Kris Wood
Department of Mechanical Engineering
California Institute of Technology
Mailcode: 104-44
Pasadena, California 91125

Michael Wozny
CICG
Rensselaer Polytechnic Institute
7023 CII
Troy, New York 12180
(518)-276-6751

Chris Yessios
Department of Architecture
Ohio State University
190 West 17th Avenue
Columbus, Ohio 43210
(614)-292-2918

Fang Zhao
Department of Civil Engineering
Carnegie-Mellon University
Pittsburgh, Pennsylvania 15213

Thomas Zimmie
Department of Civi Engineering
Rensselaer Polytechnic Institute
4038 JEC
Troy, New York 12180
(518)-274-6939